MODELLING WATER AND NUTRIENT DYNAMICS IN SOIL–CROP SYSTEMS

COST-the acronym for European **CO**operation in the field of **S**cientific and **T**echnical Research-is the oldest and widest European intergovernmental network for cooperation in research. Established by the Ministerial Conference in November 1971, COST is presently used by the scientific communities of 35 European countries to cooperate in common research projects supported by national funds. The funds provided by COST – less than 1% of the total value of the projects – support the COST cooperation networks (COST Actions) through which, with only around €20 million per year, more than 30,000 European scientists are involved in research having a total value which exceeds €2 billion per year. This is the financial worth of the European added value which COST achieves. A "bottom-up approach" (the initiative of launching a COST Action comes from the European scientists themselves), "à la carte participation" (only countries interested in the Action participate), "equality of access" (participation is open also to the scientific communities of countries not belonging to the European Union) and "flexible structure" (easy implementation and light management of the research initiatives) are the main characteristics of COST. As precursor of advanced multidisciplinary research COST has a very important role for the realization of the European Research Area (ERA) anticipating and complementing the activities of the Framework Programmes, constituting a "bridge" towards the scientific communities of emerging countries, increasing the mobility of researchers across Europe and fostering the establishment of "Networks of Excellence" in many key scientific domains such as: Physics, Chemistry, Telecommunications and Information Science, Nanotechnologies, Meteorology, Environment, Medicine and Health, Forests, Agriculture and Social Sciences. It covers basic and more applied research and also addresses issues of pre-normative nature or of societal importance.

Modelling water and nutrient dynamics in soil–crop systems

Proceedings of the workshop on "Modelling water and nutrient dynamics in soil–crop systems" held on 14–16 June 2004 in Müncheberg, Germany

Edited by

Kurt Christian Kersebaum
Jens-Martin Hecker
Wilfried Mirschel

and

Martin Wegehenkel

Funded by European Science Foundation ESF and COST 718

A C.I.P. catalogue record for this book is available from the Library of Congress

ISBN 978-1-4020-4478-6 (HB)
ISBN 978-1-4020-4479-3 (e-book)

Published by Springer,
P.O. Box 17, 3300 AA Dordrecht, The Netherlands

www.springer.com

Printed on acid-free paper

All Rights Reserved
© 2007 Springer
No part of this work may be reproduced, stored in a retrieval system, or transmitted in any form or by any means, electronic, mechanical, photocopying, microfilming, recording or otherwise, without written permission from the Publisher, with the exception of any material supplied specifically for the purpose of being entered and executed on a computer system, for exclusive use by the purchaser of the work.

Contents

Preface vii

1. **Modelling water and nutrient dynamics in soil–crop systems: a comparison of simulation models applied on common data sets** 1
   Kurt Christian Kersebaum, Jens-Martin Hecker, Wilfried Mirschel and Martin Wegehenkel

2. **The performance of the model AMBAV for evapotranspiration and soil moisture on Müncheberg data** 19
   Hans Friesland and Franz-Josef Löpmeier

3. **Performance of the model SIMWASER in two contrasting case studies on soil water movement** 27
   Elmar Stenitzer, Heiko Diestel, Uwe Franko, Reinhild Schwartengräber and Thomas Zenker

4. **Application and validation of the models THESEUS and OPUS with two field experimental data sets** 37
   Martin Wegehenkel and Wilfried Mirschel

5. **Integrating a spatial micrometeorological model into the risk assessment for arable crops in hilly terrain** 51
   Marco Acutis, Gianfranco Rana, Patrizia Trevisiol, Luca Bechini, Mario Laudato, Ferrara R. and Goetz Michael Richter

6. **Modelling soil–crop interactions with AGROSIM model family** 59
   Wilfried Mirschel and Karl-Otto Wenkel

7. **Crop simulation model of the second and the third productivity levels** 75
   Ratmir Aleksandrovich Poluektov and Vitaly Viktorovich Terleev

8. **The NDICEA model, a tool to improve nitrogen use efficiency in cropping systems** 91
   Geert-Jan H. M. van der Burgt, Gerard J. M. Oomen, A. S. J. Habets and Walter A. H. Rossing

9. Simulation of water and nitrogen flows on field scale; application of the SWAP–ANIMO model for the Müncheberg data set 111
   Joop Kroes and Jan Roelsma

10. Evaluation of water and nutrient dynamics in soil–crop systems using the eco-hydrological catchment model SWIM 129
    Joachim Post, Anja Habeck, Fred Hattermann, Valentina Krysanova, Frank Wechsung and Felicitas Suckow

11. Modelling nitrogen dynamics in soil–crop systems with HERMES 147
    Kurt Christian Kersebaum

12. Calibration and validation of CERES model for simulating water and nutrients in Germany 161
    Ajeet Singh Nain and Kurt Christian Kersebaum

13. The impact of crop growth model choice on the simulated water and nitrogen balances 183
    Eckart Priesack, Sebastian Gayler and Hans P. Hartmann

14. Simulating trends in crop yield and soil carbon in a long-term experiment – effects of rising CO_2, N deposition and improved cultivation 197
    Jørgen Berntsen, Bjørn Molt Petersen and Jorgen E. Olesen

15. Comparison of methods for the estimation of inert carbon suitable for initialisation of the CANDY model 209
    Martina Puhlmann, Katrin Kuka and Uwe Franko

16. Müncheberg field trial data set for agro-ecosystem model validation 219
    Wilfried Mirschel, Karl-Otto Wenkel, Martin Wegehenkel, Kurt Christian Kersebaum, Uwe Schindler and Jens-Martin Hecker

17. Dynamics of water, carbon and nitrogen in an agricultural used Chernozem soil in Central Germany 245
    Uwe Franko, Martina Puhlmann, Katrin Kuka, Frank Böhme and Ines Merbach

18. The lysimeter station at Berlin-Dahlem 259
    Heiko Diestel, Thomas Zenker, Reinhild Schwartengraeber and Marco Schmidt

Index 267

Preface

Soil–crop–atmosphere interactions play a central role in the multiple functions of rural landscapes. Agricultural, environmental and economic aspects are related to this topic, and there is an increasing need to understand the complex system to develop reliable models for scenario analyses. Agro-ecosystem models are more and more used to support decision-making on different scales towards a sustainable land use and management. Nevertheless, the increasing demand of model users for model validation does not fit to the decrease of research budgets for suitable experimental research and monitoring. The increasing family of modellers is confronted with a decrease of available data for model testing.

Model workshops providing common data sets for a number of modellers are not new, but became rare during the last years. Therefore, the Leibniz Centre of Agricultural Landscape Research (ZALF) in Müncheberg/Germany organized a workshop on "Modelling water and nutrient dynamics in crop–soil systems" providing data sets from its experimental field. This was accomplished by data from the Centre for Environmental Research Leipzig-Halle (UfZ) which provided data of its long-term experiment in Bad Lauchstädt, and the Technical University of Berlin which contributed data of their lysimeter station in Berlin-Dahlem. The purpose of the workshop was to compare different agro-ecosystem models on the basis of a common data set using meteorological data as main driving forces for water and nutrient dynamics including crop growth. This idea agrees well with the interest of agro-meteorologists, who look for new applications and for customers of their products. Therefore, the workshop was gratefully supported by the COST Action 718 "Meteorological Applications in Agriculture" under the umbrella of the European Science Foundation (ESF). In June 2004, the workshop was held in Müncheberg with the participation of 38 scientists from nine different countries. Twenty presentations were given for the blocks experimental site description, water dynamic modelling, soil and crop interactions, nutrient and water dynamics in soil–crop systems and long-term nutrient and carbon dynamics.

The organizers wish to acknowledge the financial contribution of the European Science Foundation (ESF) and COST 718. We also thank our director Prof. H. Wiggering and the head of our department Prof. K.-O. Wenkel for their support and our technicians for their help in the organization.

Finally, we want to dedicate this conference proceeding book to our colleague Ernst-Walter Reiche, who participated in the workshop, but died unexpectedly in March 2005. We will remember him as an inspiring and competent colleague and as a nice and helpful fellow. We will miss him very much in the future.

K.C. Kersebaum
J.-M. Hecker
W. Mirschel
M. Wegehenkel

CHAPTER ONE

Modelling water and nutrient dynamics in soil–crop systems: a comparison of simulation models applied on common data sets

Kurt Christian Kersebaum[1*], Jens-Martin Hecker[2], Wilfried Mirschel[1] and Martin Wegehenkel[1]

Abstract The paper summarizes the results of various simulation models applied to common data sets from Germany given for the workshop "Modelling water and nutrient dynamics in crop–soil systems", held in Müncheberg, Germany in June 2004. A brief overview is given on the main characteristics of the participating 18 models. Results of the models are compared with observed data of selected plots using standard indices for model performance evaluation. Results of such comparisons have to be seen carefully keeping in mind that model agreement is often dependent on the individual extend of model calibration on a specific data set. More insight about the differences between models is gained by comparing the model output for different processes. Although, model agreement to different state variables is similar for some models, there are distinct differences in the quantification of single processes.

Keywords Crop growth, Model comparison, Performance indicators, Soil processes

Kurt Christian Kersebaum, Jens-Martin Hecker,
Wilfried Mirschel and Martin Wegehenkel
[1]Institute of Landscape Systems Analysis, Leibniz-Centre for Agricultural Landscape Research (ZALF) e. V. Müncheberg, Eberswalder Strasse 84, D-15374 Müncheberg, Germany
[2]Institute of Socio-Economics, Leibniz-Centre for Agricultural Landscape Research (ZALF) e. V. Müncheberg, Eberswalder Strasse 84, D-15374 Müncheberg, Germany.
*Author for correspondence:
tel.: +49 33432 82394; fax: +49 33432 82334;
e-mail: ckersebaum@zalf.de

Introduction

Agro-ecosystem models are more and more applied to support decision-making at different scales. Such applications can range from fertilizer recommendations for farmers on a field scale and water suppliers on a catchment scale up to a landscape or regional scale for strategic policy decision support. With an increasing size of the investigation area, input data are more and more scarce and uncertain and data of relevant state variables for calibration are not available. This requires robust models, which are able to generate their input requirements from basic standardized soil data without the necessity of parameter calibration.

On the other hand, there is an increasing demand of model users and decision-makers that the validity of models used for decision support has been proved comparing uncalibrated modelling results with field observations outside the range of model development. This should enhance the reliability of model calculations used as the basis for decision-making. However, the question of model validation is contradictorily discussed in literature. Oreskes et al. (1994) generally disavowed the possibility to validate models for natural systems arguing that natural systems are generally open systems with always uncertain inputs. Rykiel (1996) concluded that validation is neither a method to test model theory nor a certificate of truth. He mentioned that validation is just an indication for a model to give acceptable results for a specific purpose. He also stated that this might not be scientifically necessary, but leads to a higher confidence and acceptance of model results. Holling (1978) proposed to investigate the non-validity of models applying them without calibration to a large number of sites and conditions accepting the risk not to reflect the observations very well. This is suitable to demonstrate the degree of confidence, respectively the limitations of a model approach.

Under this aspect, the Centre of Agricultural Landscape Research (ZALF) in Müncheberg, Germany organized a workshop on "Modelling water and nutrient dynamics in soil–crop systems", which was held in June 2004, sponsored by the European Science Foundation under the umbrella of COST action 718 ("Meteorological Applications in Agriculture"). To compare different models, we followed the idea to provide common data sets from different sites to the participants. This has been previously realized, e.g. by Groot and Verberne (1991) and McVoy et al. (1995) for soil–crop interactions and Powlson et al. (1996) for long-term carbon dynamics. The data sets provided are focused on different subjects: the data set of three plots from the Müncheberg experimental site (Mirschel et al. 2007) and the cropped and fallow plots of Bad Lauchstädt (Franko et al. 2007) are aimed to look at model performance for short- to medium-term water and nitrogen dynamics in soils and on crop growth. The long-term experimental data set of Bad Lauchstädt (Franko et al. 2007) mainly aims at the long-term carbon dynamics under different nutrient management and the data from the Berlin-Dahlem lysimeters (Diestel et al. 2007) provide data on soil water balances under grass cover.

Due to the fact that we only received few results from uncalibrated model runs, we decided to summarize more or less adopted models within this paper. Similar comparisons have been published by de Willigen (1991) and Diekkrüger et al. (1995) for soil–crop systems.

All in all, 18 different model approaches have been applied on individually selected data sets resulting in 14 papers of this issue. Results of 14 model approaches were submitted for the comparison on plot 1 of Müncheberg. Some modellers submitted also results for the other plots and sites. Only for the Berlin-Dahlem data set, we received only simulation results of one model. Therefore, a comparison of this data set was not possible within this paper.

Model characteristics

Table 1 gives an overview of the 18 different model approaches, which were applied to at least one data set of the workshop. More information about these models can be obtained from the main references in Table 1 or from the papers included in this issue. Although the application of the model WASMOD (Reiche 1994) is not documented in an own paper due to the tragic death of Ernst-Walter Reiche, we include the model for comparison on the basis of his uncalibrated results, which were provided at the workshop.

Tables 2–4 give an overview on some general characteristics of the models (Table 2) and the processes and approaches, which are considered for soil water dynamics (Table 3), nitrogen (and C) dynamics (Table 4) and crop growth (Table 2), respectively. Some models (AMBAV, SIMWASER and THESEUS) focus mainly on soil water dynamics. They use potential driven approaches and more physically based evaporation formula. Nevertheless, some models (THESEUS, Expert-N, SWAP) are designed as a toolkit and users have the choice between different approaches, e.g. for soil water, evapotranspiration and crop growth. Two models (FASSET, CANDY) combine a capacity approach with pore space fractionation for mobile and immobile water.

Most of the models consider crop cover, but model approaches range from empirical functions (AMBAV, NDICEA) over simplified temperature-driven approaches (SWIM, WASMOD, CANDY, OPUS) to more or less complex dynamic models including photosynthesis, biomass partitioning and root development. Some of these approaches are generic (SWAP and THESEUS [option for WOFOST], Expert-N, HERMES, STAMINA, AGROTOOL, FASSET), others use different submodels for different crops (CERES, AGROSIM). Depending on their capabilities to simulate multiple years and crops, models were run continuously through the whole crop rotation or were started separately for every application year (CERES, AGROSIM).

Nitrogen is considered by 13 model approaches. While AGROSIM uses a simple nitrogen balance

Table 1 List of participating models

Model code	Model name	Model reference	Results	Paper (this issue)
1	AMBAV	Löpmeier (1994)	–	Friesland & Löpmeier
2	SIMWASER	Stenitzer & Murer (2003)	B	Stenitzer et al.
3	THESEUS	Wegehenkel (2000)	L	Wegehenkel & Mirschel
4	OPUS	Smith (1992)	L	Wegehenkel & Mirschel
5	STAMINA	Acutis et al. (2007)	M	Acutis et al.
6	AGROSIM	Mirschel et al. (2001)	B, M	Mirschel & Wenkel
7	AGROTOOL	Poluektov et al. (2002)	B, L, M	Poluektov & Terleev
8	NDICEA	Koopmans & Bokhorst (2002)	M	Van der Burght et al.
9	SWAP/ANIMO	Kroes & van Dam (2003); Renaud et al. (2004)	M	Kroes & Roelsma
10	SWIM	Krysanova et al. (1998)	M, L	Post et al.
11	HERMES	Kersebaum (1995)	M, L	Kersebaum
12	WASMOD	Reiche (1994)	M	–
13	CERES	Ritchie & Otter (1985)	M	Nain & Kersebaum
14	Expert-N (CERES)	Engel & Priesack (1993)	M, L	Priesack et al.
15	Expert-N (SPASS)	Engel & Priesack (1993)	M, L	Priesack et al.
16	Expert-N (Sucros)	Engel & Priesack (1993)	M, L	Priesack et al.
17	FASSET	Berntsen et al. (2003)	M, L	Berntsen & Petersen
18	CANDY	Franko et al. (1995)	M, L	Puhlmann et al.

B = Berlin-Dahlem, L = Bad Lauchstädt, M = Müncheberg

Table 2 Overview on general characteristics of participating models

Model code	Model name	Timeline	Timestep	Scale	Type	Simulates*	Crop model
1	AMBAV	Cont.	Day	–	1D	w	–
2	SIMWASER	Cont.	Day	Field	1D	w, p	Simple
3	THESEUS	Cont.	Day	Field–meso	1D	w, p	Opt. simple/dynamic
4	OPUS	Cont.	Day	Field	1D	w, n, c, p	Dynamic
5	STAMINA	Cont.	Hour–day	Field	1D	w, p	Generic, dynamic
6	AGROSIM	Discon.	Day	Field	1D	w, p	Crop-specific, dynamic
7	AGROTOOL	Discon.	Day	Field	1D	w, p	Generic, dynamic
8	NDICEA	Cont.	Week	Field	1D	w, n, c	Simple
9	SWAP/ANIMO	Cont.	Day	Field	1D	w, n, c, p	Opt. simple/dynamic
10	SWIM	Cont.	Day	River basin	2D	w, n, c, p	Simple
11	HERMES	Cont.	Day	Field–meso	1D	w, n, p	Opt. simple/dynamic
12	WASMOD	Cont.	Day	Catchment	2D	w, n, c	Simple
13	CERES	Discon.	Day	Field	1D	w, n, p	Crop-specific, dynamic
14	Expert-N (CERES)	Cont.	Day	Field	1D	w, n, c, p	Crop-specific, dynamic
15	Expert-N (SPASS)	Cont.	Day	Field	1D	w, n, c, p	Dynamic
16	Expert-N (SUCROS)	Cont.	Day	Field	1D	w, n, c, p	Dynamic
17	FASSET	Cont.	Day	Farm	1D	w, n, c, p	Generic, dynamic
18	CANDY	Cont.	Day	Field	1D	w, n, c	Simple

Cont.: continuous; discont.: discontinuous; opt.: optional

*Model simulates – w: water; n: nitrogen; c: carbon dynamics, p: plant growth

approach and a zero-order mineralization kinetic, HERMES describes net mineralization of nitrogen from two pools using first-order kinetics. To consider net immobilization, some models use simply C/N ratios of added organic substances (CERES, SWIM), while all the others simulate C and N turnover explicitly. Soil moisture and temperature are the main driving factors for nitrogen and carbon turnover. Denitrification is considered by nearly all models, which simulate nitrogen except AGROSIM.

Table 3 Overview on characteristics of participating models for water dynamics

Model code	Model name	Type	Potential evapotranspiration	Specific
1	AMBAV	Richards	Penman–Monteith	
2	SIMWASER	Richards	Penman–Monteith	
3	THESEUS	Richards	Penman	
4	OPUS	Richards	Penman	
5	STAMINA	Capacity	Micrometeorological approach	
6	AGROSIM	Capacity	Wendling	
7	AGROTOOL	Richards	Penman–Monteith	
8	NDICEA	Capacity	Makkink	2 layers
9	SWAP/ ANIMO	Richards	Penman–Monteith	
10	SWIM	Capacity	Priestley–Taylor	
11	HERMES	Capacity	Opt. Turc-Wendling/Haude	
12	WASMOD	Capacity	Haude	
13	CERES	Capacity	Priestley–Taylor	
14	Expert-N (CERES)	Richards	Penman–Monteith	
15	Expert-N (SPASS)	Richards	Penman–Monteith	
16	Expert-N (SUCROS)	Richards	Penman–Monteith	
17	FASSET	Capacity (mobile/immobile)	Makkink	
18	CANDY	Capacity (mobile/immobile)	Wendling	

Opt.: optional

Table 4 Overview on characteristics of participating models for N dynamics

Model code	Model name	Type	Processes*	Specific
4	OPUS	C/N	MIDNFV	
6	AGROSIM	N	M	
8	NDICEA	C/N	MDF	Organic farming + phosphorus
9	SWAP/ANIMO	C/N	MIDNFV	
10	SWIM	(C)/N	MIDN	
11	HERMES	N	MDNV	
12	WASMOD	C/N	MIDNV	
13	CERES	(C)/N	MIDN	
14	Expert-N (CERES)	C/N	MIDNV	
15	Expert-N (SPASS)	C/N	MIDNV	
16	Expert-N (SUCROS)	C/N	MIDNV	
17	FASSET	C/N	MIDNV	
18	CANDY	C/N	MIDN	

*M: mineralization; I: immobilization; D: denitrification; N: Nitrification; F: N_2 fixation; V: ammonium volatilization

Model results and performance

Figures 1–4 give an overview of the scatter plots of the four most relevant measured state variables against the corresponding model outputs for Müncheberg plot 1. Moreover, scatter plots showing the relations of model results obtained from different models can be seen. Figure 1 shows the scatter matrix of measured gravimetric and simulated soil water contents in 0–90 cm. The base row represents the scatter plots of the model outputs against the observed data. The second row, for example, shows the scatter plots of the results of the model HERMES against the results obtained from the other models. Model results differ from each other, but some linear relations between the different models can be observed. Differences occur mainly through different soil parameterization and evapotranspiration formula. The closest relation exists between the different versions of the Expert-N modelling package, where differences were only generated through the use of different crop growth models. Capacity and potential driven approaches show similar pictures.

Figure 2 is organized in the same way as Fig. 1 and displays the relations for the aboveground biomass.

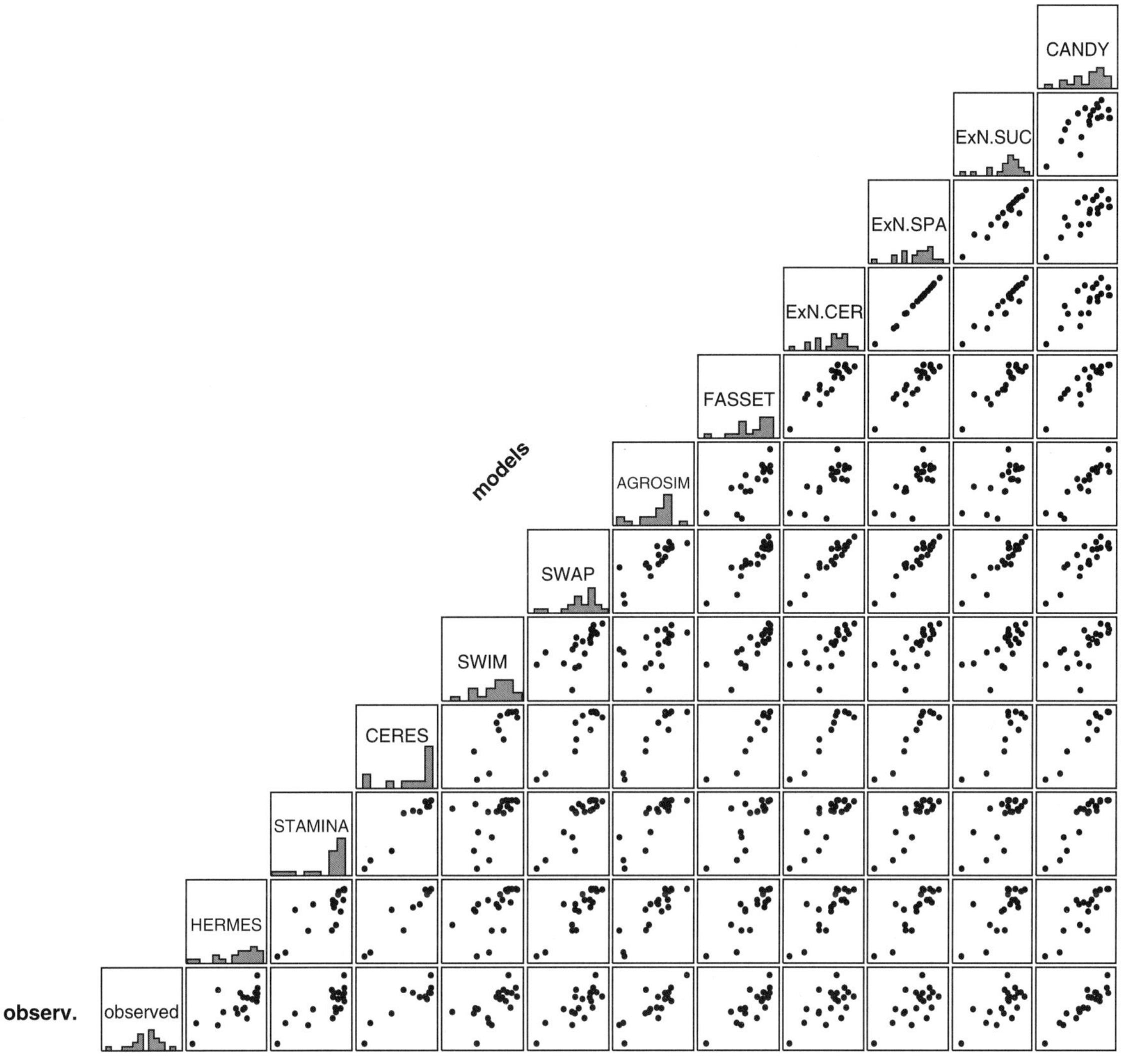

Fig. 1 Water contents in 0–90 cm of Müncheberg plot 1: scatter matrix on the relation between different models and gravimetric observations (ExN.xxx = Expert-N variants)

Here, we find more differentiated pictures. Similarities between models, e.g. between AGROSIM and FASSET or HERMES and CERES, can be observed, while surprisingly CERES and EXPERT-N-CERES are less similar. Although, different crop growth models were used within the EXPERT-N model family, the common basis led to close relationships between these approaches. This demonstrates the relative strong influence of the common basic submodels describing soil processes.

N uptake was simulated in a more variable way by the models because it is influenced directly from both, the crop growth approach and the soil nitrogen approach. Nevertheless, linear relations can be observed between HERMES, CANDY, SWAP, FASSET and AGROSIM, while the EXPERT-N approaches and SWIM show only minor relations to other models and to the observations (Fig. 3).

Figure 4 demonstrates that soil mineral nitrogen shows the poorest relation between models and observations. Clear linear relations can only be identified for FASSET and CANDY, which also show a linear relation to each other. The high spatial and temporal variability of soil mineral nitrogen and the multiple processes and interactions influencing the soil nitrogen dynamic make it difficult to achieve

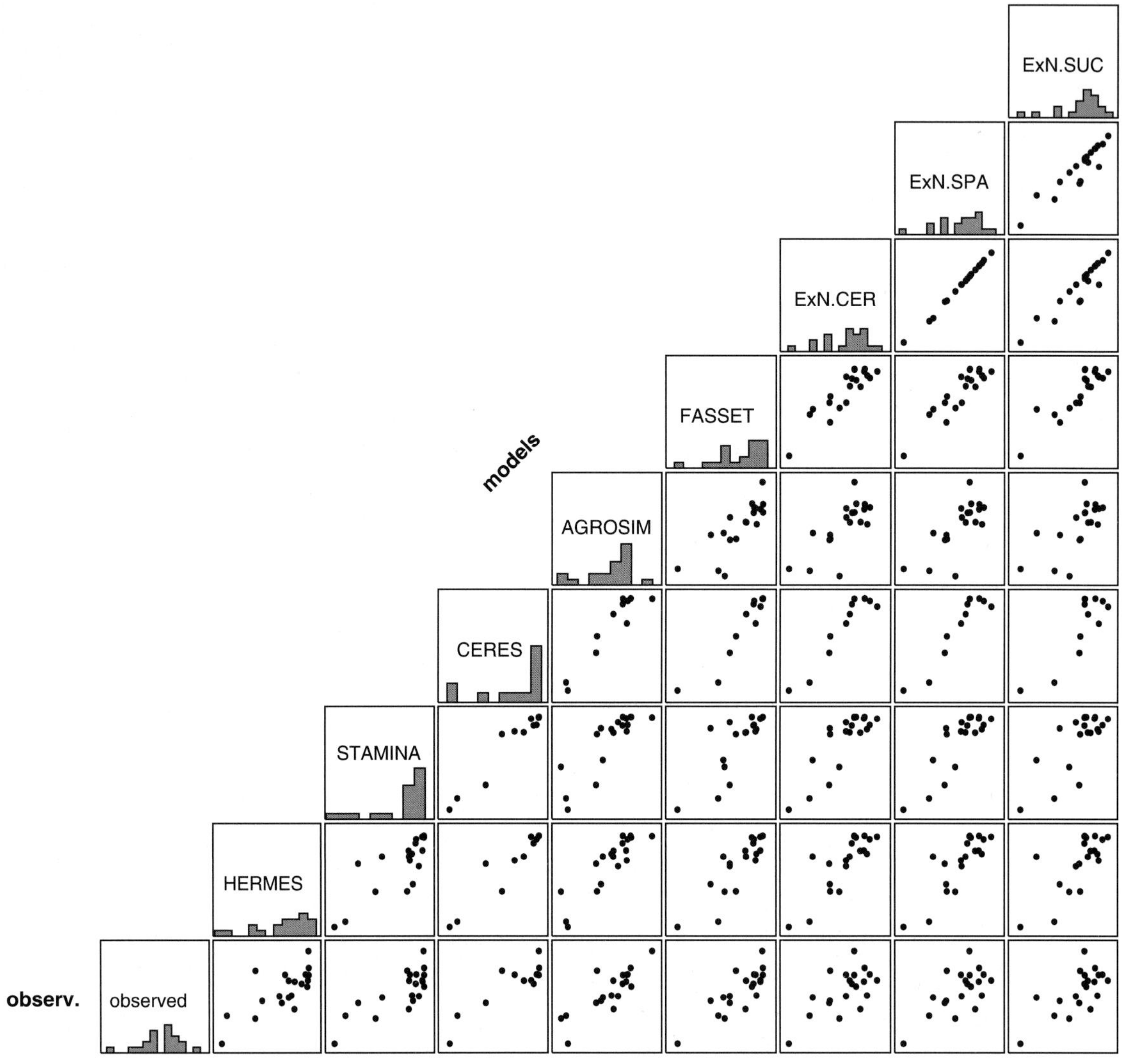

Fig. 2 Aboveground biomass production of Müncheberg plot 1: scatter matrix on the relation between different models and observations (ExN.xxx = Expert-N variants)

good agreements and still seem to be a challenge for models.

Comparison of processes

The performance of different models, which will be described below, is only one part of the comparison. Moreover, it is interesting to compare the output for different processes to analyse and demonstrate differences between the models. Therefore, we collected simulated daily rates for various processes for Müncheberg plot 1 and compared cumulative values of various models for an identical 5-year period (1 April 1993–31 March 1998).

Figure 5 shows the main parts of the water balance, the cumulative actual evapotranspiration (including interception) and the cumulative percolation in 90 cm depth. For the evapotranspiration, simulated values ranged between 1793 (EXPERT-N-CERES) and 2729 mm (WASMOD). The majority of the models calculated values between 2400 and 2500 mm. Differences occurred not only from different evapotranspiration formula, but also from the correction of precipitation (Richter 1995), which was used by HERMES and

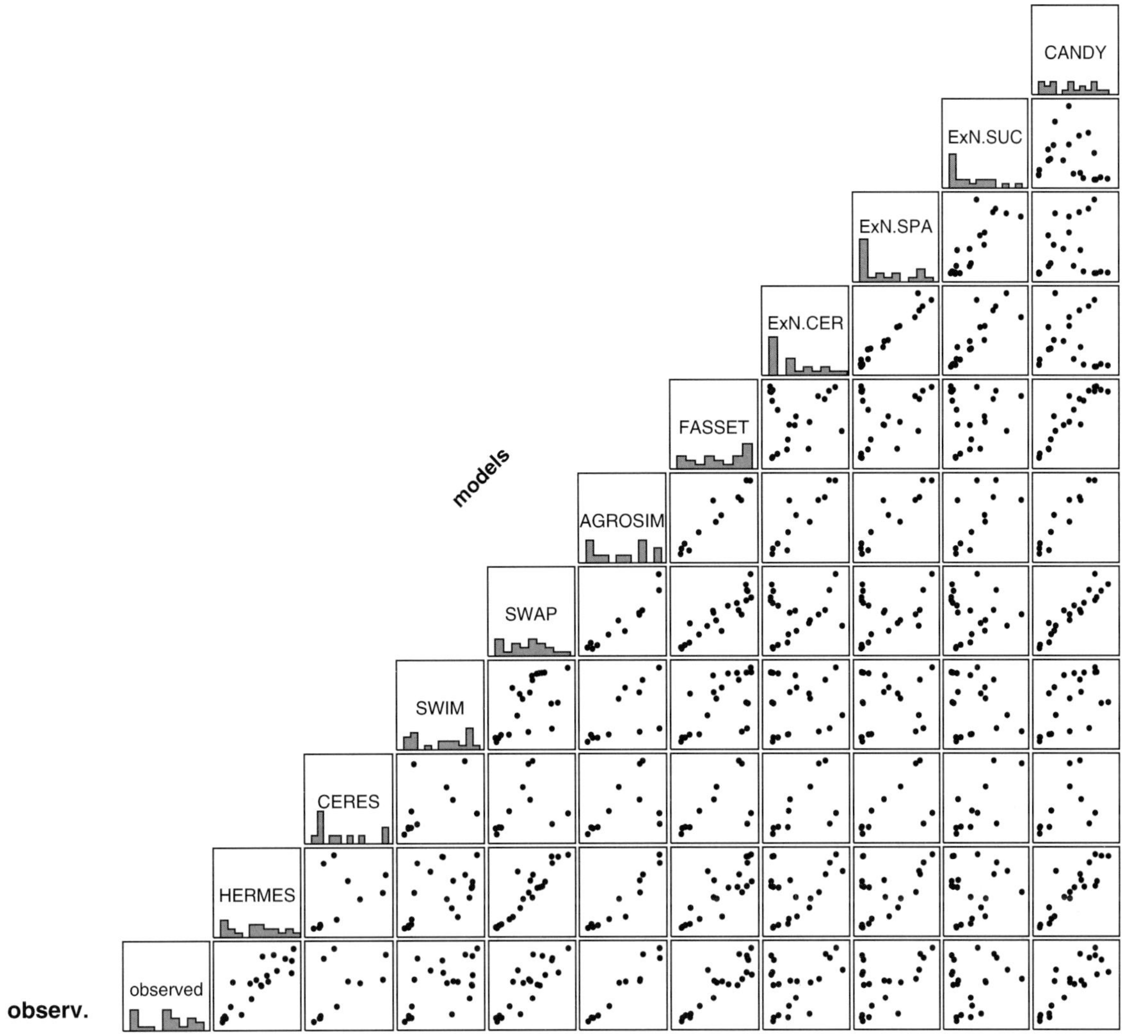

Fig. 3 N uptake by crops for Müncheberg plot 1: scatter matrix on the relation between different models and observations (ExN.xxx = Expert-N variants)

CANDY leading to an increase of about 12% for the precipitation sum, a higher percolation, and actual evapotranspiration. For the not corrected simulations, low actual evapotranspiration corresponded to high percolation (EXPERT-N variants and THESEUS). The overall range for percolation was between 500 (WASMOD) and 1114 mm (EXPERT-N-CERES).

Figure 6 shows some processes of the nitrogen dynamics accumulated for the time period between 1 April 1993 and 31 March 1998. There were large differences in the cumulative mineralization of nitrogen ranging from 98 (WASMOD) to 508 kg N ha^{-1} (HERMES). N uptake varied between 453 (SWIM) and 1159 kg N ha^{-1} (FASSET). Five models calculated values higher than 1000 kg N ha^{-1}. The nitrogen uptake is the only process, which can be roughly estimated (because not all crops were measured at harvest) from the measurements, which indicate an uptake of at least 800 kg N ha^{-1} during that period.

Denitrification plays a minor role on the sandy site of Müncheberg. Therefore, most of the models calculated only small amounts of gaseous nitrogen losses. Only CANDY showed extraordinary high losses of more than 50 kg N ha^{-1} during the simulation period. The different processes of water and nitrogen dynamics resulted in a wide range (27–189 kg N ha^{-1}) of simulated nitrogen leaching losses at the bottom of the root zone in 90 cm. The high amount simulated by

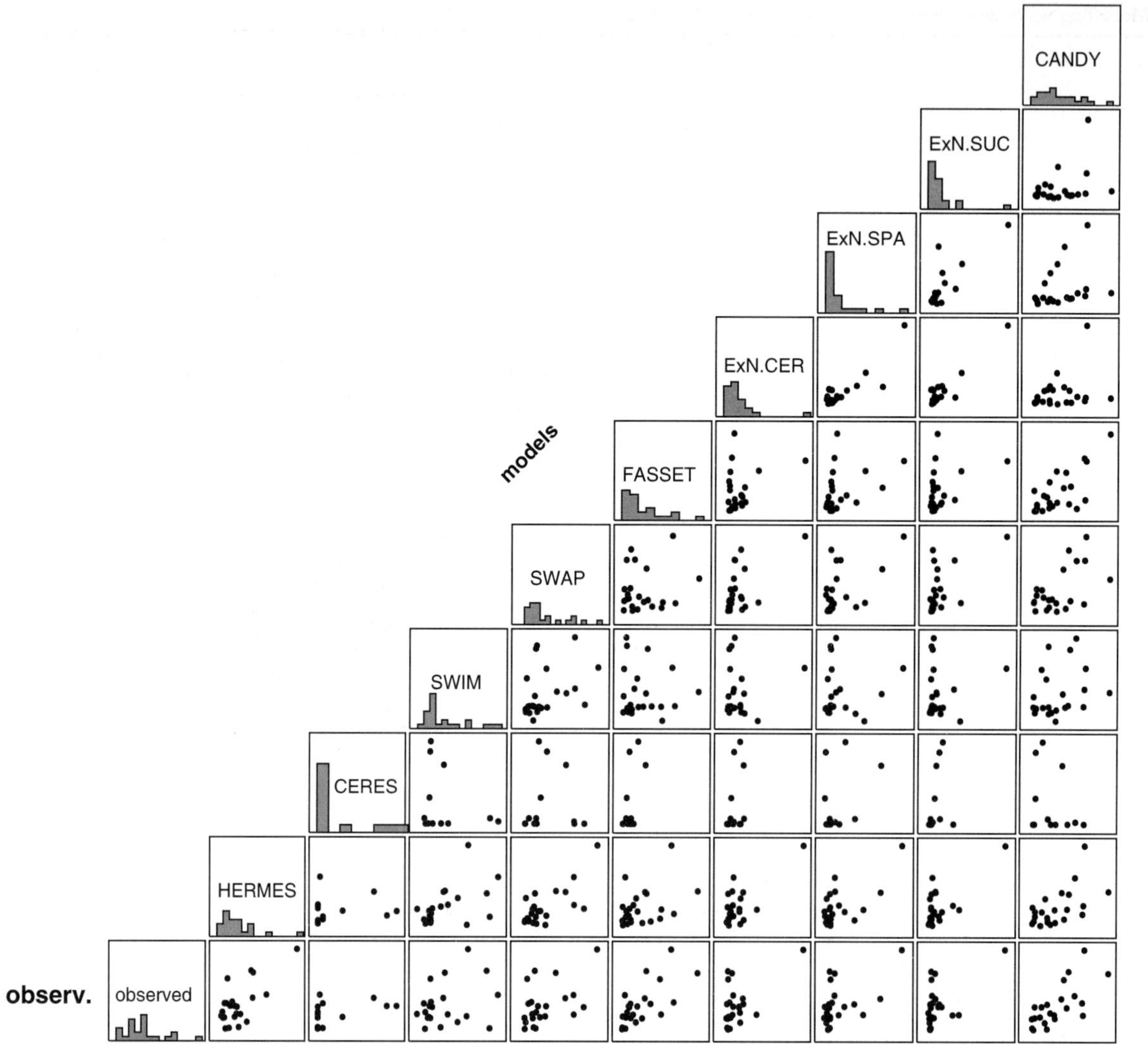

Fig. 4 Soil mineral nitrogen in 0–90 cm at Müncheberg plot 1: scatter matrix on the relation between different models and observations (ExN.xxx = Expert-N variants)

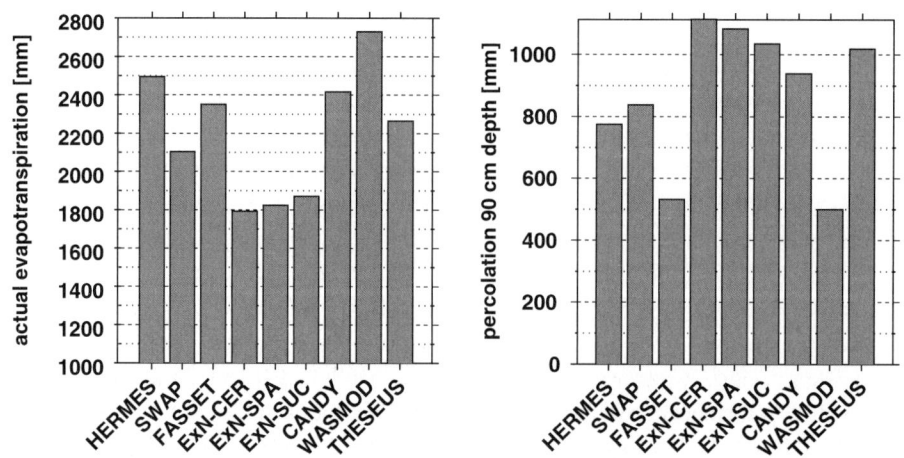

Fig. 5 Comparison of cumulative actual evapotranspiration and percolation in 90 cm depth of different models applied on Müncheberg plot 1 (period: 1.4.1993–31.3.1998)

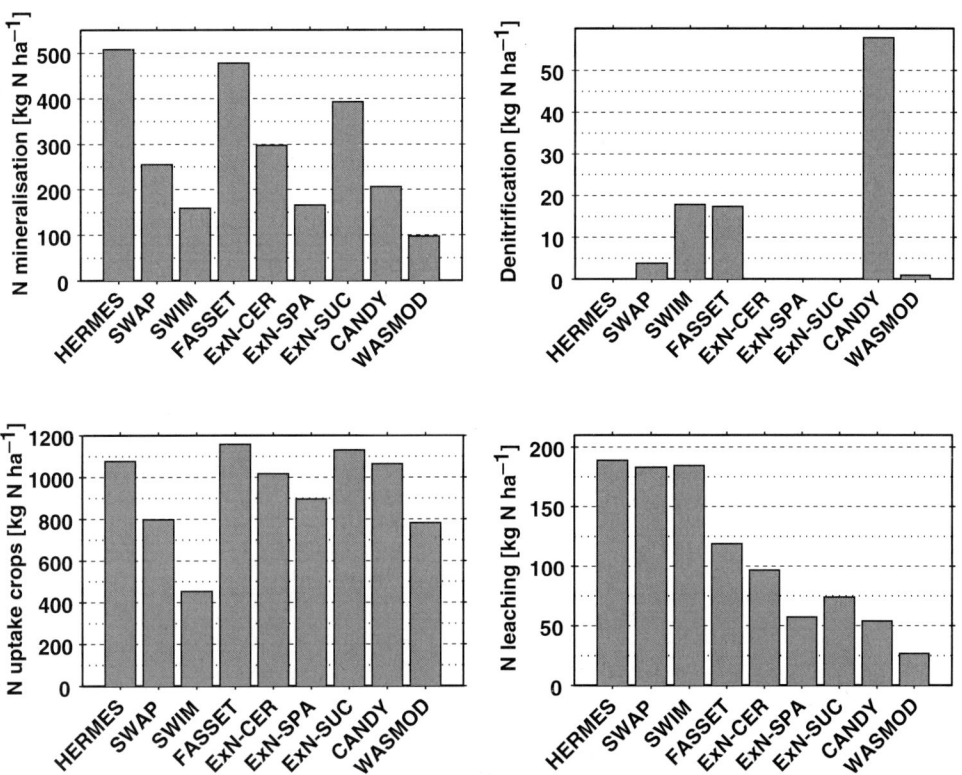

Fig. 6 Comparison of cumulative nitrogen mineralization, denitrification, nitrogen uptake by plants and nitrogen leaching in 90 cm depth of different models applied on Müncheberg plot 1 (period: 1.4.1993–31.3.1998)

HERMES was mainly a result of the high mineralization in combination with a relative high percolation, which was partly balanced through a high N uptake by crops. In contrast, the lowest simulation of nitrogen leaching by WASMOD corresponds well to the calculated low mineralization combined with low percolation. Nevertheless, other models are in the same order of magnitude forming two different groups of low (WASMOD, CANDY, Expert-N-SPASS, EXPERT-N-SUCROS) and high (HERMES, SWAP, SWIM) nitrogen leaching. FASSET and EXPERT-N-CERES are in between both groups.

Evaluation of model performance

In order to evaluate the model performance, we applied the following four statistical measures and indices based on the comparison of simulated with observed model outputs: the mean bias error (MBE; Addiscott and Whitmore 1987)

$$\text{MBE} = \sum_{i=1}^{n} \frac{S_i - O_i}{n} \quad (1)$$

the root mean square error (RMSE; Fox 1981):

$$\text{RMSE} = \sqrt{\frac{\sum_{i=1}^{n}(S_i - O_i)^2}{n}} \quad (2)$$

Index of agreement (IA) according to Willmott (1982)

$$\text{IA} = 1 - \frac{\sum_{i=1}^{n}(S_i - O_i)^2}{\sum_{i=1}^{n}\left(|S_i - \overline{O}| + |O_i - \overline{O}|\right)^2} \quad (3)$$

Modelling efficiency (ME) according to Nash and Sutcliffe (1970):

$$\text{ME} = 1 - \frac{\sum_{i=1}^{n}(S_i - O_i)^2}{\sum_{i=1}^{n}(O_i - \overline{O})^2} \quad (4)$$

where n is the number of samples, S_i and O_i are the simulated and the observed values, and $\bar{O}$ is the mean of the observed data.

The MBE considers positive and negative deviations, which make it suitable to indicate the bias of the model error, while RSME describes the average absolute deviation between simulated and observed values. The lower the RSME, the more accurate the simulation. The IA as an additional method for the evaluation of modelling performance results in a range between 0 and 1. The closer IA is to 1, the better the simulation quality, similar to the coefficient of determination. In contrast to IA, the ME also allows negative values and compares the deviation between simulated and observed state variables with the variance of the observed values.

The estimation of model performance was mainly concentrated to the data set obtained from Müncheberg plot 1. For this data set, most of the models provided their final result files.

The performance indices MBE, RMSE, IA, and ME for the modelling results calculated for plot 1 at the location Müncheberg are summarized in Figs. 7–11. For the soil water content of the rooting zone (millimetre in 0–90 cm), MBE ranged within −3.9 and 39 mm, RMSE between 14 and 45 mm, IA between 0.50 and 0.93, as well as ME within −3.1 and 0.66 (Fig. 7), whereas the best performance values were achieved by those models, which simulated only a limited part of the whole period.

Concerning the simulation quality of the aboveground biomass, MBE ranged within −1355 and 452

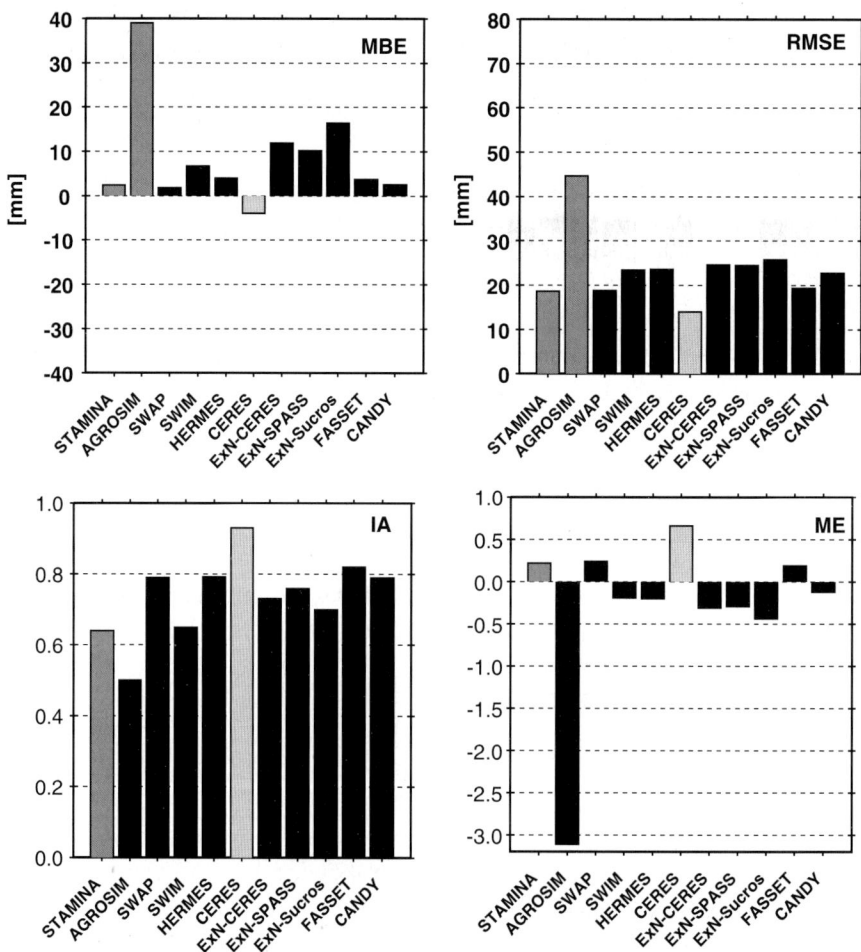

Fig. 7 Model performance indices for soil water content in 0–90 cm (against TDR measurements) of different models applied on Müncheberg plot 1 (dark grey bars: ~80% of values, light grey: ~50% of values)

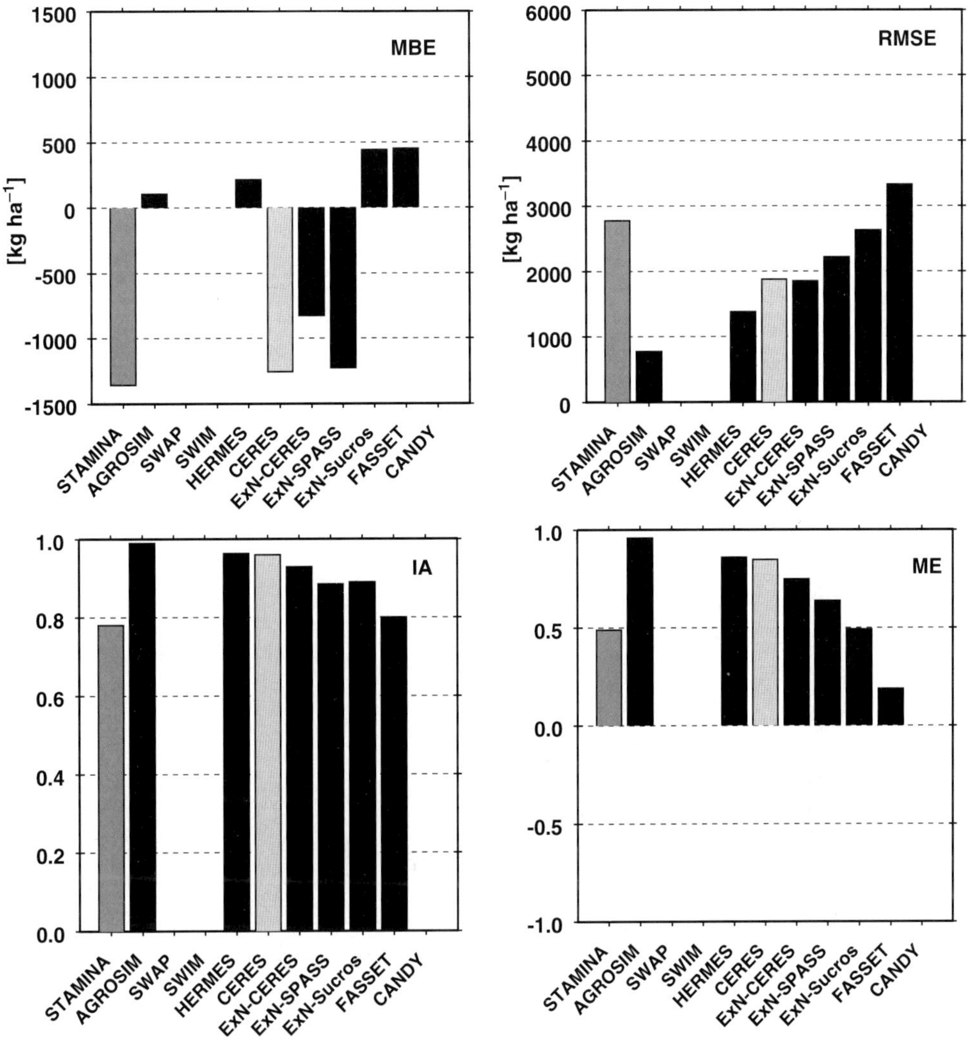

Fig. 8 Model performance indices for aboveground biomass of different models applied on Müncheberg plot 1 (dark grey bars: ~80% of values, light grey: ~50% of values)

kg ha^{-1}, RMSE between 773 and 3329 kg ha^{-1}, IA between 0.78 and 0.99, as well as ME within 0.19 and 0.96 (Fig. 8).

Taking into account only the simulated dry matter of the crop storage organ as an indicator for the performance of yield estimation, MBE ranged within −2613 and 1186 kg ha^{-1}, RMSE between 1015 and 2764 kg ha^{-1}, IA between 0.90 and 0.99, as well as ME within 0.66 and 0.95 (Fig. 9).

Looking at the simulated nitrogen uptake of crops, MBE ranged within −35 and 33 kg N ha^{-1}, RMSE between 25 and 77 kg N ha^{-1}, IA between 0.5 and 0.94, as well as ME within −1.1 and 0.78 (Fig. 10).

For the simulated soil mineral nitrogen contents (kg N ha^{-1} in 0–90 cm), MBE ranged from −26 to 7 kg N ha^{-1}, RMSE between 22 and 42 kg N ha^{-1}, IA between 0.56 and 0.83, as well as ME within −0.8 and 0.43 (Fig. 11). For nearly all models, the MBEs of nitrogen uptake by crops and soil mineral nitrogen content show the same bias for both outputs, except for HERMES, where the overestimation of nitrogen uptake corresponds to an underestimation of soil mineral nitrogen in the same order of magnitude.

The statistical measures and indices of performance for those models, which also provided data for the other data sets, are summarized in Table 5 for

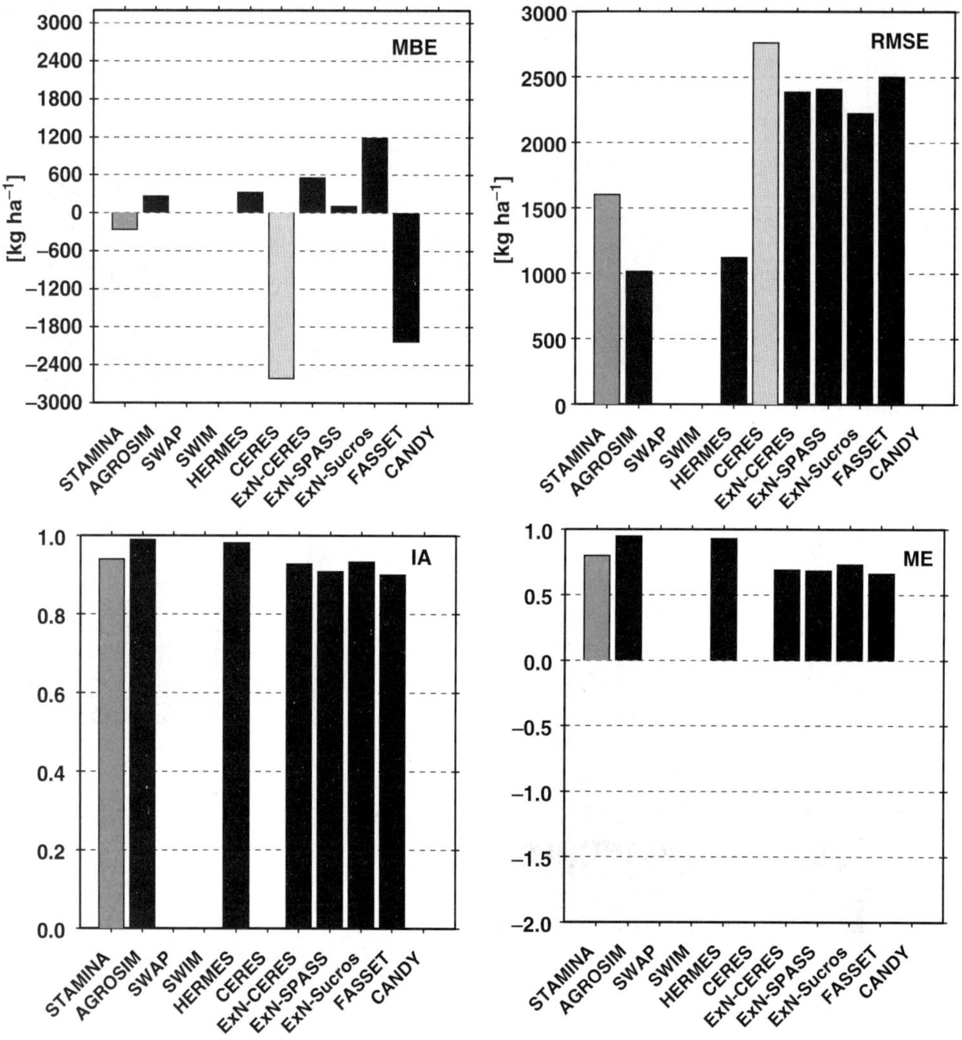

Fig. 9 Model performance indices for crop storage organ (DM) of different models applied on Müncheberg plot 1 (dark grey bars: ~80% of values, light grey: ~50% of values)

the Müncheberg plots 2 and 3, and in Table 6 for the Bad Lauchstädt short-term crop rotation. Especially, the performance indicator ME for the soil mineral nitrogen showed an interesting effect on plot 2 in Müncheberg and for the Bad Lauchstädt crop rotation plot. On Müncheberg plot 2, the temporal variation of soil mineral nitrogen was low due to the limited nitrogen supply by organic fertilizer. On the Bad Lauchstädt plot, soil mineral nitrogen observations in 0–20 cm varied only by 24 kg N ha^{-1}, which is relatively small compared to the fluctuations indicated by the model results and the input and output of nitrogen by fertilization and crop uptake (Table 5). In both cases, the indicator showed negative values for all models because even small deviations were related to the very low fluctuation of the measurements (Table 5). Therefore, the coefficient ME can also be seen as an indication that the temporal resolution of measurements might be insufficient to reflect adequately the temporal dynamic of a state variable.

Taking into account the broad range of the results, it is obvious that each model has its specific range of application and therefore, its own strengths and weaknesses. For example, the model AGROSIM showed the strongest performance concerning the simulation quality of aboveground biomass, storage organs and N uptake by crops taking into account all four statistical quality measures (Figs. 8–10). In contrast to that, the

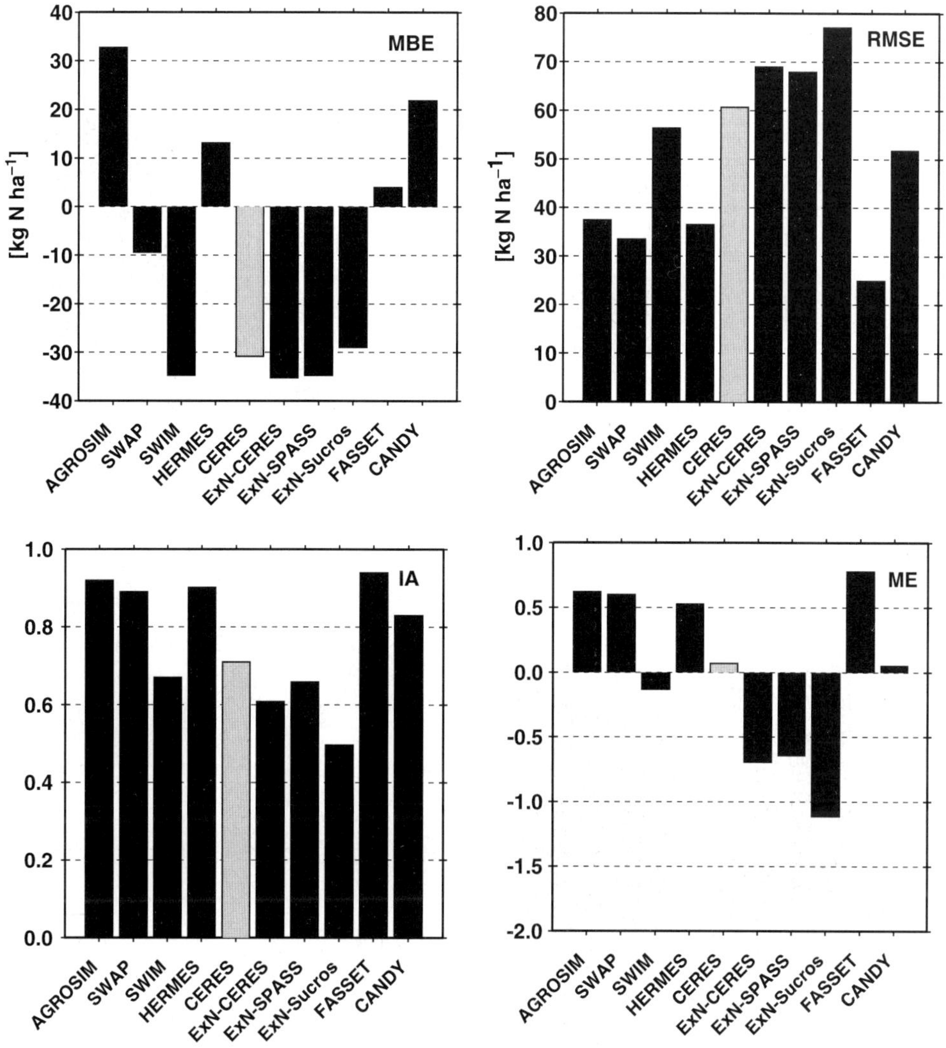

Fig. 10 Model performance indices for nitrogen in aboveground biomass of different models applied on Müncheberg plot 1 (dark grey bars: ~80% of values, light grey: ~50% of values)

simulation quality of soil water contents calculated by AGROSIM was low (Fig. 7). The strongest performance concerning soil mineral nitrogen content could be observed for the models CANDY, HERMES and NDICEA (Fig. 11). The latter one uses a model internal optimization procedure for parameters and crop yield data as input, which reduces the probability of errors for soil mineral nitrogen. The other two models achieved a similar result for soil mineral nitrogen in a very different way, which was shown in the comparison of processes (see Fig. 6).

The FASSET model performs well in N uptake, soil mineral nitrogen, and water, but was less successful for crop biomass and storage organ. For the soil water contents, the model CERES showed the best performance (Fig. 7). However, CERES was only applied for cereal crop growth periods from seeding to harvest date (Fig. 7). Therefore, the comparison of the model performance indicators was restricted due to the following reasons:

- Some models ran over the whole time period from 1993 to 1998 continuously, while others were reinitialized every year using the first observation of the year. Some models were restricted to cereals and left out years with other crops and/or bare soil periods.

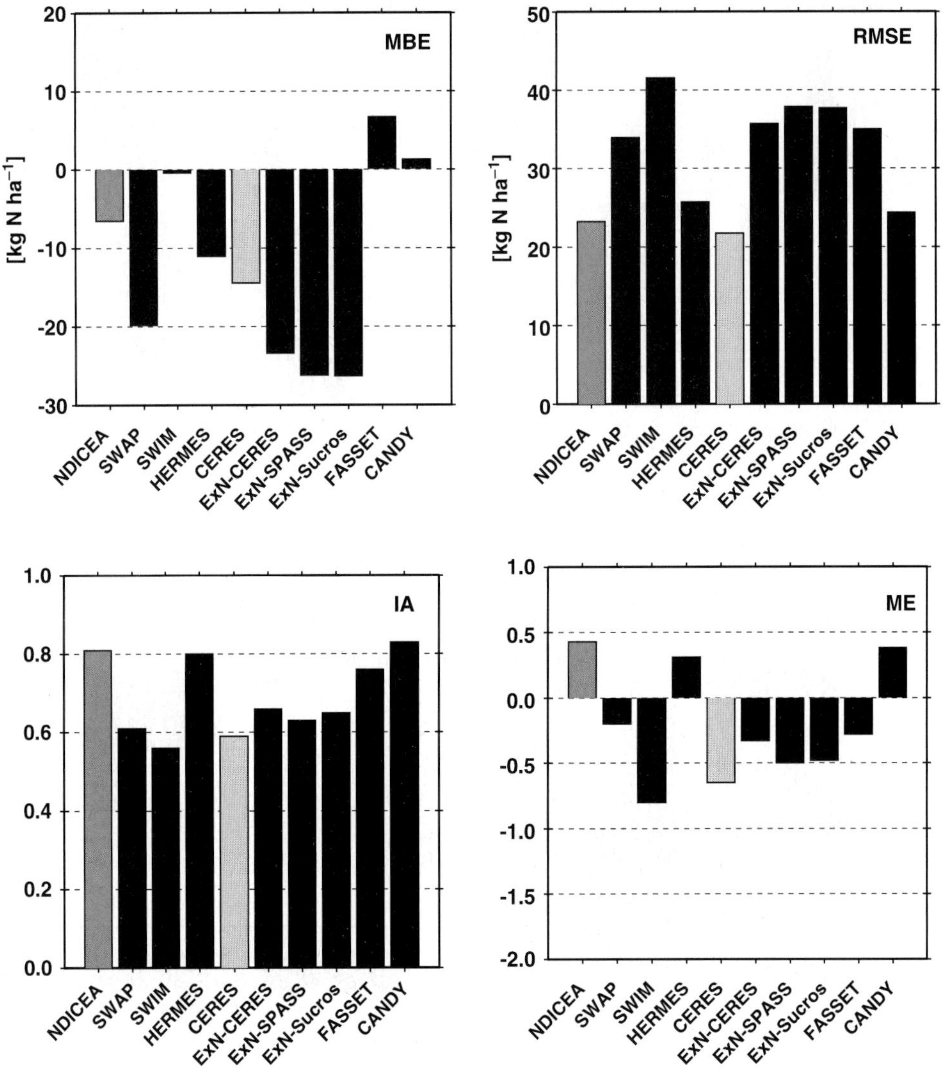

Fig. 11 Model performance indices for soil mineral nitrogen content in 0–90 cm of different models applied on Müncheberg plot 1 (dark grey bars: ~80% of values, light grey: ~50% of values)

- Efforts for model calibration were different. Some models used just basic information, like texture and organic matter content, and standard procedures or parameters without further calibration, others used more specified soil parameters given by the data providers. In some cases, models used their own optimized parameters from an internal or external calibration procedure.
- Some models used observed data directly as input, e.g. yield or N uptake data if a dynamic crop growth model was not included. Therefore, the full dynamic approaches were more at risk to fail also on other state variables, e.g. soil mineral nitrogen.

More details about the application, calibration and parameter estimation procedures are given in the specific model papers within this issue.

Conclusions

Applications of agro-ecosystem models on a field or regional scale are mostly characterized by a high uncertainty of input data, especially regarding soil and management information. This means that empirical relations have to be used to fill gaps and to create the required model input data. If applied on a specific

Table 5 Summary of model performance of participating models for water, soil mineral nitrogen (N_{min}), crop growth and N uptake by crops on Müncheberg plots 2 and 3

Model		Water (TDR) in 0–90 cm [mm]		N_{min} in 0–90 cm [kg N ha^{-1}]		Above ground biomass [kg ha^{-1}]		Storage organ [kg ha^{-1}]		N in crop [kg N ha^{-1}]	
		Plot 2	Plot 3	Plot 2	Plot 3	Plot 2	Plot 3	Plot 2	Plot 3	Plot 2	Plot 3
SWIM		−12.9	−6	−0.7	−8.4	–	–	–	–	–	–
HERMES		0.1	10.8	−8.3	3.5	291	−336	−1097	−1830	3.1	12.6
ExN-SPA	M	−1.3	−3.7	−11.6	−6.2	−758	−1157	−868	−438	−20.4	−31.8
ExN-CER	B	−2.8	−3.6	−11.5	−3.5	−356	−854	−232	67	−15.6	−30.6
ExN-SUC	E	4.7	1.0	−12.4	−7.8	406	434	−100	676	−17.0	−26.6
NDICEA		–	–	5.1	−8.4	–	–	–	–	–	–
SWIM		27.4	22.4	29.1	31.2	–	–	–	–	–	–
HERMES	R	26.6	24.6	19.1	25.8	1735	1424	2768	2965	31.2	50.4
ExN-SPA	M	26.9	20.7	22.2	27.3	1583	2149	1407	1914	44.2	56.4
ExN-CER	S	27.8	20.4	20.4	26.1	1697	1719	1740	2002	41.0	57.9
ExN-SUC	E	24.1	18.4	20.3	27.4	2327	2726	1483	1803	46.1	59.9
NDICEA		–	–	22.8	23.4	–	–	–	–	–	–
SWIM		0.60	0.61	0.52	0.57	–	–	–	–	–	–
HERMES		0.66	0.71	0.69	0.82	0.89	0.96	0.87	0.90	0.75	0.80
ExN-SPA	I	0.68	0.73	0.64	0.77	0.89	0.89	0.97	0.96	0.50	0.66
ExN-CER	A	0.67	0.72	0.69	0.80	0.89	0.94	0.96	0.97	0.46	0.60
ExN-SUC		0.70	0.76	0.70	0.80	0.83	0.88	0.97	0.97	0.37	0.56
NDICEA		–	–	0.65	0.83	–	–	–	–	–	–
SWIM		−0.28	−0.47	−1.85	−0.53	–	–	–	–	–	–
HERMES		−0.21	−0.78	−0.23	−0.04	0.52	0.85	0.61	0.68	−0.04	−0.23
ExN-SPA	M	−0.23	−0.26	−0.65	−0.17	0.60	0.66	0.90	0.87	−1.09	−0.53
ExN-CER	E	−0.32	−0.22	−0.40	−0.07	0.54	0.79	0.85	0.85	−0.79	−0.61
ExN-SUC		0.01	0.01	−0.39	−0.18	0.14	0.46	0.89	0.88	−1.26	−0.73
NDICEA		–	–	−0.76	0.14	–	–	–	–	–	–

site, the comparison of simulated with measured data often shows unsatisfactory results due to deviating conditions and parameters. This does not automatically mean that these parameters are wrong. It might also be that the site or the specific location of measurement is not representative for the whole field or soil map unit. The comparison of different models applied on the same data set is not suitable to serve as a model contest or to find the best model. Although, the application of different indices for model performance helps to identify strengths and weaknesses of each model, an objective comparison is nearly impossible due to different levels of input requirements, calibration efforts and last but not least the uncertainties and errors within the measured data themselves. Moreover, the analysis of model behaviour concerning different processes demonstrated differences between the approaches and discrepancies within the model results. The wide differences between different model approaches for various processes show that there is still a need for further research and model improvement, although model performance seems to be sufficiently good in some cases. The results indicate that the conclusions of the two previous workshops (de Willigen 1991; Diekkrüger et al. 1995) are still valid, that especially the microbial turnover processes of carbon and nitrogen and the interaction with crops are still not sufficiently understood. At last, the specific user has to decide which model is suitable for his specific needs and data availability.

Acknowledgements This contribution was supported by the German Federal Ministry of Consumer Protection, Food and Agriculture and the Ministry of Agriculture, Environmental Protection and Regional Planning of the Federal State of Brandenburg (Germany). We also thank the European Science Foundation and COST 718 for sponsoring the model workshop. A special thanks is given to the data providers from Bad Lauchstädt and Berlin-Dahlem and to the technical staff of the Müncheberg experimental stations for sampling and laboratory work and our colleagues Petra Eisermann, Rena Dühnelt, Karin Luzi, Michael Bähr and Heidi Wegehenkel helping to organize

Table 6 Summary of model performance of participating models for water, soil mineral nitrogen (N_{min}), crop growth and N uptake by crops on Bad Lauchstädt crop rotation plot. (italics = only growing periods of cereals)

		Water (FDR) [vol%]			N_{min} [kg N ha^{-1}]	Above ground biomass [kg ha^{-1}]	Storage organ [kg ha^{-1}]	N in crop [kg N ha^{-1}]
Depth (cm)		5	45	90	0–20			
Model								
AGROSIM		–	4.9	6.3	–	–87	148	24.8
FASSET	M	–9.9	4.4	1.5	5.8	1031	–2419	–4.4
HERMES	B	8.3	4.4	–2.1	21.3	736	–830	30.1
SWIM	E	–6.3	0.9	–1.5	–2.2	–	–	–
THESEUS		4.9	5.3	6.3	–	–401	–	–
OPUS		–1.3	–5.0	–2.2	–14.4	–1379	–	–7.4
AGROSIM		–	5.4	8.1	–	660	414	36.4
FASSET	R	11.0	6.6	4.7	12.7	3979	3775	24.2
HERMES	M	10.5	9.1	4.8	27.3	1735	2854	63.4
SWIM	S	8.9	5.6	5.3	6.4	–	–	–
THESEUS	E	7.5	7.6	7.1	–	4123	–	–
OPUS		6.5	7.5	4.7	20.7	4376	–	41.6
AGROSIM		–	0.64	0.13	–	0.99	1.00	0.51
FASSET		0.25	0.55	0.74	0.75	0.66	0.88	0.91
HERMES	I	0.52	0.50	0.83	0.45	0.93	0.94	0.71
SWIM	A	0.39	0.54	0.81	0.72	–	–	–
THESEUS		0.50	0.46	0.47	–	0.89	–	–
OPUS		0.64	0.34	0.51	0.16	0.83	–	0.57
AGROSIM		–	–2.07	–1.70	–	0.96	1.00	–2.56
FASSET		–3.42	–3.35	0.06	–1.44	–1.00	0.69	0.62
HERMES	M	–3.03	–7.14	0.02	–10.17	0.62	0.83	–1.60
SWIM	E	–1.85	–2.06	–0.20	0.37	–	–	–
THESEUS		–1.03	–4.64	–1.19	–	0.59	–	–
OPUS		–0.53	–4.51	0.05	–5.43	0.59	–	–0.12

the workshop at Müncheberg. Finally, we thank all participants of the workshop for their kind cooperation especially providing their model results for this comparison.

References

Addiscott TM, Whitmore AP (1987) Computer simulation of changes in soil mineral nitrogen and crop nitrogen during autumn, winter and spring. J Agric Sci (Cambridge) 109:141–157

Acutis M, Rana G, Trevisiol P, Bechini L, Laudato M, Ferrara R, Richter GM (2007) Integrating a spatial micrometeorological model into the risk assessment for arable crops in hilly terrain. In: Kersebaum KC, Hecker J-M, Mirschel W, Wegehenkel M (eds) Modelling water and nutrient dynamics in soil-crop systems. Springer, Dordrecht, pp 51–57

Berntsen J, Petersen BM, Jacobsen BH, Olesen JE, Hutchings NJ (2003) Evaluating nitrogen taxation scenarios using the dynamic whole farm simulation model FASSET. Agr Syst 76:817–839

Diekkrüger B, Söndgerath D, Kersebaum KC, McVoy CW (1995) Validity of agroecosystem models – a comparison of results of different models applied to the same data set. Ecol Model 81:3–29

Diestel H, Zenker T, Schwartengräber R, Schmidt M (2007) The lysimeter station at Berlin-Dahlem. In: Kersebaum KC, Hecker J-M, Mirschel W, Wegehenkel M (eds) Modelling water and nutrient dynamics in soil-crop systems. Springer, Dordrecht, pp 259–266

Engel T, Priesack E (1993) Expert-N, a building block system of nitrogen models as resource for advice, research, water management and policy. In: Eijsackers HJP, Hamers T (eds) Integrated soil and sediment research: a basis for proper protection, Kluwer Academic, Dordrecht, The Netherlands, pp 503–507

Fox DG (1981) Judging air quality model performance: a summary of the AMS workshop on dispersion model performance. Bull Am Meteorol Soc 62:599–609

Franko U, Oelschlägel B, Schenk S (1995) Simulation of temperature-, water- and nitrogen dynamics using the model CANDY. Ecol Model 81:213–222

Franko U, Puhlmann M, Kuka K, Böhme F, Merbach I (2007) Dynamics of water, carbon and nitrogen in an agricultural used Chernozem soil in Central Germany. In: Kersebaum KC, Hecker J-M, Mirschel W, Wegehenkel M (eds)

Modelling water and nutrient dynamics in soil–crop systems. Springer, Dordrecht, pp 245–258

Groot JJR, Verberne ELJ (1991) Response of wheat to nitrogen fertilization, a data set to validate simulation models for nitrogen dynamics in crop and soil. Fert Res 27: 349–383

Holling CS (1978) Adaptive environmental assessment and management. Wiley, New York.

Kersebaum KC (1995) Application of a simple management model to simulate water and nitrogen dynamics. Ecol Model 81:145–156

Koopmans CJ, Bokhorst J (2002) Nitrogen mineralisation in organic farming systems: a test of the NDICEA model. Agronomie 22:855–862

Kroes JG, van Dam JC (eds) (2003) Reference Manual SWAP version 3.0.4. Wageningen, Alterra, Green World Research. Alterra-report 773. Wageningen, The Netherlands, 211 pp

Krysanova V, Müller-Wohlfeil D-I, Becker A (1998) Development and test of a spatially distributed hydrological/water quality model for mesoscale watersheds. Ecol Model 106:263–289

Löpmeier F-J (1994) The calculation of soil moisture and evapotranspiration with agrometeorological models (in German). Zeitschrift f Bewaesserungswirtschaft 29:157–167

McVoy CW, Kersebaum KC, Arning M, Kleeberg P, Othmer H, Schröder U (1995) A data set from north Germany for the validation of agroecosystem models: documentation and evaluation. Ecol Model 81:265–300

Mirschel W, Schultz A, Wenkel K-O (2001) Assessing the impact of land use intensity and climate change on ontogenesis, biomass production, and yield of Northeast German agro-landscapes. In: Tenhunen JD, Lenz R, Hantschel R (eds) Ecosystem approaches to landscape management in Central Europe, Ecological Studies, Vol. 147. Springer-Verlag, Berlin, Heidelberg, New York, pp 299–313

Mirschel W, Wenkel K-O, Wegehenkel M, Kersebaum KC, Schindler U, Hecker J-M (2007) Müncheberg field trial data set for agro-ecosystem model validation. In: Kersebaum KC, Hecker J-M, Mirschel W, Wegehenkel M (eds) Modelling water and nutrient dynamics in soil–crop systems. Springer, Dordrecht, pp 219–243

Nash JE, Sutcliffe IV (1970) Riverflow forcasting through conceptual model. J Hydrol 273:282–290

Oreskes N, Shrader-Fechette K, Belitz K (1994) Verification, validation and confirmation of numerical models in earth sciences. Science 263:641–646

Poluektov RA, Fintushal SM, Oparina IV, Shatskikh DV, Terleev VV, Zakharova ET (2002) Agrotool – a system for crop simulation. Arch Acker- Pfl Boden 48:609–635

Powlson DS, Smith P, Smith JU (eds) (1996) Evaluation of soil organic matter models using existing long-term datasets. NATO ASI series, Series I, Vol. 38. Springer-Verlag, Heidelberg, pp 237–246

Renaud LV, Roelsma J, Groenendijk P (2004) User's guide of the ANIMO 4.0 nutrient leaching model. Wageningen, Alterra Report 224. Wageningen, The Netherlands, 154 pp

Reiche E-W (1994) Modelling water and nitrogen dynamics on catchment scale. Ecol Model 75/76:371–384

Richter D (1995) Ergebnisse methodischer Untersuchungen zur Korrektur des systematischen Messfehlers des Hellmann-Niederschlagsmessers. Ber Dtsch Wetterd 159:93

Ritchie JT, Otter S (1985) Description and performance of CERES-Wheat: a user oriented wheat yield model. In: ARS Wheat Yield Project, ARS–38. National Technical Information Service, Springfield, VA, pp 159–175

Rykiel EJ jr (1996) Testing ecological models: the meaning of validation. Ecol Model 90:229–244

Smith R (1992) OPUS: an integrated simulation model for transport of non-point-source pollutants at field scale. Vol. 1, Documentation. USDA-ARS, ARS-98. USDA, Washington, DC

Stenitzer E, Murer E (2003) Impact of soil compaction upon soil water balance and maize yield estimated by the SIMWASER model. Soil Till Res, 73:43–56

Wegehenkel M (2000) Test of a modelling system for simulating water balances and plant growth using various different complex approaches. Ecol Model 129:39–64

Willigen P de (1991) Nitrogen turnover in the soil–crop system; comparison of fourteen simulation models. Fert Res 27:141–149

Willmott CJ (1982) Some comments on the evaluation of model performance. Bull Am Meteorol Soc 64:1309–1313

CHAPTER TWO

The performance of the model AMBAV for evapotranspiration and soil moisture on Müncheberg data

Hans Friesland and Franz-Josef Löpmeier

Abstract A model for the calculation of evapotranspiration and soil moisture, called by the German abbreviation AMBAV, is used within an agrometeorological advisory routine. AMBAV shows realistic soil moisture results compared with the Müncheberg data set of measurements from 1993 to 1998, even though the soil texture, biometric and phenological data are only roughly adjusted. Weather during the vegetation period greatly influences the number of irrigations to a crop: between zero and three or more for the considered years in Müncheberg. The model serves as an agrometeorological tool to find the optimum date for a crop-specific irrigation, in routine use under consideration of forecast rain amounts, too.

Keywords Agrometeorological model, Evapotranspiration, Irrigation advice; Soil water content

Introduction

The model AMBAV (Agrarmeteorologisches Modell zur Berechnung der aktuellen Verdunstung = agrometeorological model for calculating the actual evapotranspiration) is part of the complex agrometeorological model toolbox AMBER developed by the Agrometeorological Research Braunschweig (German Weather Service, Deutscher Wetterdienst [DWD]).

Material and methods

The model AMBAV calculates the potential and real evapotranspiration and the soil water balance for different crop covers. It is disposed for producing irrigation recommendations which are disseminated by the DWD via fax service for different soil types using hourly data from the meteorological station network of the weather service including weather forecast up to 5 days. The model is designed to be used by local meteorological advisory services.

AMBAV simulates the water balance in the crop–soil system using the Penman-Monteith formula on an hourly basis. The model calculates separately soil evaporation, transpiration and interception for different crop covers considering the relevant processes of heat, water and vapour transport in the soil–crop–atmosphere interface including water losses during irrigation. Soil water dynamics are simulated using a mechanistic model based on the Richards equation. Soil water characteristics and hydraulic conductivity functions have been described by pedotransfer functions (PTF) (Vereecken et al. 1989, 1990). The coefficients have

---

Hans Friesland and Franz-Josef Löpmeier
German Weather Service, Agro-meteorological Research,
Bundesallee 50, D-38116 Braunschweig/Germany
tel.: +49-531-25205-0; fax: +49-531-25205-45;
e-mail: hans.friesland@dwd.de

been recalculated in order to get field capacities and wilting points in accordance with the German soil evaluation (Ad-hoc-Arbeitsgruppe Boden 1996). The reduction of evaporation as well as transpiration is calculated from soil water potentials and resistances representing the plant roots.

The model considers 13 different crops: winter wheat, spring wheat, winter barley, rye, oats, maize, sugar beets, potatoes, oilseed rape, grassland, fruit trees, coniferous and deciduous forest.

Input files are separated into starting file, parameter files and meteorological files which should be filled with adequate data.

(a) The starting file pre-defines as a batch the model runs for different stations, input and output path, simulation period, crop, soil type and irrigation settings
(b) The parameter files contain:
    - The plant parameters for the crops (height, leaf area index [LAI], etc.)
    - Parameters for soil-hydraulic properties for specific soil type
    - List of station coordinates
    - Five phenological development stages during the season for each crop
    - Crop- and soil-specific initial / boundary conditions for a given location
    - Intermediate results of soil water budget and rooting which is automatically generated after a model run to be used for the next run
(c) The meteorological files contain hourly data of cloud cover (octas), relative humidity (2 m), air temperature (2 m), global radiation, precipitation amount and wind speed

The output data files are partly standard, partly the content can be at choice, e.g. actual and potential evapotranspiration (AET, PET), available soil water (in percentage, 0–60 cm depth, soil water budget and rooting percentage in layers.

For this model test the Müncheberg data for plot 2 with loamy sand (Sl2 classification in AMBAV) have been taken. Details of the Müncheberg field data are supplied in this volume by Mirschel et al. (2007). Some meteorological data gaps have been filled by interpolation with the nearby DWD station at Lindenberg. A detailed model description is given by Löpmeier (1994) and Braden (1995).

Results and discussion

AMBAV results for the experimental years at Müncheberg

For 1993 sugar beet season, Fig. 1 shows the soil water content (SWC) of 0–100 cm and measurements of 0–90 cm in millimetre together with the calculated daily AET in millimetre. Cardinal values are 170 mm for field capacity and 30 mm for wilting point for this soil. The period from April 1st to September 30th covers the main vegetation period of sugar beet. Both curves, measured and calculated water content, go widely in parallel beginning with saturated conditions in spring. Periods without or small rain events correspond well with decreasing soil moisture, whereas considerable rain amounts cause sharp rises in the soil water curve as in the end of June 1993. At the end of August the lowest soil moisture content is reached, and consequently the water balance of rain and evapotranspiration becomes positive again.

In the third graph of Fig. 1 the available water (%) for the rooting zone and the same for 0–60 cm are added as the thicker curve. On May 17th both curves match when the sugar beet rooting depth of 60 cm is reached. In this simulation the maximum rooting has been set to 100 cm. It is well to discern that all curves of water content reveal their peaks and minima widely in parallel. During dry periods the volume of 100 cm soil shows a higher water content due to the deeper profile depth, while after a heavier rain the 60 cm soil volume indicates a more distinct water recharge. Figure 1 further shows in the lowest graph the course of actual evapotranspiration together with calculated daily drainage at 100 cm soil depth for the whole year (bare soil is applied to the model before sowing and after beet harvest). Calculated AET reaches maximum daily values of more than 5 mm on single days in June and July, whereas later in the season the lower soil moisture together with actual weather depresses high daily AET amounts. In July and August 1993 slight capillary rise of 0.1 mm water per day is the soil reaction to the crop and weather induced vertical moisture gradient. As expected in the winter months larger peaks of percolating water occur during phases of the water-saturated soil column.

In 1994 winter wheat was grown at the Müncheberg plot 2, and the comparison of measurements and

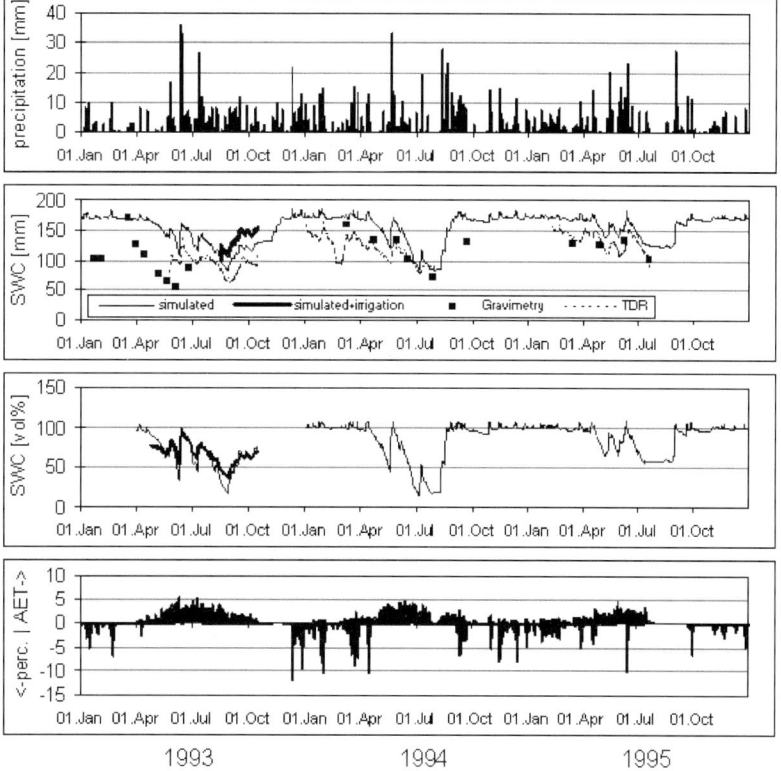

Fig. 1 Müncheberg plot 2, 1993–1995 (sugar beet, winter wheat, winter barley), daily measurements and AMBAV results from top to bottom: precipitation, soil water content (SWC, simulated 0–100 cm, gravimetry and TDR measurement 0–90 cm), SWC as available soil water (vol%) 0–100 cm (short thicker line: in root zone), percolation and actual evapotranspiration (AET)

calculations by AMBAV for the 0–100 cm soil volume is given in Fig. 1. For this vegetation period, in the graph beginning in January, the AET curve reflects very well the reaction to meteorological conditions. The two curves for water content (calculated 0–100 cm, measured 0–90 cm) show a reasonably good correspondence, although some disfunction of the time domain reflectometry (TDR) sensor at the beginning of March appears. In that year a cool May and enough rain led to a late soil water reduction. But afterwards a very hot July gave rise to high AET rates. At last for 1995 Fig. 1 presents the curves of the daily soil water (calculated, measured) and AET of winter barley. Only a short summer vegetation time is needed by the crop, and soil water calculations and TDR measurements are in good agreement. The data show the reaction to a rather cool, rainy season with lower AET than normal and in consequence high soil moisture for spring and early summer. The reaching of field capacity in early summer for full-grown cereals is a rare event.

For the 1993–1995 period Fig. 2 presents the selected crop data, LAI and rooting depth calculated by AMBAV together with actual evapotranspiration for the crops sugar beet, winter wheat, and winter barley. The curves of these crop data depend on the preset phenological data and remain constant at their maximum level unless meaningful changes in LAI occur. The increasing LAI in spring is reflected in the rising daily AET rates, as well as LAI and AET decrease in late summer.

In 1996 a relatively short summer vegetation time of winter rye shows calculated and measured soil water curves, which go widely in parallel and have rather high SWC (Fig. 3). 1996 was the coolest summer with enough rain and often low to moderate AET rates except at the end of May to the beginning of June. As in the year before the SWC several times approached near the maximum (170 mm) field capacity.

The sugar beet crop of 1997 met a rather warm season, especially in August, and the SWC graph shows two distinct dry periods with a strongly decreasing SWC. The calculated AET rates typically took low values when soil moisture was down, as the plant roots failed to get their full water supply. Two main events of considerable rainfall in summer caused sharp rises in the soil moisture curve.

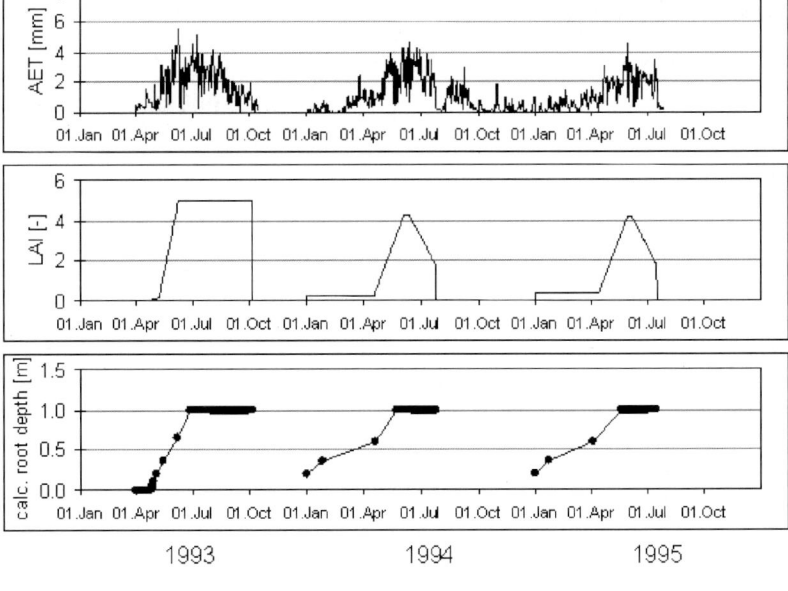

Fig. 2 Evapotranspiration and crop data for Müncheberg plot 2, 1993–1995, AMBAV results: actual evapotranspiration (AET), leaf area index (LAI), root depth

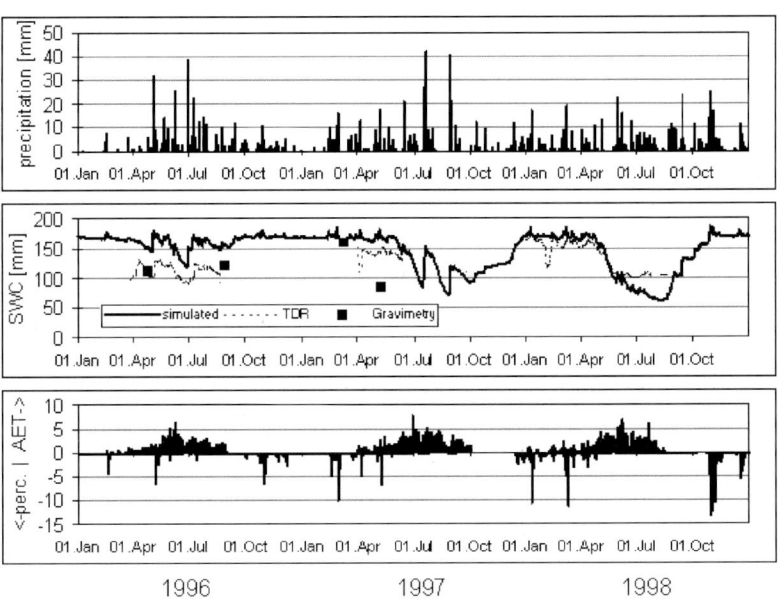

Fig. 3 Müncheberg plot 2, 1996–1998 (winter rye, sugar beet, winter wheat), daily measurements and AMBAV results from top to bottom: precipitation, soil water content (SWC, simulated 0–100 cm, gravimetry and TDR measurement 0–90 cm), percolation and actual evapotranspiration (AET)

Winter wheat again was grown in 1998, the year with the hottest June and long periods of missing considerable rain amounts. This year presents the curves (calculated, measured) with the lowest water content of all experimental years at the end of season. The daily AET plot shows a few extreme high values, but considerable lower ones in summer because of lack of water together with quick maturation of the wheat. Harvest is assumed on the 8 August and accounted for by the model. Nevertheless considerable percolation occurs in three events in spring and late autumn due to high rain amounts during soil water saturation.

Figure 4 presents AET again in a better resolution together with data of the 1996–1998 crops. As expected the highest evapotranspiration rates occured in midsummer during the period of maximum LAI and rooting. Due to atmospheric conditions some peak AET rose higher than in the previous 3-year period.

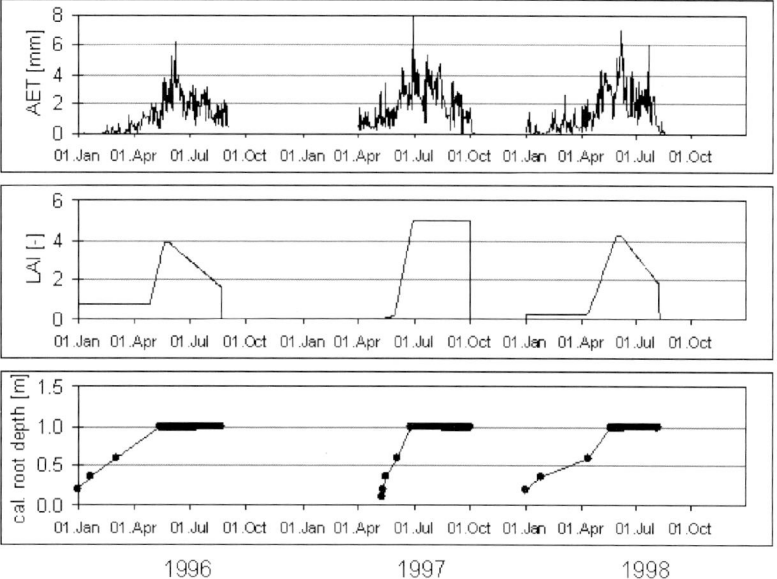

Fig. 4 Evapotranspiration and crop data for Müncheberg plot 2, 1996–1998, AMBAV results: actual evapotranspiration (AET), leaf area index (LAI), root depth

In the synopsis of all the years, the AMBAV model SWC of the 0–100 cm depth shows in every winter a full recharge of soil water up to field capacity. For the summer water supply only 1995 and 1996 were characterized by sufficient soil moisture, whereas the remaining four years contained distinct dry periods. Here the SWC reached or felt below 80 mm water in the 100 cm soil volume.

For April to September Table 1 presents the monthly mean temperatures, as well as precipitation and AET sums of the experimental years at Müncheberg. The temperature maximum values are underlined whereas minimum values are printed in italic. Data show that the 1994, 1997 and 1998 seasons had been the warmest ones and contribute to higher AET and lower soil moisture than the remaining seasons. The sums for rain and AET (April to September) astonishingly were not very different. The AET sum did not only result from the direct meteorological conditions, but also from soil moisture in the rooting

Table 1 Meteorological conditions at Müncheberg for the 1993–1998 period

Monthly mean temperature				[°C]				
Year	A.-S.	Apr.	May	June	July	Aug.	Sept.	
93	14.1	<u>9.9</u>	<u>15.7</u>	15.0	16.0	15.6	12.0	
94	14.8	8.4	12.2	15.5	<u>21.5</u>	17.9	13.4	
95	14.4	7.9	12.1	*14.6*	20.2	18.6	13.0	
96	13.1	8.1	*11.3*	15.4	*15.5*	17.6	*10.5*	
97	14.6	*5.9*	12.8	16.6	18.0	<u>20.3</u>	13.5	
98	14.7	9.5	14.5	<u>17.0</u>	16.8	16.4	<u>13.6</u>	
Precipitation sum (Apr.–Sept. and monthly)				[mm]				AET
Year	A.-S.	Apr.	May	June	July	Aug.	Sept.	A.-S. [mm]
93	370	19	43	97	96	41	73	339
94	411	48	89	36	45	109	84	301
95	312	40	62	92	21	46	51	229
96	374	12	112	81	94	33	42	292
97	368	39	53	59	130	63	24	371
98	339	53	52	71	61	60	40	331

zone and from plant phenology. The drying effect of the hottest July in 1994 was somewhat attenuated by higher rain amounts before the ripening wheat ceases its water use. The soil moisture in the 1995 and 1996 season remained relatively high, not only due to the low temperature conditions, but partly due to high precipitation sums. On the other hand, the dry period years 1997 and 1998 were mainly attributable to the very warm June (1997) and August (1998). Of course the meteorological data of August and September were irrelevant for the winter wheat and winter barley seasons, as ripeness happened before.

Irrigation recommendations

As possible irrigation settings in AMBAV date and amount can be given to the model after an irrigation, or the model automatically recommends irrigations in consideration of the users wishes: available water (%) to start irrigation, water amount (mm), hour of irrigation start, and duration of irrigation (h). For example, the 1993 sugar beet SWC has been calculated with and without irrigation (see left side in Fig. 1). Only one irrigation was needed in the simulation when 50% available water in the rooting zone was set as a limit. The irrigation amount of 30 mm is included in the precipitation bars. As may be expected the two soil water curves behave quite similar in their further course through time, and astonishingly the difference between them remains rather stable until autumn (day 288 = 15 October).

The year 1998 represents an example for three irrigations (each 30 mm) needed for the winter wheat season. Figure 5 shows that from May onwards there was a strong tendency to soil drying down due to evapotranspiration. The calculated soil water curves have a tendency to converge and only after the third irrigation on day 181 (30 June) stay apart, also due to the increasing ripeness.

Table 2 indicates the number of recommendable irrigations for 1993–1998 at Müncheberg for the different cultures. As expected already, 1995 and 1996 remained free of irrigation, taking 50% available water in the rooting zone as a threshold in the model AMBAV. But 30 mm each for 1993 and 1994 and 90 mm as a sum each for 1997 and 1998 would have had to be expended to keep the crop within the soil water boundaries for a positive crop development. It can be concluded from these Müncheberg 6 years results that it is more the year characteristic (i.e. the weather) than the type of crop that rules the number of irrigations.

Comparing nowadays irrigation conditions (last 40 years) with future irrigation conditions (40-year scenario) Table 3 gives the AMBAV results for a light and a heavy soil site and irrigation starts at certain soil water levels. The figures for sandy soil near Soltau (northern Germany) show that in the mean the numbers are reduced by half an irrigation due to the CO_2 effect on the stomata of the leaves. In addition, it can be seen that an early onset of irrigation at 60% available water needs more irrigation water. The comparison with the sugar

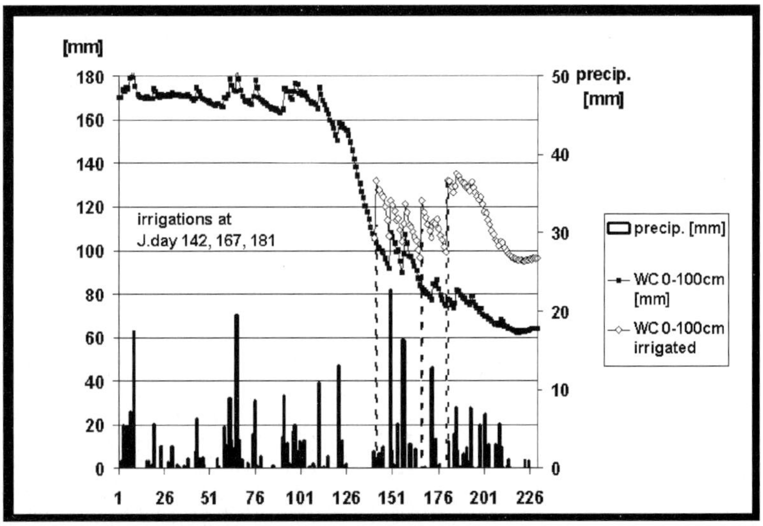

Fig. 5 AMBAV results: soil water content, precipitation and irrigation, Müncheberg 1998 (winter wheat)

Table 2 Recommended irrigations by model AMBAV for Müncheberg (JD = julian day)

	1993 Sugar beets	1994 W. wheat	1995 W. barley	1996 Rye	1997 Sugar beets	1998 W. wheat
30 mm Recomm. at:	JD 229	JD 178	No	No	JD 191 JD 230 JD 237	JD 142 JD 167 JD 181
Irrigation Sum [mm]	30	30	0	0	90	90

Table 3 Mean number of irrigations per year, Soltau (sandy soil, winter cereals) and Magdeburg (loamy soil, sugar beet), northern Germany

Irrigation	Starting at 40% avail. water		Starting at 50% avail. water		Starting at 60% avail. water	
	Climate today	2010–2050	Climate today	2010–2050	Climate today	2010–2050
Soltau	1.8	1.3	2.2	1.6	2.7	2.2
Magdeburg	3.3	2.9	3.8	3.4	4.3	4.0

beet crop on a heavy soil demonstrates very similar effects.

Extreme dry year example 2003

A cross section from west to east is taken by using meteorological data of the stations Braunschweig, Magdeburg and Lindenberg (nearby Müncheberg) for an AMBAV run with 2003 data. All three places were calculated with the same sandy soil (Sl2) and the same crop winter wheat assumed. At the eastern place Lindenberg, the soil water depletion tendency was greatest, so that the number of irrigations and the time (from day 148 to 175 = end of May to end of June) was the largest, while the other two places needed only three irrigations. Figure 6 shows the soil water curves with very similar water losses until the first irrigation, even afterwards comparable water contents occur, but in autumn a different recharge and in winter a field capacity water level can be seen at all stations. Irrigation recommendations of model AMBAV end some time before yellow ripeness in cereals. The passage of frontal systems, often from west to east, is discernible in these curves from the

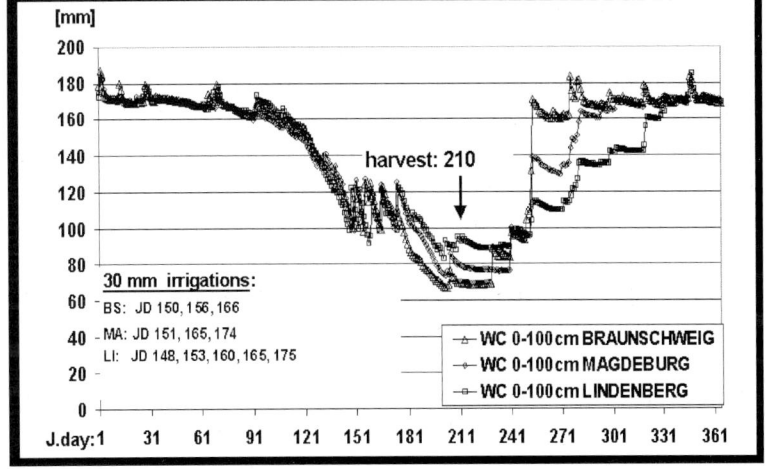

Fig. 6 AMBAV results: soil water content 2003, sandy soil, winter wheat at three stations

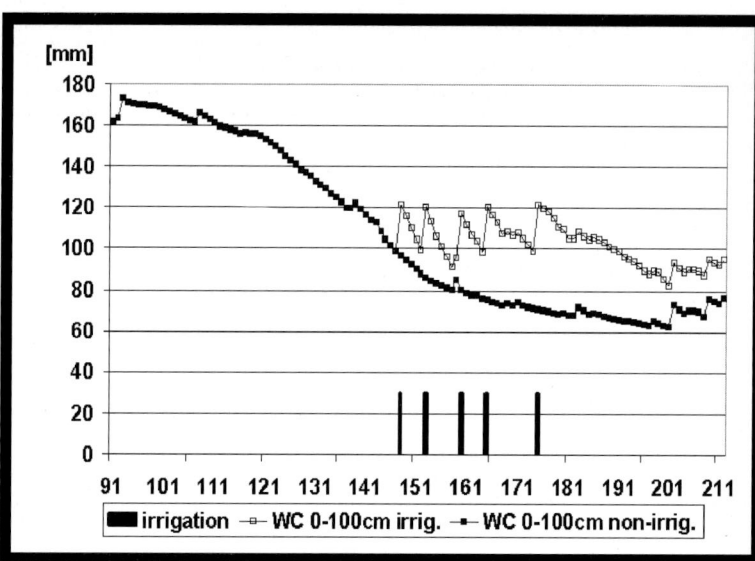

Fig. 7 AMBAV results: soil water content 2003 Lindenberg, with and without irrigation, sandy soil, winter wheat

almost simultaneous rain events and according soil water recharges.

At last Fig. 7 demonstrates the above 2003 Lindenberg soil water curves with five calculated irrigations (dates indicated within the graph) and without. Here only the most important vegetation period for wheat from April to end of July is displayed. Near mid of July (about day 196) both water curves tend to merge again, as the irrigated version has larger water losses from crop transpiration.

Conclusions

The measured and calculated values for Müncheberg plot 2 crops show reasonably good accordance, although not fully comparable concerning the depth of the SWC. The time course of constant measurements is susceptible to periodical or single errors. Here a daily model output can produce a better result over a season or a year. The variety of field crops and meteorological conditions through 6 years leads to rather different soil water curves. Consequently, the number of needed irrigations varied between zero and three and even more considering the hot, dry year 2003. The model AMBAV, in use for the routine agrometeorological advisory service in Germany, has proved as a useful means to calculate evapotranspiration, SWC for defined soil columns, and amount and date of crop- and soil-specific irrigations. The model can serve as a tool for the calculation of historical soil water conditions of a crop season as well as for prognostic irrigation recommendations in agrometeorological advisory schemes, including the numerical weather forecast.

References

Ad-hoc-Arbeitsgruppe Boden (1996) Bodenkundliche Kartieranleitung, 4th edn. 1994 reprint, E.Schweizerbart'sche Verlagsbuchhandlung, Stuttgart

Braden H (1995) The model AMBETI – a detailed description of a soil-plant-atmosphere model. Berichte des Deutschen Wetterdienstes 195:117

Löpmeier F-J (1994) The calculation of soil moisture and evapotranspiration with agrometeorological models (in German). Zeitschrift f. Bewaesserungswirtschaft 29:157–167

Mirschel W, Wenkel K-O, Wegehenkel M, Kersebaum K-C, Schindler U, Hecker J-M (2007) Müncheberg field trial data set for agro-ecosystem model validation. In: Kersebaum KC, Hecker J-M, Mirschel W, Wegehenkel M (eds) Modelling water and nutrient dynamics in soilcrop systems. Springer, Dordrecht, pp 219–243

Vereecken H, Maes J, Feyen J, Darius P (1989) Estimating the soil moisture retention characteristic from texture, bulk density, and carbon content. Soil Sci 148:390–403

Vereecken H, Maes J, Feyen J (1990) Estimating unsaturated soil hydraulic conductivity from easily measured soil properties. Soil Sci 149:1–12

CHAPTER THREE

Performance of the model SIMWASER in two contrasting case studies on soil water movement

Elmar Stenitzer[1*], Heiko Diestel[2], Uwe Franko[3], Reinhild Schwartengräber[2] and Thomas Zenker[2]

Abstract The performance of the water balance and crop growth model SIMWASER to estimate either the capillary rise from shallow groundwater or the N leaching from a deep soil profile was demonstrated using measurements on the one hand from the Berlin-Dahlem lysimeter station and on the other hand from the Bad Lauchstädt experimental field. In case of the Berlin-Dahlem data, the robustness of the model was evaluated by running it either with measured hydraulic soil parameters or with estimated pedotransfer functions (PTF) derived from texture class, bulk density class, and humus content class. Simulations were performed for the period from March to September 1997. Simulated capillary rise and actual evapotranspiration (AET) deviated by less than 10% for both the measured and the estimated hydraulic soil characteristics, showing somewhat better results for the sandy soil.

Nitrogen leaching from the deep Chernozem soil at Bad Lauchstädt is estimated for a bare and a cropped experimental plot from measured Nitrogen concentration of the soil solution and simulated percolation flux at 170 cm depth. Because of low N concentration at both plots, the amount of N leached down to groundwater during the investigated period from the beginning of 1999 to the end of 2002 was limited to 7 kg/ha at the bare plot and to 0.5 kg/ha at the cropped site.

Keywords Capillary conductivity, Capillary rise, Lysimeter, Moisture characteristic, Nitrogen leaching, Simulation, Suction cups

Introduction

Capillary rise from shallow groundwater may be important for crop growth during rainless periods: even in areas with more than 1000 mm of annual precipitation, essential yield reductions from grassland on sandy soils at the valley floor of the Drau river in Austria (Quendler et al. 1998) were observed, when shallow groundwater was lowered by deepening the river bed. In Swedish experiments with "controlled drainage", water table was kept higher than usual during summer period and capillary rise caused better crop growth, and therefore, less surplus nutrients were left in soil and leached to the groundwater (Wesström 2002). In dry regions, uptake of shallow groundwater by capillary rise may reduce irrigation demand (Ayars and Hutmacher

---

Elmar Stenitzer, Heiko Diestel, Uwe Franko, Reinhild Schwartengräber and Thomas Zenker
[1]Institute for Soil Water Management Research, A-3252 Petzenkirchen, Austria
[2]Institute for Environmental Planning and Landscape Architecture, TU Berlin, Germany
[3]UFZ-Centre for Environmental Research Leipzig-Halle, Halle/Saale, Germany;
*Author for correspondence (e-mail: elmar.stenitzer@baw.at; fax: ++43 7416 52108 90)

1994; Guitjens 1990; Ragab and Amer 1986; Saini and Ghildyal 1977), but on the other hand may lead to salination problems (Rhoades and Loveday 1990). In all cases mentioned above, quantification of the relation between ground water depth and capillary rise is necessary for finding sustainable solutions. The most powerful but most expensive tool in this respect is the "groundwater" or "compensation" lysimeter (Aboukhaled 1982), by which a constant water level within the lysimeter is established and the capillary flux is measured by the amount of water necessary to hold the water table at constant height. Although a constant water table will not reflect the natural conditions in most cases, lysimeter measurements may be used to calibrate simulation models, by which ground water uptake and plant growth are estimated (Zhang et al. 1999). Lysimeter findings may be extrapolated by such models in time and space, but when dealing with regional studies, measuring the soil physical characteristics necessary to run the models for all different soil profiles will be impossible. An alternative would be to use soil characteristics inferred from readily available soil data by class pedo transfer functions (PTF). In this study the soil water and plant growth model SIMWASER (Stenitzer and Murer 2003) was run for a sandy and a clayey soil using soil physical characteristics from either laboratory measurements or from standard tables as input. Simulated capillary rise with both input data sets were compared with the respective measurements at the Berlin-Dahlem groundwater lysimeters (Diestel et al. 2007).

One of the goals of sustainable agriculture is protecting groundwater from contamination by nutrients and pesticides. Deep percolation as the transport medium of these contaminants depends on weather, soil, and agronomic measures and must be known for a classification of the field site with regard to the vulnerability of the groundwater quality. Depending on this taxation, appropriate cultivation may be chosen, thus securing high quality of the groundwater. The deep loessian Black Earth soil of the long-term experimental field at Bad Lauchstädt near Halle/Saale in Germany is supposed to be very favourable regarding nutrient utilization because of its rather high water storage capacity, which together with low precipitation at the site should show insignificant leaching of nutrients into groundwater (Körschens et al. 2002). The aim of this part of the present work is to estimate the amount of nitrogen losses from measured nitrogen concentration of the soil solution and the simulated percolation.

The result of both simulation studies were already presented at the workshop *Modelling Water and Nutrient Dynamics in Soil–Crop Systems* in July 2004 at Müncheberg/Germany.

Material and methods

Experimental sites

All information on both experimental sites are given in detail by Diestel et al. (2007) for the Lysimeter station Berlin-Dahlem and by Franko et al. (2007) for the experimental field in Bad Lauchstädt.

Simulation model

The deterministic simulation model SIMWASER on soil water balance and plant growth describes the one-dimensional vertical water flux within a soil profile and at the same time the growth of a vegetation cover. The water balance and the growth of plants are connected by the physiological interaction between transpiration and assimilation: accumulation of plant material depends on the amount of CO_2 incorporated via the stomata, by which at the same time water vapour is lost from the saturated vacuole into the unsaturated ambient air. Potential assimilation and therefore potential growth is only possible as long as the water supply towards the stomata can meet the potential transpiration loss. If this is not the case, stomata will close and formation of plant material will be restricted. All these processes depend on the respective plant development stage as, for example, the partition of the daily assimilates between leaves, stem and roots. SIMWASER calculates the actual development stage by dividing the currently accumulated growing degree days by the sum of growing degree days necessary for ripeness of the respective crop: a growing degree day corresponds to the mean daily temperature minus a base temperature which is specific to that crop.

Water flux between the soil layers is calculated according to DARCY's law as function of the capillary conductivity and the gradient of the matric potential using small but variable time steps, which restrict changes of water content to 0.1 vol%. Filter velocity at the lower end of a soil layer V_i is calculated according to the following equation:

$$V_i = \frac{(K_i + K_{i+1})}{2} \cdot \left(\frac{\Psi_{i+1} - \Psi_i}{Z_i} + 1\right)$$

V_i filter velocity (mm day^{-1})
K_i, K_{i+1} capillary conductivity of layers i, $i + 1$ (mm day^{-1})
Ψ_i, Ψ_{i+1} matric potential of layers i, $i + 1$ (dm)
Z_i distance from centre of layer i to centre of layer $i + 1$ (dm)

The filter velocity V_i of the water flowing out of the bottom of layer i is at the same time the filter velocity of the water flowing at the top of the next layer $i + 1$ into it. The soil profile model may be divided into 50 layers maximum, each 5–10 cm thick and must reach down to a depth which is outside the range of plant roots. When calculating the soil water flux within the soil profile, one must take into account if it may be influenced by the ground water level or not: in latter case, it may be assumed that there exists no capillary rise from the coarse aquifer, whereas in the former case the variable ground water level will form the lower boundary of the profile.

Input data needed are daily weather data (maximum and minimum of air temperature and of relative humidity, wind speed at 2 m height, global radiation and precipitation), soil parameters (pF-, Ku- and penetration resistance curve) of each soil layer and water content at begin of simulation. In case of a soil profile influenced by groundwater, daily groundwater depth must also be given. The simulation is controlled by specifying the cropping sequence by giving a list of the names of each crop and their dates of sowing and harvesting. Model output covers the daily values of soil water storage, accumulated amounts and daily fluxes of precipitation, evapotranspiration, transpiration, deep percolation, capillary rise and surface runoff, as well as daily water contents and matric potentials of each soil layer.

The model mainly has been used in practical work of the Institute for Soil Water Management Research experts in assessing influence of capillary rise upon plant growth, influence of land use change upon ground water recharge and estimating irrigation demand. Some more details on the model and its application may be found in Stenitzer (2004).

Soil parameters of lysimeters in Berlin-Dahlem

For simulation of the capillary rise within the lysimeters, two different sets of soil parameters were used: "measured (LAB)" curves were constructed graphically by eye-fitting from measurements on undisturbed soil samples in the laboratory, and "estimated (PTF)" curves were derived according to the soil class of each soil horizon. Soil texture, total pore volume and water contents at specific soil suctions, as well as the saturated hydraulic conductivity for both soils are given in Diestel et al. (2007). The LAB-pF curves were constructed from these data; the LAB-Ku curves were extrapolated from measured Ksat using the procedure of Millington and Quirk (Bouwer and Jackson 1974), by which the shape of the Ku curve is estimated from the shape of the pF curve. The PTF-pF curves were interpolated according to the measured soil class and to the estimated total porosity from tabulated PTFs, which have been developed for the SIMWASER model. These PTF are based on standard tables of air capacity, field capacity and usable field capacity, as well as the saturated hydraulic conductivity of the different soil classes as function of the bulk density (DIN 4220, 1998). The total pore volume estimated by these tables was corrected for the influence of the humus content by the respective factors, which are also given in the DIN tables. The PTF-Ku curves were deduced from PTF-pF curves with the Millington-Quirk method using an estimated Ksat, which was interpolated according to the estimated total pore volume from the above-mentioned standard tables. Both types of pF- and Ku curves for the different layers of the Wildeshausen and the Weckesheim soil are shown in Figs. 1 and 2.

Soil parameters of experimental field in Bad Lauchstädt

For simulation of the soil water balance at Bad Lauchstädt, measured soil physical parameters of the different soil layers down to 200 cm depth of the so called "Intensivmessfeld" were derived from a detailed laboratory analysis on undisturbed soil samples (Franko et al. 2007). Capillary conductivity was deduced from measured hydraulic conductivity according to the method of Millington and Quirk as described above.

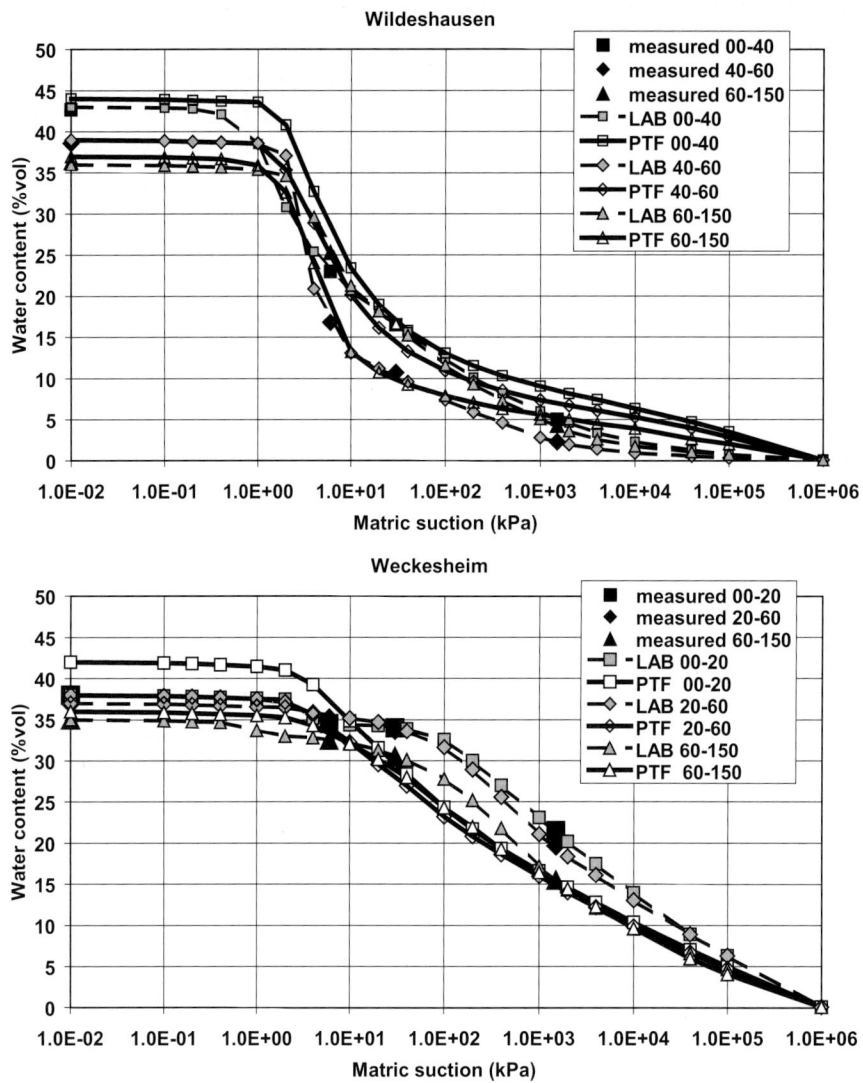

Fig. 1 Comparison of measured water content at saturation, pF 1.8, 2.5 and 4.2 with eye-fitted (LAB) and estimated (PTF) pF curves at Wildeshausen and Weckesheim

Results and discussion

Lysimeter Berlin-Dahlem

Comparison of simulated and measured accumulated percolation and capillary rise, as well as evapotranspiration and rain in the Wildeshausen and in the Weckesheim soil are shown in Figs. 3 and 4. Negative percolation means capillary rise.

Deviation of simulated capillary rise using LAB curves from measurements were rather low for both sites: at the sandy soil simulated accumulated capillary rise was 122 mm, the measured value was 115 mm, while for the clayey soil, simulation yielded 16 mm and was also close to the measured 22 mm. Using PTF curves produced similar results for the sandy soil (108 mm simulated versus 115 mm measured), but for the clayey soil, the difference between 10 mm simulated and 22 mm measured capillary rise was higher.

Simulated and measured actual evapotranspiration (AET) were also in acceptable agreement, regardless of the type of hydraulic parameter curves: for the sandy soil simulated AET using LAB curves amounted to 460 mm, which is about the same as the

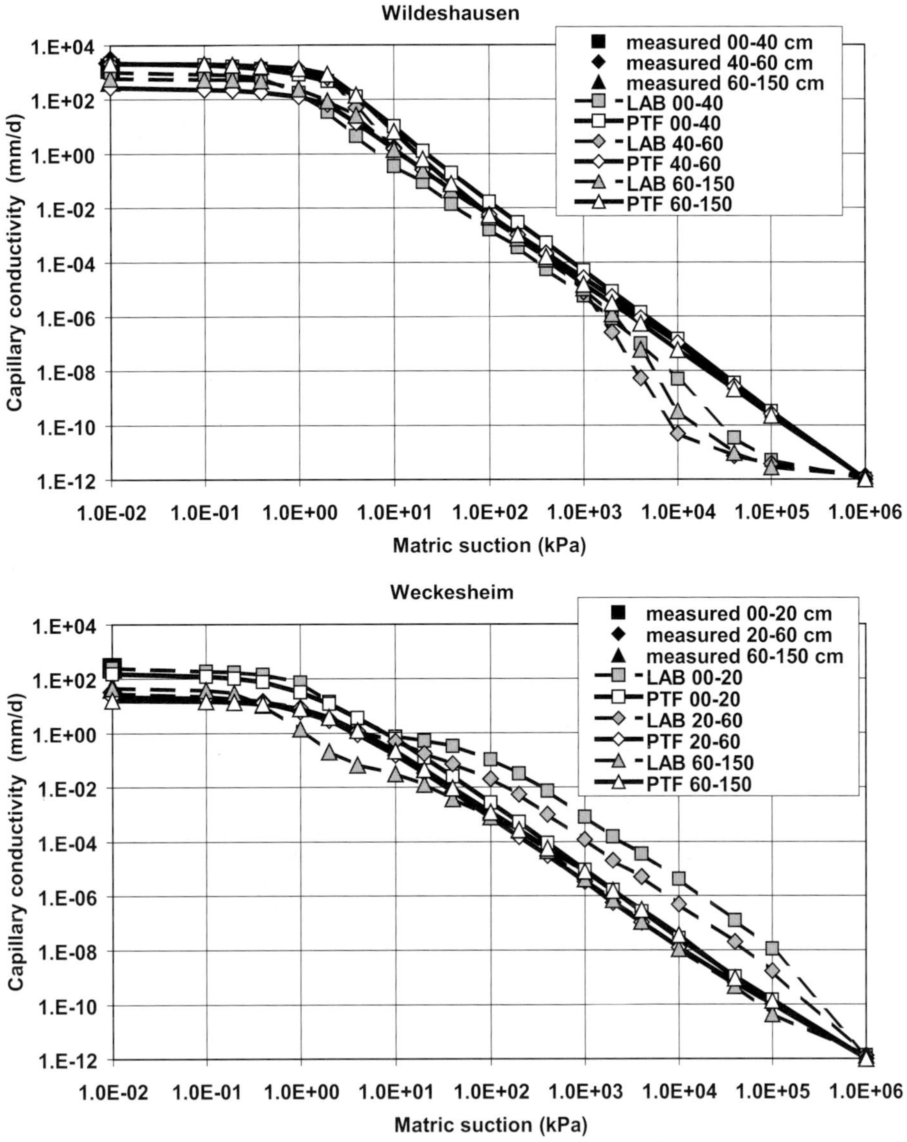

Fig. 2 Comparison of "measured" (LAB) and estimated (PTF) Ku curves at Wildeshausen and Weckesheim

measured value of 454 mm. For the clayey soil, the LAB curves yielded 382 mm, which is distinct lower than the measured evapotranspiration of 412 mm. Using PTF curves, simulated evapotranspiration of the sandy soil was 483 mm, which is nearly 30 mm more than the measured value, but this difference is still well below 10% of the measured 454 mm and seems to be acceptable. With the clayey soil, simulation based on the PTF curves resulted in 386 mm evapotranspiration, which is about the same as for the LAB curves and is also acceptable.

Experimental field Bad Lauchstädt

Simulated and measured water contents at 5, 45, 90 and 170 cm are compared in Figs. 5 and 6, showing extremely differences in the uppermost layer. These measurements, therefore, were not taken into account in estimating the water storage of the whole soil profile.

Comparison of estimated soil water storage and the simulated one is shown in Fig. 7. There generally exists conformity for both plots. Although there are some deviations from measured soil water storage during

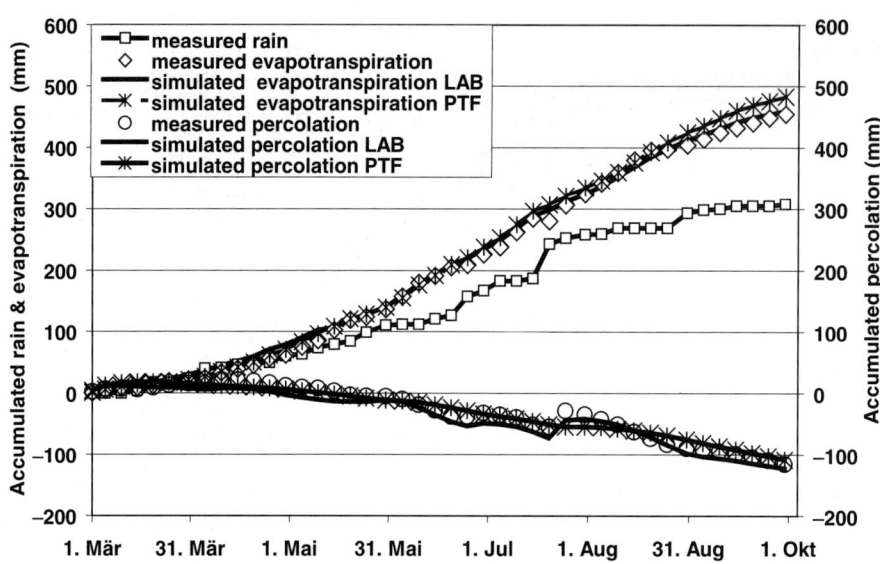

Fig. 3 Simulated and measured evapotranspiration and percolation/capillary rise at the sandy Wildeshausen soil. Measured rain is also shown

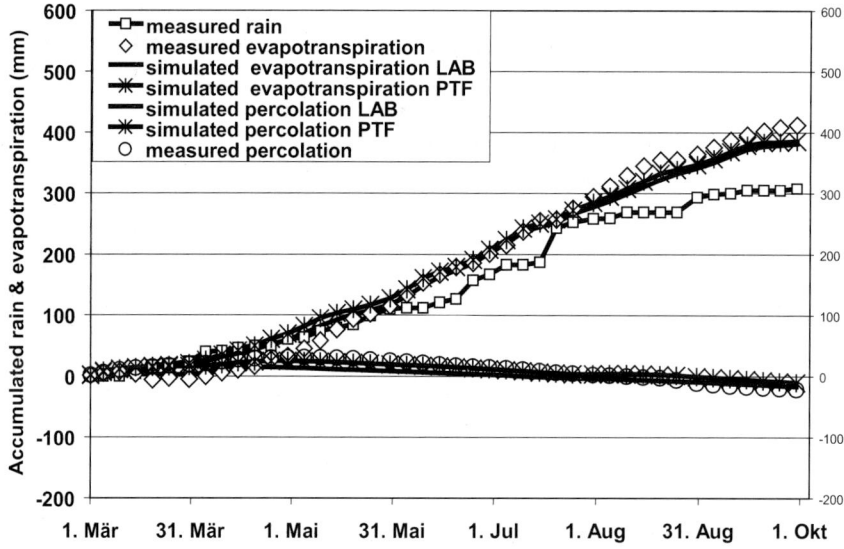

Fig. 4 Simulated and measured evapotranspiration and percolation/capillary rise at the clayey Weckesheim soil. Measured rain is also shown

short periods, the distinct difference of both plots is clearly worked out. Taking into account the limited database, one might state, that SIMWASER was able to estimate the soil water balance at both sites sufficiently.

Simulated AET and deep percolation Perc at both sites are shown together with calculated potential evapotranspiration (PET) and measured precipitation R in Fig. 8: at the cropped plot, AET (1780 mm) corresponds roughly to the amount of rain (1850 mm), so only 130 mm are left for deep drainage. At the black fallow soil evaporation was restricted to 1320 mm and deep percolation amounted to 580 mm.

Accumulated nitrogen losses calculated for the simulated percolation and measured N concentrations at 170 cm depth are shown in Fig. 9: because of rather low concentrations in both cases, the total amount of N losses was very low compared to less favourable soils.

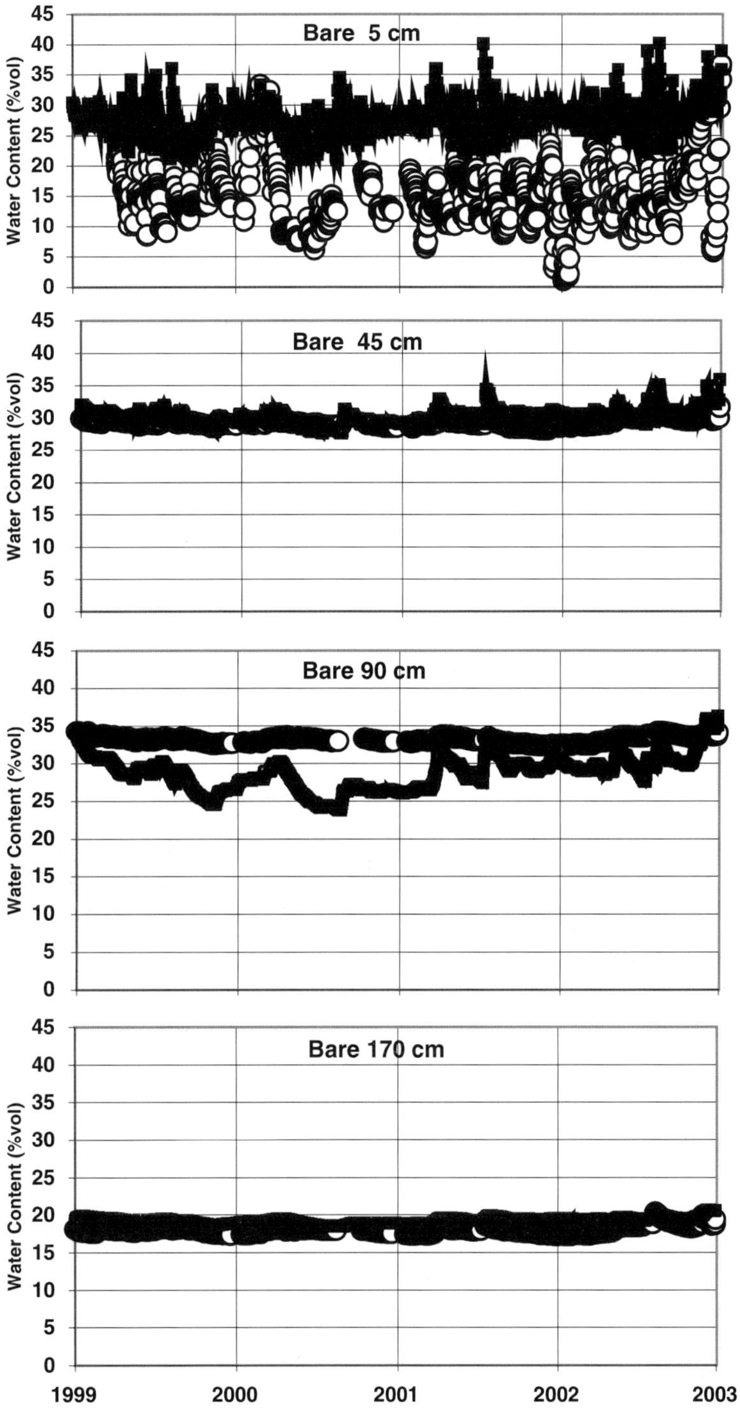

Fig. 5 Comparison of measured (open dots) and simulated (full line) water content at the black fallow plot

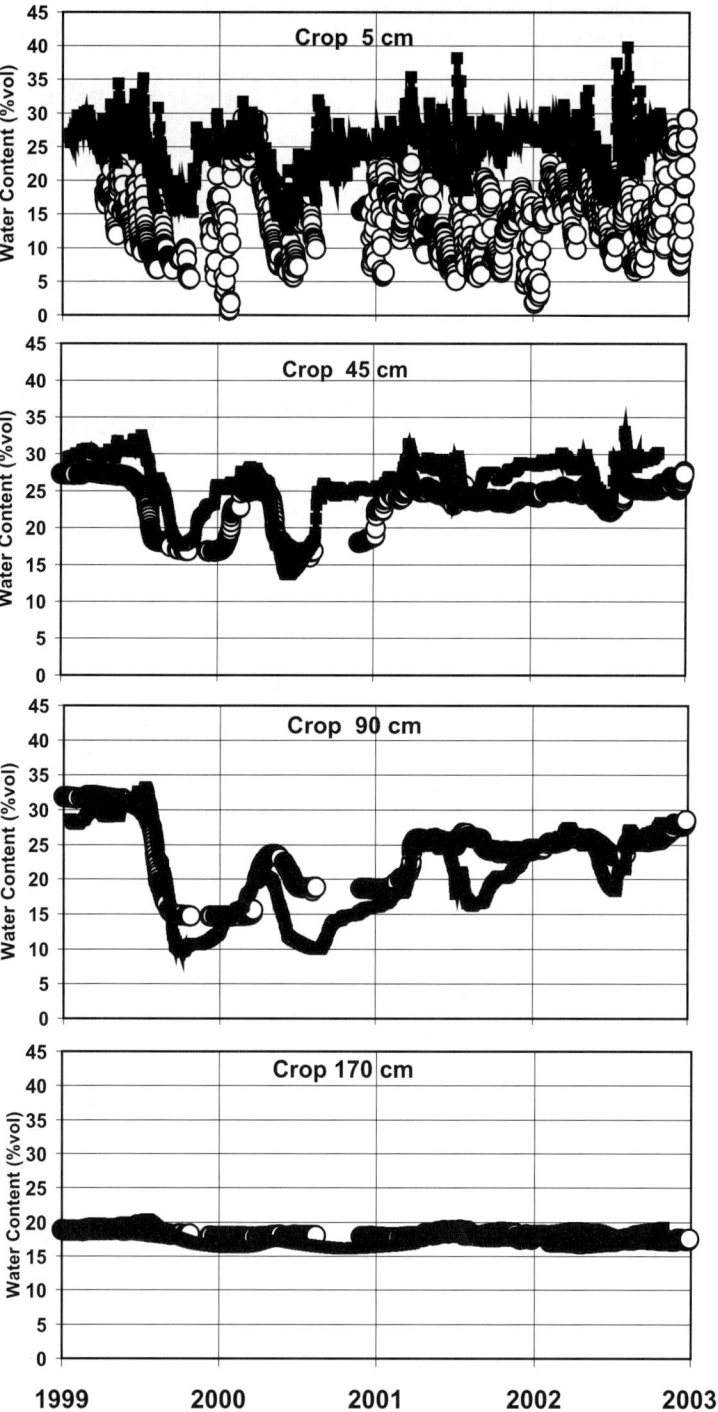

Fig. 6 Comparison of measured (open dots) and simulated (full line) soil water content at the crop rotation plot

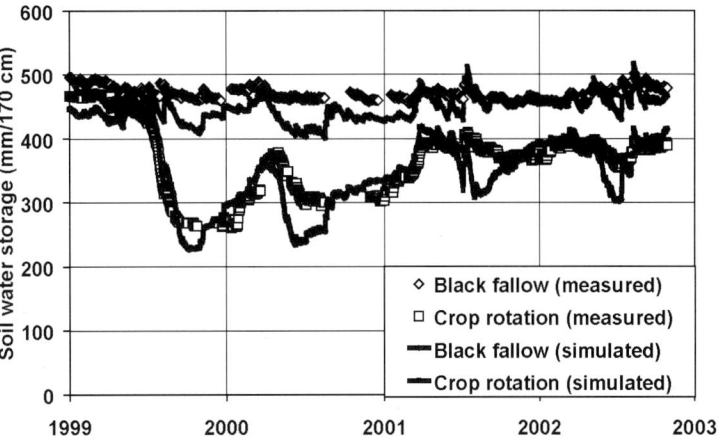

Fig. 7 Comparison of simulated and measured soil water storage at the black fallow and the crop rotation plot

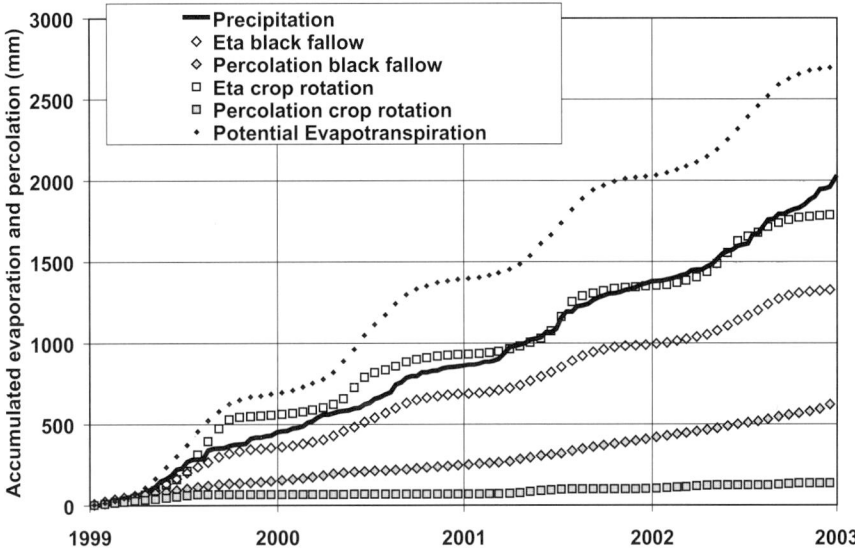

Fig. 8 Simulated evapotranspiration and percolation (accumulated values)

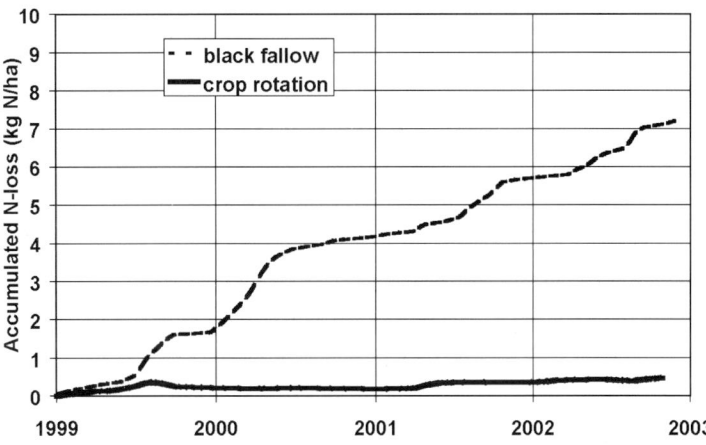

Fig. 9 Estimated nitrogen loss due to percolation at the fallow and the crop rotation plot

Conclusions

Capillary rise at Lysimeter station Berlin-Dahlem

Although using only one vegetation period and confining to a constant ground water level, the results of this study show that the model SIMWASER is capable to simulate capillary water supply of a lawn from shallow groundwater on two contrasting soils with acceptable degree of accuracy. Simulation results using estimated PTF as input data deviated within 10% from the measured values. Although these findings are encouraging to use SIMWASER in combination with PTF curves in working on practical problems concerned to soil water balance and solution transport, further investigations on the model performance with variable soils and with variable ground water depths are needed.

Nitrate leaching at experimental field in Bad Lauchstädt

Combination of measured concentrations of nutrients in the soil solution with simulated deep percolation and capillary rise may be used to quantify the leaching losses due to different soil use. In case of the deep soil and low rainfall as presented here, the amount of nitrogen leaching may be neglected for the selected simulation period. For the cropped soil, the low leaching losses correspond quite well to the aboveground nitrogen balance, where an input of 98 kg N/ha from fertilizer is compensated by an uptake of 127 kg N/ha by the crop products (beets, grain and tubers). The low leaching losses of the bare soil are probably not characteristic for this type of land use and result from relative short time period of fallow treatment. A much higher leaching loss is expected in the future, when all mineralized nitrogen will be transported to the deeper soil layers.

References

Aboukhaled A (1982) Lysimeters. FAO irrigation and drainage paper 39. Food and Agricultural Organization of the United Nations, Rome

Ayars JE, Hutmacher RB (1994) Crop coefficients for irrigating cotton in the presence of groundwater. Irrig Sci 15:45–52

Bouwer H, Jackson RD (1974) Determining soil properties. In: Schilfgaarde J van (ed.) Drainage for agriculture. Agronomy Series 17. American Society of Agronomy, Madison, WI, pp 611–672

DIN 4220 (1998) Pedologic site assessment – designation, classification and deduction of soil parameters (in German). Beuth Verlag GmbH, Berlin

Diestel H, Zenker T, Schwartengräber R, Schmidt M (2007) The lysimeter station at Berlin Dahlem. In: Kersebaum KC, Hecker J-M, Mirschel W, Wegehenkel M (eds) Modelling water and nutrient dynamics in soil-crop systems. Springer, Dordrecht, pp 259–266

Franko U, Puhlmann M, Kuka K, Böhme F, Merbach I (2007) Dynamics of water, carbon and nitrogen in an agricultural used Chernozem soil in Central Germany. In: Kersebaum KC, Hecker J-M, Mirschel W, Wegehenkel M (eds) Modelling water and nutrient dynamics in soil-crop systems. Springer, Dordrecht, pp 245–258

Guitjens JC (1990) Alfalfa. In: Stewart BA, Nielsen DR (eds) Irrigation of agricultural crops. Agronomy Series 30. American Society of Agronomy, Madison, WI, pp 537–568

Körschens M, Schulz E, Weigel A, Knappe S (2002) Influence of high doses of farmyard manure on enrichment and transfer of nutrients on Loess Black Earth. 12th ISCO (International Soil Conservation) Congress, Beijing, pp 300–303

Quendler Th, Gritsch H, Stenitzer E (1998) Impact of lowered ground water on grassland yields in the Upper Drau Valley (In German). Der Förderungsdienst-Spezial, Heft 2:10–16

Ragab RA, Amer F (1986) Estimating water table contribution to the water supply of maize. Agric Water Manage 11:221–230

Rhoades JD, Loveday J (1990) Salinity in irrigated agriculture. In: Stewart BA, Nielsen DR (eds) Irrigation of agricultural crops. Agronomy Series 30. American Society of Agronomy, Madison, WI, pp 1089–1142

Saini BC, Ghildyal BP (1977) Seasonal water use by winter wheat grown under shallow water table conditions. Agric Water Manage 1:201–298

Stenitzer E (2004) SIMWASER – a numerical model on soil water balance and plant growth. IKT-Report 5/2004. Institut für Kulturtechnik und Bodenwasserhaushalt, A-3252 Petzenkirchen, Austria

Stenitzer E, Murer E (2003) Impact of soil compaction upon soil water balance and maize yield estimated by the SIMWASER model. Soil Till Res 73:43–56

Wesström I (2002) Controlled drainage. Effects of subsurface runoff and nitrogen flows. Acta Universitatis Agriculturae Sueciae, Agraria 350

Zhang L, Dawes WR, Slavich PG, Meyer WS, Thorburn PJ, Smith DJ, Walker GR (1999) Growth and ground water uptake responses of lucerne to changes in groundwater levels and salinity: lysimeter, isotope and modelling studies. Agric Water Manage 39 (2–3):265–282

CHAPTER FOUR

Application and validation of the models THESEUS and OPUS with two field experimental data sets

Martin Wegehenkel and Wilfried Mirschel

Abstract The agro-ecosystem models THESEUS and OPUS were tested with data obtained from an agricultural experimental field plot located at Bad Lauchstädt and a data set obtained from 12 weighable lysimeters located at Berlin-Dahlem. The comparisons of simulated with measured model outputs were analysed using the modelling efficiency index (IA) and the root mean squared error (RMSE). According to this analysis, at Bad Lauchstädt, both models simulated adequately the time courses of volumetric soil water contents in 5 cm depth, but in deeper soil layers, the simulation quality was insufficient. At this agricultural experimental field, the time courses of pressure heads were also predicted with a low quality by both models. OPUS underestimated the development of aboveground biomass and yields. THESEUS also underestimated the development of aboveground biomass, but in contrast to OPUS, the yields were simulated quite well by THESEUS at the test site Bad Lauchstädt. The discrepancies between the measured aboveground biomass accumulation rates and yields with those simulated by OPUS is due to an overestimation of the impact of the limitations in nitrogen availability on the aboveground biomass accumulation and yield by the model OPUS.

Martin Wegehenkel and Wilfried Mirschel
Institute of Landscape Systems Analysis, Leibniz-Centre for Agricultural Landscape Research (ZALF) e. V. Müncheberg, Eberswalder Strasse 84, D-15374 Müncheberg, Germany;
e-mail: mwegehenkel@zalf.de

For the lysimeter data set, THESEUS simulated adequately daily evapotranspiration rates, but underestimated the flow peaks of ground water recharge (GWR) in comparison with the lysimeter data. The model OPUS was not applied to this data set.

The most likely reason for the discrepancies between simulated and measured model outputs regarding the soil water contents, pressure heads and GWR rates at both test locations is the uncalibrated application of THESEUS and OPUS using parameters of the soil water retention functions obtained from a tabular database and not from field or laboratory measurements using soil core samples from the field. This indicates that a realistic, physically based simulation of water movement in the unsaturated soil zone requires reasonable estimates of the water retention and unsaturated hydraulic conductivity functions. The results of our study indicated also the need of a site-specific parameter calibration of the crop growth modules especially for that included in OPUS.

Keywords Agro-ecosystem modelling, Crop growth, Nitrogen balance, Soil water balance, Validation, Yield

Introduction

Agro-ecosystem models are widely used to estimate or predict the impact of climate conditions and management practices on crop biomass development and yield. Some of these models were developed for selected crops like winter wheat between seeding and harvest such as CERES-WHEAT (Godwin et al. 1989) or AGROSIM-WINTERWHEAT (Schultz and Mirschel 1995). Other models like OPUS (Smith 1992a, 1995) and WOFOST (Supit et al. 1994) are able to simulate complete crop rotations and, therefore, more appropriate tools for the simulation of long-term periods. All these models have to be validated by comparing simulated model outputs with the corresponding data measured in the field. Such validation studies with different models have been carried out at different locations (e.g. Diekkrüger et al. 1995; Jamieson et al. 1998; Eitzinger et al. 2004). The main objective of our study is a report about the application of the modelling system THESEUS (Wegehenkel 2000) and the agroecosystem model OPUS (Smith 1992a) for the simulation of soil water balance and crop growth including a validation based on the comparison of simulated with measured model outputs such as soil water contents, pressure heads, soil nitrate contents, aboveground biomass, yield, actual evapotranspiration (AET) rates and ground water recharge (GWR) rates obtained from two field data sets. One data set is obtained from measurements of soil water balance, crop growth and nitrogen balance at an agricultural experimental field and the other data set is obtained from weighable lysimeter measurements of daily AET and GWR rates. Both data sets and the results of the application, as well as the validation of THESEUS and OPUS were presented at an international modelling workshop "Modelling water and nutrient dynamics in soil crops systems" carried out at the Centre of Agricultural Landscape and Landuse Research, Müncheberg from 14 June 2004 to 16 June 2004, organized by the Institute of Landscape Systems Analysis.

Material and methods

Description of the test sites and the data

In our study, we used the two following data sets:

- Bad Lauchstädt, 4 years crop rotation (Franko et al. 2007)
- Lysimeter data, Berlin-Dahlem (Diestel et al. 2007; Zenker 2003)

More information about these data sets can be obtained from the above-mentioned references.

Simulation models

For the Bad Lauchstädt data set, we used the models THESEUS1 and OPUS. For the lysimeter data set, we used the model THESEUS2. THESEUS1 and THESEUS2 were obtained from the modular modelling system THESEUS, which integrates various different complex model components for the purposes of water balance modelling and crop growth simulation. These components are modules for calculating potential and actual evapotranspiration (AET), crop growth, and soil water balance, which are stored in a module library. This system enables the user to create a suitable simulation model for the desired application, taking into account the goals of the project and the actual availability of data by selectively coupling appropriate modules from the module library within a simple, non-graphical user interface (Wegehenkel 2000).

Model THESEUS1

The crop growth module in Theseus1 was obtained from the Wofost6.0 model. It simulates crop growth based on ecophysiological processes and calculates phenological development, CO_2 assimilation, transpiration, growth and maintenance respiration, distribution of assimilates on stem, leaf, fruit and root, as well as dry matter formation (Supit et al. 1994).

The module uses a modified Penman approach for calculating potential evapotranspiration, stated as follows:

$$\text{PET} = \frac{\Delta R_{na} + \gamma \cdot [0.26 \cdot (e_s - e_a)] \cdot [f + c \cdot u(2)]}{\Delta + \gamma} \quad (1)$$

where Δ is the slope of the saturation vapour pressure curve in kPa °C^{-1}; R_{na} is the net radiation defined as evapotranspiration equivalent in mm day^{-1}; γ is the psychrometer constant (=0.65) in kPa °C^{-1}; e_s is the saturated vapour pressure and e_a is the actual vapour pressure, both in kPa; f is an empirical constant (=1) and c is an empirical coefficient calculated from the difference between daily maximum and minimum air temperature; $u(2)$ is the mean wind speed at 2 m height in m s^{-1}. The crop growth is limited by the availability of water and the nutrients nitrogen (N), phosphorus (P) and potassium (K) in the root zone, as well as oxygen shortage caused by stagnant water in the root zone (Supit et al. 1994). The influence of the availability of nutrients on the yield is calculated on yearly basis using a simple empirical mass balance approach taking into account the potential natural supply of N, P, and K in the soil depending on soil properties such as organic carbon or pH plus the extra supply of N, P, and K generated by the use of fertilizer per year and the total nutrient uptake of the crops, which is determined by the water-limited crop biomass accumulation at harvest date. Within this approach, fertilizer applications are defined only as simple yearly N, P, or K supply without taking into account fertilizer type or application date (Jansen et al. 1990; Supit et al. 1994). If the nutrient supply of the soil is lower than the potential nutrient uptake of the crop, the resulting yield is reduced (Jansen et al. 1990; Supit et al. 1994). Therefore, within the dynamic simulation of the development of aboveground crop biomass, no interaction between crop growth and nutrient availability and none of the nitrogen balance components such as mineralization and denitrification or any nitrogen transformations are calculated by this model. This crop growth module was combined with the multilayer soil water balance model Sawah (Ten Berge et al. 1995), which simulates saturated and unsaturated water fluxes in soil profiles and is based on a numerical solution of the flux density and continuity equations. The flux density q in cm day^{-1} is calculated as follows

$$q = -k(h) \cdot \frac{\partial H}{\partial z} \quad \text{with} \quad H = h + z \quad (2)$$

where H is the hydraulic head composed of soil water potential h, including gravitational potential z, both in cm; $k(h)$ is the soil hydraulic conductivity in cm day^{-1}; and z is the depth in cm. The changes in soil water contents per time step Δt are obtained from the continuity equation

$$\frac{\partial \theta}{\partial t} = -\frac{\partial q}{\partial z} + s \quad (3)$$

where s is the sink term in cm day^{-1}, and θ is the soil water content in cm^3 cm^{-3}. For $k(h)$ and $\theta(h)$, the parametrizations according to van Genuchten (1980) were used. The crop growth and the soil water balance module are parts of the modelling system Theseus, which is described in Wegehenkel (2000, 2005). More details about the crop growth model Wofost6.0 included in Theseus can be obtained from Supit et al. (1994).

THESEUS2

The crop growth module in THESEUS1 was not suitable for the vegetation grassland at the 12 weighable lysimeters in Berlin-Dahlem in the time period from 1996 to 1999. Therefore, we had to use a modified vegetation model to calculate transpiration, interception and evaporation of the lysimeters. This model was also obtained from the modelling system THESEUS and used Eq. 1 for calculating potential evapotranspiration (PET) and the soil water balance model SAWAH described in Eqs. 2 and 3. In THESEUS2, the impact of the vegetation on soil water balance is simulated with a semiempirical plant modelling approach according to Koitzsch and Günther (1990). This approach is based on simple empirical two-dimensional table functions with linear interpolation between neighbouring table values for rooting depth (RD), for plant height, and for soil coverage of a given crop for the calculation of transpiration, interception, and evaporation (Koitzsch and Günther 1990). These table functions consist of a certain number of data pairs (crop parameter; Julian day of the year), which correspond to significant phenological stages of the crop. The data were obtained from measurements in long-term field experiments, which were carried out for a sustainable irrigation management (Wenkel et al. 1989). Functions are available for winter and summer cereals, sugar beet, maize, potatoes, peas, faba beans, winter and summer rape, pasture, oats, alfalfa, grassland, as well as coniferous, mixed and deciduous forests. This semiempirical model was extended with a simple module, which calculates crop emergence based on daily heat accumulation. Daily heat accumulation is calculated by totalling the daily mean air temperature beginning at the sowing date of the crop. Crop emergence will occur if a certain heat accumulation is reached. Based on these calculations, the starting points of the functions can be adjusted to the actual climatic boundary conditions (Wegehenkel 2000).

Interception Int (mm day^{-1}) is calculated as a function of maximum interception capacity k (2.5 mm for agricultural crops and grassland, 4–7 mm for forests), plant height (PLH) (in m), and relative crop soil cover days (SCD) (0–1).

$$\text{Int} = k \cdot \text{PLH} \cdot \text{SCD} \qquad (4)$$

Potential evapotranspiration of the vegetated fraction of the soil surface ETPOT (mm day^{-1}) and potential evaporation PE, (mm day^{-1}) for uncovered soil are given by

$$\text{ETPOT} = \text{PET} \cdot F(t) \qquad (5)$$

$$\text{PE} = \text{PET}$$

where $F(t)$ is a conversion factor. For agricultural crops, $F(t)$ is calculated empirically from PLH. For grassland and forests, $F(t)$ is directly obtained from the table function with linear interpolation between neighbouring table values ($0 \leq F(t) \leq 1.4...1.6$). AET is determined by ETPOT and PE, by the relative SCD, RD, and by a root density function, which establishes the potential water extraction from various soil layers. Depending upon available water, the contribution of water from each soil layer to AET decreases linearly from a given threshold to zero at the wilting point (Mirschel et al. 1995; Wegehenkel 2000).

Model OPUS

OPUS is an integrated agro-ecosystem model (Smith 1992a, 1995). Evapotranspiration (E_t in mm day^{-1}) calculations are made as follows:

$$E_t = \frac{(1 + c_w) \cdot \Delta \cdot \dfrac{R_i \cdot (1 - \xi)}{58.3}}{\Delta + 0.68} \qquad (6)$$

where c_w is a coefficient = 0.28 expressing effects of wind and humidity; Δ is the slope of curve for saturation vapour pressure at mean air temperature in mbar K^{-1}; R_i is the incoming solar radiation in langley day^{-1}, ξ is the albedo of the field surface. The model simulates one-dimensional movement of soil water using the well-known Richards equation:

$$C_c(h) \cdot \frac{\partial h}{\partial t} = \frac{\partial \left(k(h) \dfrac{\partial h}{\partial z} + k(h) \right)}{\partial z} + q_e \qquad (7)$$

where $C_c(h)$ is the slope of the curve $\partial \theta / \partial h$, h is soil water pressure head in mm, t is time, z is depth in mm, $k(h)$ is the hydraulic conductivity, and q_e is a sink due to the removal of water by plants and/or losses of water by evaporation from soil surface layers, both in mm day^{-1}. A modification of the Brooks–Corey functions (Brooks and Corey 1964) describes the soil water retention and hydraulic characteristics. The dynamics of the soil microbial system and the carbon, nitrogen and phosphorus cycles in the soil are simulated using an organic residue decomposition model. Processes of nitrification and denitrification are simulated similar to the soil water movement linked to local soil water contents, temperatures, and soil nitrogen contents. The nitrate movement is simulated with a linear equilibrium isotherm model.

Plant growth is simulated with a simple mechanistic model, which relates daily dry matter production to leaf area index (LAI) and solar radiation.

$$dM = c_e \cdot R_i \cdot f_g(DM) \cdot f_L(DM) \cdot f_t \cdot S \qquad (8)$$

in which dM is the daily rate of biomass growth and DM presents total accumulated biomass, both in kg dry matter ha^{-1}; c_e is a plant-specific biomass conversion factor in kg dry matter ha^{-1} langley^{-1}; R_i is daily global radiation in langley (cal cm^{-2} day^{-1}); f_g is a growth limit coefficient, which goes to zero if the plant nears its physiological size limit; f_L is a photosynthetic area factor which is calculated from the LAI, in turn a function of DM; f_t is a senescence coefficient which goes to zero if the plant ontogenesis reaches maturity in thermal time (°C day^{-1}); S is a stress factor, representing the most critical of independently calculated stresses from water, temperature, and nutrients. Following emergence, daily plant material production is allocated among root, stem and leaf, as well as fruit material (Smith 1992a).

Modelling procedures

For the agricultural experimental field Bad Lauchstädt, the simulation period ranges from 1 January 1999 up to 31 December 2002. At this test site, the models OPUS and THESEUS1 were applied. The soil profile of Bad Lauchstädt was discretized by OPUS in 11 numerical layers with a variable thickness down to a depth of 120 cm. This discretization is an internal procedure in OPUS and is determined by the maximum RD of the crops defined by the model user (Smith 1992a, b). In our case, maximum RD for all crops in OPUS was set at 110 cm according to Smith (1992b). The initial nitrate concentrations in the root zone for the calculations of the OPUS model were estimated from field data (Franko et al. 2007). Within the THESEUS1 model, the soil profile

of the Bad Lauchstädt field plot was discretized in 20 layers each with a thickness of 10 cm. The parameters for the crop growth module in OPUS were obtained from Smith (1992a) and those for the crop module included in THESEUS1 from Boons-Prins et al. (1993).

For the lysimeter data sets of the test site Berlin-Dahlem with the time period from 1 January 1996 up to 31 December 1999, the model THESEUS2 was applied using the table functions of RD, SCD, and $F(t)$ of grassland, which was grown at all lysimeters in this time period. The soils of the lysimeters were discretized in 15 layers each with a thickness of 10 cm due to the construction of the lysimeters (Diestel et al. 2007; Zenker 2003).

Within the data sets for Bad Lauchstädt and for the lysimeters at Berlin-Dahlem, only a limited amount of 3–4 measured data pairs soil water content versus pressure head were available (Franko et al. 2007; Diestel et al. 2007; Zenker 2003). With this amount of data pairs, it is difficult to estimate reliable parameters for the soil hydraulic functions, first, water content versus pressure head $\theta(h)$ and, second, hydraulic conductivity versus pressure head $k(h)$ using such programmes like SHYPFIT (Durner 2000). Therefore, for the simulation runs of THESEUS1 and THESEUS2 with the data sets Bad Lauchstädt and Berlin-Dahlem, the parameters of the soil hydraulic functions according to van Genuchten (1980) were estimated using type of soil horizon, bulk density and soil texture using a tabular database published by Bohne et al. (1993). In OPUS, the soil hydraulic functions were parametrised using a modified Brooks–Corey approach (Smith 1992a). The parameters for this approach were calculated using the corresponding parameters obtained from the van Genuchten method.

The saturated soil hydraulic conductivities for Bad Lauchstädt and the lysimeter soils were estimated using soil texture and bulk density according to a tabular data set in AG Boden (1994). Furthermore, the initial soil water contents of all soil layers were set equal to the corresponding field capacity obtained from the corresponding $\theta(h)$ function.

The simulation quality of all models was evaluated using the IA according to Willmot (1982), stated as follows:

$$\mathrm{IA} = 1 - \frac{\sum_{i=1}^{n}(\theta_{\mathrm{sim}} - \theta_{\mathrm{obs}})^2}{\sum_{i=1}^{n}\left[\left|\theta_{\mathrm{sim}} - \theta_{\mathrm{obs-mean}}\right| + \left|\theta_{\mathrm{obs}} - \theta_{\mathrm{obs-mean}}\right|\right]^2} \quad (9)$$

where θ_{sim} and θ_{obs} are the simulated and measured values, and $\theta_{\mathrm{sim-mean}}$ as well as $\theta_{\mathrm{obs-mean}}$ present the corresponding mean values. IA results in a range between 0 and 1, the closer it is to 1, the better the simulation quality (Willmot 1982).

Results and discussion

Bad Lauchstädt

The soil water contents in 5 and 45 cm depths are simulated sufficiently by THESEUS1 (Figs. 1 and 2). This is also illustrated by the corresponding values of IA between 0.60 and 0.62 and a RMSE between 0.080 and 0.081 cm^3 cm^{-3} (Table 1). The model OPUS showed a similar simulation quality only in 5 cm depth (Fig. 2 and Table 2). However, both models overestimated the temporal dynamics of soil water contents in soil layers deeper than 5 cm in comparison with those measured by frequency domain reflectometry (FDR) (Figs. 1 and 2). However, THESEUS1 calculated higher soil water contents and therefore higher soil water storage than OPUS (Figs. 1 and 2). Moreover, both models showed a distinctly lower simulation quality for the pressure heads with a range of IA between 0.10 and 0.44 and a RMSE between 750 and 9923 hPa (Tables 1 and 2). Only in the depth of 90 cm, OPUS showed an IA = 0.60 and a RMSE of 118 hPa for the pressure heads in 90 cm depth (Table 2).

At an international workshop on the validity of 18 different agro-ecosystem models held at Brunswick, Germany, in 1993, the simulation quality were evaluated using also RMSE obtained from simulated and observed soil water contents, as well as pressure heads. At this workshop, RMSE was within 0.020 and 0.126 cm^3 cm^{-3} for the soil water contents and within 100 and 555 hPa for the pressure heads (Diekkrüger et al. 1995). In the study of Jacques et al. (2002), a model for the calculation of field-scale water flow was validated, using also the time domain reflectometry (TDR) method and tensiometer measurements. Here, RMSE for simulated and measured soil water contents showed a range of 0.038 and 0.125 cm^3 cm^{-3} and RMSE for calculated and measured pressure heads was within 31 and 371 hPa (Jacques et al. 2002). In a similar study dealing with the validation of different

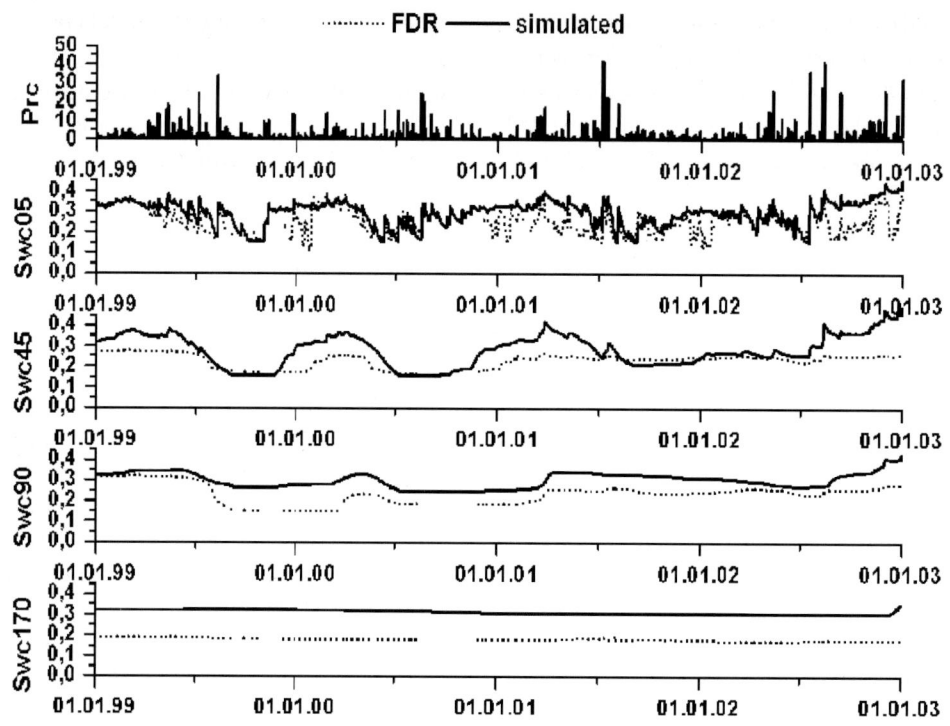

Fig. 1 Daily rates of precipitation (Prc) in mm day^{-1} and observed soil water contents in cm^3 cm^{-3} in 5 cm (Swc05), 45 cm (Swc45), 90 cm (Swc90) and 170 cm depth (Swc170) compared with those simulated by THESEUS1, Bad Lauchstädt, 1999–2002, crop rotation

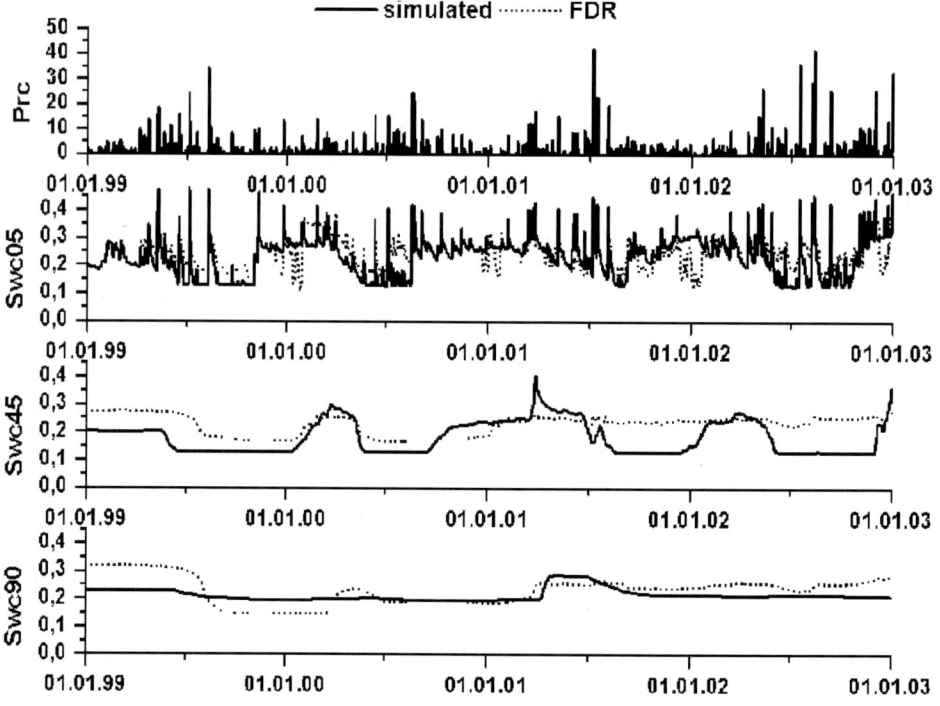

Fig. 2 Daily rates of precipitation (Prc) and observed soil water contents in cm^3 cm^{-3} in 5 cm (Swc05), 45 cm (8Swc45) and 90 cm depth (Swc90) compared with those simulated by OPUS, Bad Lauchstädt, 1999–2002, crop rotation

Table 1 IA and RMSE obtained from the comparison of soil water contents (IA-Swc, RMSE-Swc) and pressure heads (IA-Prh, RMSE-Prh) measured with FDR and tensiometers with those simulated by Theseus 1, Bad Lauchstädt, crop rotation

Depth in cm FDR	IA-Swc	Swc-RMSE (cm³ cm⁻³)	N-Swc	Depth in cm Tensiometer	IA-Prh	RMSE-Prh (hPa)	N-Prh
5	0.62	0.080	1135	–	–	–	–
45	0.60	0.081	1225	45	0.10	9000	771
90	0.55	0.074	1225	90	0.44	750	706
170	0.10	0.133	1215	170	0.28	588	588

Table 2 IA and RMSE obtained from the comparison of soil water contents (IA-Swc, RMSE-Swc) and pressure heads (IA-Prh, RMSE-Prh) measured with FDR and tensiometers with those simulated by Opus, Bad Lauchstädt, crop rotation

Depth in cm FDR	IA-Swc	RMSE-Swc (cm³ cm⁻³)	N-Swc	Depth in cm Tensiometer	IA-Prh	RMSE-Prh (hPa)	N-Prh
5	0.69	0.071	1135	–	–	–	–
45	0.45	0.081	1225	45	0.10	9923	771
90	0.57	0.052	1225	90	0.61	112	706

pesticide leaching models, IA and RMSE obtained from the comparison of soil water contents measured with TDR with those simulated with the pesticide leaching models ranged within 0.65 and 0.91 and between 0.01 and 0.023 cm³ cm⁻³ (Vanclooster and Boesten 2000). In the validation study of Heidmann et al. (2000) with the agroecoystem model Soiln using also TDR measurements, RMSE obtained from simulated and observed soil water contents ranged between 0.01 and 0.035 cm³ cm⁻³. In a similar study with the models Ceres and Swap, the corresponding IA for simulated and observed soil water contents was in a range of 0.58–0.93 and RMSE between 0.007 and 0.068 cm³ cm⁻³ (Eitzinger et al. 2004). Therefore, the simulation quality of Theseus1 and Opus is only partly comparable with those obtained from the above-mentioned references. One likely reason for this insufficient model performance of both models might be the fact, that at Bad Lauchstädt, the results of the FDR probes were not calibrated in the field using corresponding gravimetrically determined soil water contents. In contrast to that, some studies recommended such a field calibration of FDR and TDR probes (e.g. Jacques et al. 2002; Malicki et al. 1996; Wegehenkel 2005). However, the model performance is also insufficient taking into account simulated and measured pressure heads and tensiometers need no field calibration such as the FDR probes. The parameters of the soil water retention function and unsaturated hydraulic conductivity function were estimated from tabular data and not measured directly in the field or in the laboratory using soil core samples from the field. Therefore, this is the most likely reason for these discrepancies between simulated and observed soil water contents and pressure heads, as well as the resulting low simulation quality of both models.

Whilst the annual rates of precipitation range within 409 and 640 mm year⁻¹, simulated annual rates of evapotranspiration, run-off and, GWR obtained from Theseus1 ranged within 416 and 507 mm year⁻¹, no run-off and between −1 and −3 mm year⁻¹ (Table 3). The corresponding rates of ETr, run-off and GWR obtained from Opus ranged within 386 and 609 mm year⁻¹, 0 and 8 mm year⁻¹, as well as 0 and −34 mm year⁻¹ (Table 3). The annual rate of ETr in 2002 simulated by Theseus1 with 416 mm year⁻¹ lead to a higher soil water storage in comparison with Opus (Table 3, Figs. 1 and 2). Opus calculated an annual ETr rate of 609 mm year⁻¹ and a significantly lower soil water storage at the end of the simulation period in 2002 (Table 3, Figs. 1 and 2).

The comparison of the simulated and observed nitrate contents showed, that Opus underestimated nitrate contents in the upper soil compartment 0–20 cm especially in the years 2000–2002 (Fig. 3). During 2000–2002, the time course of the simulated nitrate

Table 3 Annual rates of precipitation (Prc), simulated actual evapotranspiration (ETr), actual transpiration (Trans), run-off and ground water recharge (GWR) in mm year^{-1} obtained from THESEUS1 and OPUS, Bad Lauchstädt, crop rotation

Year	Prc	THESEUS1				OPUS			
		ETr	Trans	Run-off	GWR	ETr	Trans	Run-off	GWR
1999	452	490	255	0	−3	481	288	1	0
2000	409	446	185	0	−2	386	214	0	0
2001	518	507	284	0	−1	433	173	8	−34
2002	640	416	102	0	−1	609	491	1	−2

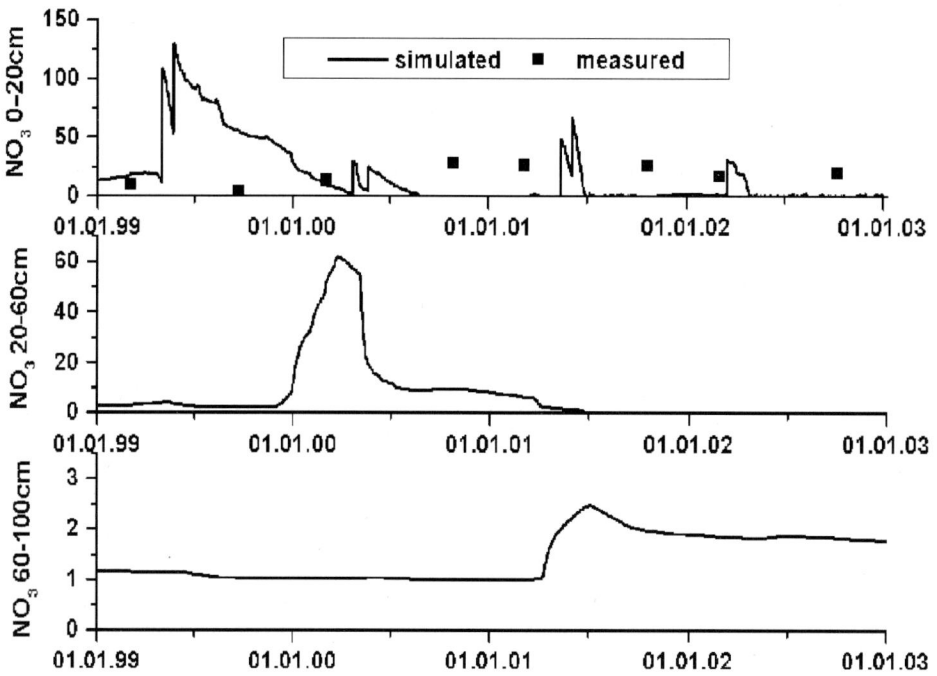

Fig. 3 Simulated (by OPUS model) and observed nitrate contents in kg ha^{-1} in 0–20 cm depth (NO$_3$ 0–20 cm), 20–60 cm depth (NO$_3$ 20–60 cm) and 60–100 cm depth (NO$_3$ 60–100 cm) at Bad Lauchstädt, 1999–2002, crop rotation

contents in 0–20 cm showed mostly values near zero and only the peaks due to the fertilizer application dates could be observed (Fig. 3). The simulated decrease in nitrate contents in 0–20 cm soil depth beginning in the year 1999 was caused by the nitrogen uptake by crops and a nitrate transport into the lower soil compartment 20–60 cm calculated by OPUS, which is indicated also by a simulated strong increase of the nitrate contents in 20–60 cm depth in the first months of the year 2000 (Fig. 3, Table 4). The simulated annual rates of nitrogen-balance components showed only low rates for nitrogen net mineralization and nitrogen leaching despite a high amount of fertilizer application in the first year 1999 with an amount of 344 Kg N ha^{-1} year^{-1} (Table 4). The low annual rates of N leaching were due to the low amount of the simulated annual rates of GWR and the calculated high N uptake rates by the crops (Tables 3 and 4). These N uptake rates ranged between 95 and 188 kg ha^{-1} year^{-1} (Table 4).

In comparison with THESEUS1, OPUS simulated a lower accumulation of the aboveground biomass and calculated also lower yields, especially in the year 2001 for potatoes and in the year 2002 for winter wheat (Fig. 4, Table 5). The crop growth simulated by THESEUS1 is limited mainly by the soil water availability for crop transpiration. OPUS takes also into account the impact of limitations of nitrogen

Table 4 Simulated nitrogen (N)-balance components in kg ha^{-1}, Opus, Bad Lauchstädt, crop rotation

	N fertilizer	N Net mineralization	Denitrification	N leaching	N uptake by crops
1999	344	3	0	0	156
2000	64	−7.5	0	0	95
2001	100	−3.8	0	2.9	103
2002	59	1.6	0	0	188

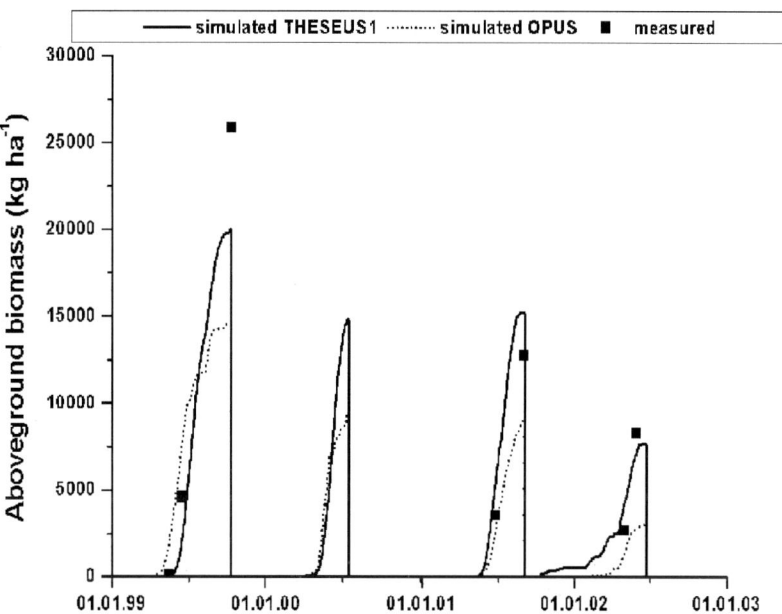

Fig. 4 Measured aboveground biomass accumulation in comparison with those simulated by Theseus1 and Opus, Bad Lauchstädt, 1999–2002, crop rotation

Table 5 Simulated and observed yields in kg ha^{-1} at Bad Lauchstädt

Year	Crop	Theseus 1	Opus	Measured
1999	Sugar beet	12,806	11,658	19,440
2000	Spring barley	2,616	2,908	2,886
2001	Potatoes	10,588	6,895	12,750
2002	Winter wheat	5,045	218	5,094

availability in the soil on the development of aboveground biomass. One likely reason for these differences in the simulated aboveground biomass accumulation and the yields might be the higher soil water contents in the soil profile calculated by Theseus1 in comparison with those simulated by Opus (Figs. 1 and 2). In comparison with Theseus1, this might be an indication for a higher crop water stress simulated by Opus at Bad Lauchstädt. However, except in the year 2001, Opus calculated higher annual rates of transpiration with a range of 173 and 491 mm year^{-1} in comparison with Theseus1 with a range of 185 and 284 mm year^{-1} (Table 3). Furthermore, the comparison of the sums of water stress ratios and of water stress days calculated by both models showed also no indication for a higher crop water stress calculated by Opus in comparison with Theseus1 (Table 6). In Opus and Theseus1, water stress days are defined similar as days, where the simulated ETr rates are below the

Table 6 Simulated nitrogen (N stress) and water (W stress) stresses at Bad Lauchstädt

Year	Opus				Theseus1	
	Sum of N stress ratios	Sum of N stress days	Sum of W stress ratios	Sum of W stress days	Sum of W stress ratios	Sum of W stress days
1999	0.6	12	66.9	86	58.2	78
2000	106.9	114	57.0	75	11.8	14
2001	13.9	44	8.0	13	31.2	42
2002	161.0	181	8.0	9	0.0	0

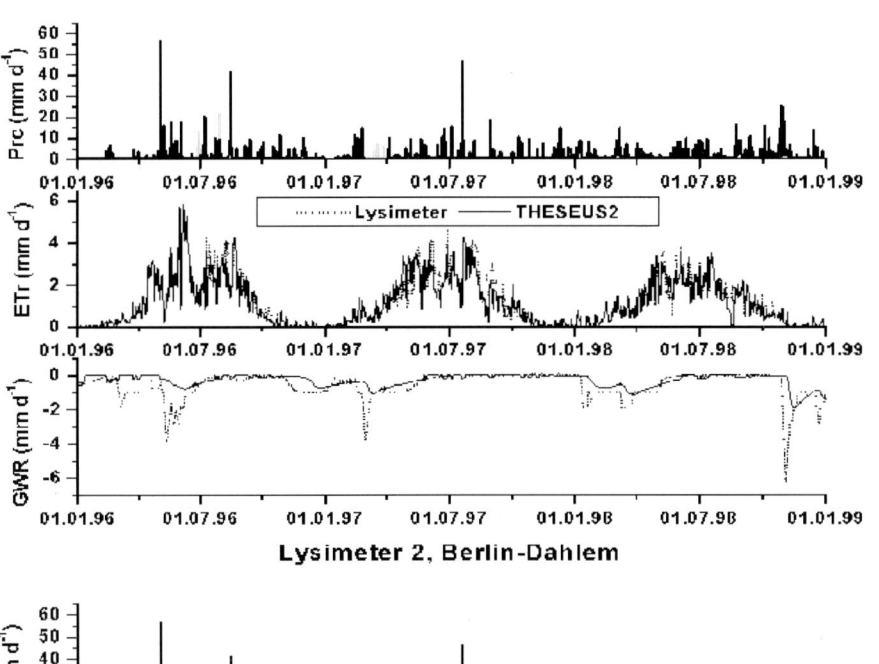

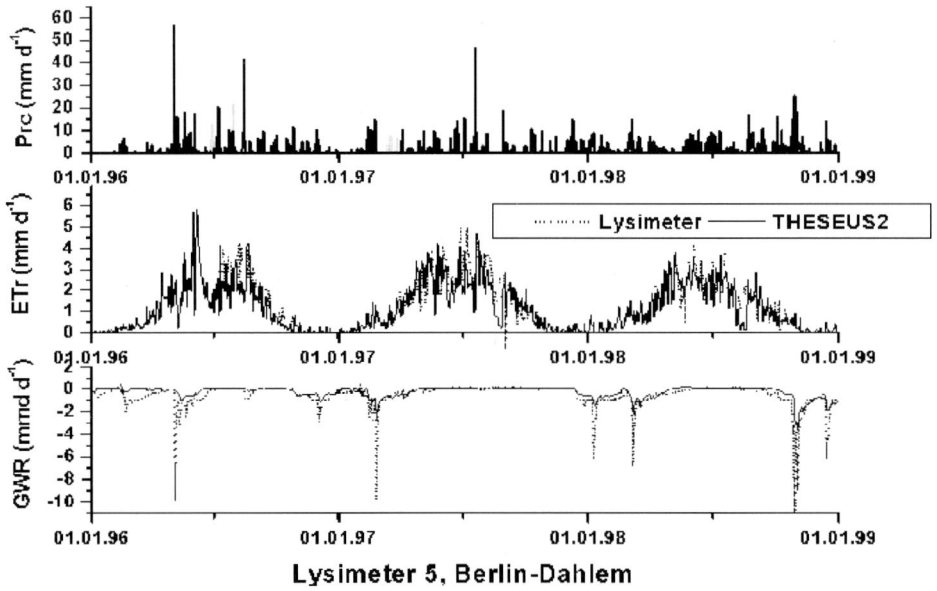

Fig. 5 Daily precipitation rates (Prc), observed actual evpotranspiration (ETr) and ground water recharge rates (GWR) at the bottom of the lysimeters 2 and 5 in comparison with those simulated by Theseus2, Berlin-Dahlem, 1996–1999

PET rates. The sums of water stress ratios are the sums of the simulated ETr: PET ratios in both models. The comparison of the results of the aboveground biomass accumulation, the crop yields, and the sums of water stress ratios, as well as water stress days calculated by OPUS with those simulated by THESEUS1 including the observed crop data indicates, that the lower aboveground biomass accumulation and the yields simulated by OPUS are due to an overestimation of the impact of the limited nitrogen availability in the root zone on the accumulation of the aboveground biomass and yields, especially in the years 2001 and 2002. This is also illustrated by the simulated high amount of nitrogen stress days and the simulated low nitrate contents in the years following 1999 (Table 6, Fig. 3). In OPUS, nitrogen stress days are defined as days, where the simulated actual nitrogen uptake (ANU) rates by crops are below the corresponding potential nitrogen uptake (PNU) rates. The sum of stress ratios is the sums of the ratios ANU:PNU. This overestimation trend is mainly due to the use of uncalibrated or insufficient calibrated crop growth parameters (e.g. Smith 1992b; Smith, 1995).

Lysimeter Berlin-Dahlem

As an example, simulated and observed daily rates of ETr and GWR of lysimeter 2 and lysimeter 5 are shown in Fig. 5. Whilst ETr rates were simulated quite well in comparison with those measured by the lysimeters, THESEUS2 underestimated the peaks of the GWR rates in comparison with the measured ones and, therefore, showed a lower simulation quality in comparison with the ETr rates (Fig. 5). This was a general trend in the modelling results obtained from all lysimeters. This is also illustrated by the results of IA and RMSE obtained from the comparison of simulated with observed ETr and GWR rates in Table 7. For the daily ETr rates, IA and RMSE ranged within 0.82 and 0.91 and within 0.60 and 1.08 mm day^{-1} (Table 7). For the GWR rates, IA and RMSE ranged within 0.37 and 0.70 and within 0.74 and 2.14 mm day^{-1} (Table 7). For the lysimeters 9 and 10, the use of the parameters of the $\theta(h)$ and $k(h)$ functions estimated from the tabular data set of Bohne et al. (1993) lead to numerical instabilities and water balance errors in the calculations of the soil water balance model. Therefore, results of the calculations of ETr and GWR for these both lysimeters were not available and could be presented in our study. It was also difficult to analyse exactly the reasons for the discrepancies between simulated and observed GWR rates. One likely reason may be the fact, that there exists fast flow components in some lysimeters, which could not be simulated by THESEUS1. Similar to the procedure at the test site Bad Lauchstädt, the parameters of the soil hydraulic functions for the lysimeter soils were estimated using soil texture and bulk density using the tabular database from Bohne et al. (1993). Other likely reasons are the so called "Lysimeter effects" such as preferential flow paths caused by an insufficient fit of the soil monolith into the lysimeter and disturbances of the water dynamics by the bottom

Table 7 IA and RMSE for simulated and measured actual evapotranspiration (ETr) and ground water recharge (GWR) rates at the bottom of the 12 lysimeters in Berlin-Dahlem, THESEUS2

	Groundwater depth in cm	IA-ETr	RMSE-ETr (mm day^{-1})	IA-GWR	RMSE-GWR (mm day^{-1})
No. 1	210	0.91	0.61	0.39	1.16
No. 2	210	0.91	0.60	0.59	0.74
No. 3	135	0.86	0.81	0.42	1.91
No. 4	135	0.80	1.08	0.37	2.14
No. 5	210	0.86	0.78	0.70	0.82
No. 6	210	0.87	0.75	0.59	1.05
No. 7	210	0.87	0.77	0.72	0.80
No. 8	210	0.86	0.78	0.63	0.93
No. 9	135	–	–	–	–
No. 10	135	–	–	–	–
No. 11	210	0.83	1.02	0.52	1.06
No. 12	210	0.82	1.01	0.49	1.17

boundary layer of the lysimeter. These effects are addressed in detail in the paper of Diestel et al. (2007) and the study of Zenker (2003).

Conclusions

In our study, the uncalibrated use of all models resulted in more or less distinct discrepancies between the simulated and measured model outputs such as soil water contents, pressure heads, GWR rates, aboveground biomass accumulation and yields.

In our study, all models were using the Richards equation for the calculation of the soil water balance. It is well-known from other publications, that the use of physically based soil water balance models using the Richards equation with retention parameters estimated by pedotransfer functions can sometimes lead to discrepancies between simulated and measured pressure heads, as well as soil water contents (e.g. Vereecken et al. 1992; Christiaens and Feyen 2001; Minasny and McBratney 2002; Wegehenkel 2005). A realistic, physically based simulation of water movement in the unsaturated soil zone requires reasonable estimates of the water retention and unsaturated hydraulic conductivity functions. A variety of studies have revealed the importance of how these unsaturated soil parameters are assessed (e.g. Vereecken et al. 1992; Christiaens and Feyen 2001; Wegehenkel 2005). At the scale of an experimental test field or a lysimeter, a good performance of soil balance models mostly needs a fit of the soil hydraulic parameters to the retention data measured in the field or in the laboratory using soil core samples obtained from the field (e.g. Vereecken et al. 1992; Christiaens and Feyen 2001; Jacques et al. 2002; Wegehenkel 2005). In our study, this was also indicated by the observed numerical instabilities of the calculations of the soil water model in THESEUS2 at lysimeters 9 and 10 of the data set Berlin-Dahlem. This leads also to the conclusion, that a detailed and proper evaluation of soil hydraulic properties should be included in the working plans of field and lysimeter experiments, which are used for model validation.

Furthermore, parameters of crop growth models have to be calibrated at the field scale to get an optimum precision of the model calculations (e.g. Wegehenkel et al. 2004).

Acknowledgements This contribution was supported by the German Federal Ministry of Consumer Protection, Food and Agriculture and the Ministry of Agriculture, and Environmental Protection and Regional Planning of the Federal State of Brandenburg (Germany). A special thanks is given to the data providers from Bad Lauchstädt and Berlin-Dahlem and to the organizers of the modelling workshop as well.

References

AG Boden (1994) Bodenkundliche Kartieranleitung. 4. Aufl. Schweizerbart Stuttgart.

Berge HFM ten, Metselaar K, Jansen MJW, San Agustin EM, Woodhead T (1995) The SAWAH riceland hydrology model. Water Resour Res 31:2721–2731

Bohne K, Horn R, Baumgartl Th (1993) Bereitstellung von van Genuchten Parametern zur Charakterisierung der hydraulischen Bodeneigenschaften. Zeitschrift für Pflanzenernährung und Bodenkunde 56:229–233

Boons-Prins ER, Koning GHJ de, Diepen CA van, Penning de Vries FWT (1993) Crop specific simulation parameters for yield forecasting across the European Community. Simulation reports CABO-TT, No. 32. Wageningen, The Netherlands

Brooks R, Corey A (1964) Hydraulic properties of porous media. Hydrology Paper 3. Colorado State University, Fort Collins CO, pp 22–27

Christiaens K, Feyen J (2001) Analysis of uncertainties associated with different methods to determine soil hydraulic properties and their propagation in the distributed hydrological model MIKE-SHE. J Hydrol 246:63–81

Diekkrüger B, Söndgerath D, Kersebaum K-C, McVoy CW (1995) Validity of agroecosystem models – a comparison of results of different models applied to the same data set. Ecol Model 81:3–29

Diestel H, Zenker Th, Schwartengräber R, Schmidt M (2007) The lysimeter station at Berlin-Dahlem. In: Kersebaum KC, Hecker J-M, Mirschel W, Wegehenkel M (eds) Modelling water and nutrient dynamics in soil-crop systems. Springer, Dordrecht, pp 259–266

Durner W (2000) SHYPFIT 0.22 Users manual. Research report 95.1. Department of Hydrology, University of Bayreuth, Bayreuth, Germany

Eitzinger J, Trnka M, Höbsch J, Zalud Z, Dubrovsky M (2004) Comparison of CERES, WOFOST and SWAP models in simulating soil water content during growing season under different soil conditions. Ecol Model 171:223–246

Franko U, Puhlmann M, Kuka K, Böhme F, Merbach I (2007) Dynamics of water, carbon and nitrogen in an agricultural used Chernozem soil in Central Germany. In: Kersebaum KC, Hecker J-M, Mirschel W, Wegehenkel M (eds) Modelling water and nutrient dynamics in soil-crop systems. Springer, Dordrecht, pp 245–258

Genuchten M van (1980) A closed form equation for predicting the hydraulic conductivity of unsaturated soils. Soil Sci Soc Am J 44:892–898

Godwin DC, Ritchie JT, Singh U, Hunt L (1989) A users guide to Ceres-Wheat-V2.10. International Fertilizer Development Center, Muscle Shoals, AL

Heidmann T, Thomsen A, Schelde K (2000) Modelling soil water dynamics in winter wheat using different estimates of canopy development. Ecol Model 129:229–243

Jacques D, Simunek J, Timmerman A, Feyen J (2002) Calibration of Richards' and convection–dispersion equations to field-scale water flow and solute transport under rainfall conditions. J Hydrol 259:15–31

Jamieson PD, Porter JR, Goudriaan J, Ritchie JT, Keulen H van, Stol W (1998) A comparison of the models ARCFCWHEAT, CERES-WHEAT, SIRIUS, SUCROS2 and SWHEAT with measurements from wheat under drought. Field Crops Res 55:23–44

Jansen BH, Guiking FCT, Eijk D van der, Smaling EMA, Wolf J, Reuler H van (1990) A system for quantitative evaluation of the fertility of tropical soils QUEFTS. Geoderma 46:299–319

Koitzsch R, Günther R (1990) Modell zur ganzjährigen Simulation der Verdunstung und der Bodenfeuchte landwirtschaftlicher Nutzflächen. Arch. Acker-Pflanzenbau Bodenkd 24:717–725

Malicki MA, Plagge R, Roth CH (1996) Improving the calibration of dielectric TDR soil moisture determination taking into account the solid soil. Eur J Soil Sci 23:357–366

Minasny B, McBratney AB (2002) Uncertainty analysis of pedotransfer functions. Eur J Soil Sci 53:417–429

Mirschel W, Wenkel K-O, Koitzsch R (1995) Simulation of soil water and evapotranspiration using the model BOWET and data sets from Krummbach and Eisenbach, two research catchments in North Germany. Ecol Model 81:53–69

Schultz A, Mirschel W (1995) Simulating soil water balance, nitrogen behaviour and biomass components using the agroecosystem model AGROSIM-WINTERWHEAT and data from the North German Krummbach catchment. Ecol Model 81:133–145

Smith R (1992a) OPUS: an integrated simulation model for transport of non-point-source pollutants at field scale. Vol. I. Documentation. USDA ARS-98, USDA, Washington, DC

Smith R (1992b) OPUS: an integrated simulation model for transport of non-point-source pollutants at field scale. Vol. II. Users Manual. USDA ARS-98. USDA, Washington, DC

Smith R (1995) OPUS: simulation of a wheat-sugarbeet plot near Neuenkirchen, Germany. Ecol Model 81:121–132

Supit I, Hooijer AA, Diepen CA van (eds) (1994) System description of the WOFOST 6.0 crop simulation model implemented in CGMS, Vol.1. Theory and algorithms. Joint Research Centre, Commission of the European Communities, EUR 15956, European Communities Luxembourg, 146 pp

Vanclooster M, Boesten JJTI (2000) Application of pesticide simulation models to the Vredepeel data set I. Water, solute and heat transport. Agricultural Water Management 44:105–117

Vereecken H, Diels J, Orshoven J van, Feyen J, Bouma J (1992) Functional evaluation of pedotransfer functions for the estimation of soil hydraulic properties. Soil Sci Soc Am J 56:1371–1378

Wegehenkel M (2000) Test of a modelling system for simulating water balances and plant growth using various different complex approaches. Ecol Model 129:39–64

Wegehenkel M, Mirschel W, Wenkel KO (2004) Predictions of soil water and crop growth dynamics using the agroecosystem models THESEUS and OPUS. J Plant Nutr Soil Sci 167:736–744

Wegehenkel M (2005) Validation of a soil water balance model using soil water content and pressure head data. Hydrological Processes 19:1139–1164

Wenkel KO, Neumeyer M, Mirschel W, Groth R (1989) COBB-integrierte Lösungen zur Berechnung und Prognose der Bodenfeuchte sowie zur operativen Steuerung des Beregnungseinsatzes. Feldwirtschaft Berlin 29(10):456–457

Willmot CJ (1982) Some comments on the evaluation of model performance. Bull Am Meteorol Soc 64:1309–1313

Zenker Th (2003) Verdunstungswiderstände und Grasreferenzverdunstung. Dissertation TU Berlin, 161. http://edocs.tu-berlin.de/diss/2003/zenker_thomas.pdf

CHAPTER FIVE

Integrating a spatial micrometeorological model into the risk assessment for arable crops in hilly terrain

Marco Acutis[1]*, Gianfranco Rana[2], Patrizia Trevisiol[1], Luca Bechini[1], Mario Laudato[1], R Ferrara[2] and Goetz Michael Richter[3]

Abstract Arable crops in hilly terrain may experience additional abiotic stress over crops growing in the plain, which affects crop establishment and productivity. We aim to predict risks and sustainability of crops in hilly terrain for current and future climates. In the STAMINA project we include a proper micrometeorological model (MM) to simulate the effects of terrain on atmospheric variables and their impact on crop growth. A generic crop growth model (CGM) is connected to the MM model, simulating growth and development of an arable crop. The model system estimates the distributed components of the soil water and energy balance with reference to standard weather variables, crop and topographic characteristics. Finally, specific risk indicators, e.g. crop water stress index (CWSI) and thermal stress index (TSI), are calculated to characterize yield reduction and to test mitigation options. The model was successfully calibrated for cereals and sugar beet. Here, we give an overview of this new model and present some results for a 5-year simulation scenario for northern Europe, in which we quantified the effects of terrain (slope and aspect) on meteorological variables and crop yields. The effect of topography on productivity was considerable: south-facing aspects were beneficial for winter wheat in wet years (+2.3 t DM ha^{-1}), similar effects were seen for sugar beet.

Keywords Agrometeorology, Crop growth model, Energy balance, Sugar beet, Terrain, Winter wheat

Introduction

The land use in hilly terrain is currently under review in Europe; contrasting phenomena in different countries are being compared. In many situations, arable crops on slopes show reduced yield compared to those grown in the plain below (Godwin and Miller 2003) due to a range of problems. These include soil erosion and losses of fertility through the run-off, difficulty in the proper use of machinery for soil tillage and fertilization. Differences in soil characteristics between the top and bottom of slopes create different levels of water and nitrogen availability (Malhi et al. 2004). Different local meteorological conditions affect primary production in different hillsides of a catchment (Whitman et al. 1985).

Marco Acutis, Gianfranco Rana, Patrizia Trevisiol, Luca Bechini, Mario Laudato, R Ferrara and Goetz Michael Richter
[1]Department of Crop Science, University of Milano, Via Celoria 2, 20133 Milano, Italy
[2]Istituto Sperimentale Agronomico, Via Celso Ulpiani 5, 70125 Bari, Italy
[3]Rothamsted Research, Harpenden, Hertfordshire, AL5 2JQ, United Kingdom
*Corresponding author:
tel.: +02.5031.6591; fax: +02.5031.6575;
e-mail: marco.acutis@unimi.it

Slope and aspect induce effects on yields through, for example, higher radiation in southern compared to northern aspects and different partitioning between direct and diffuse radiation. Different radiation due to slope and aspect can create differences in air and soil temperatures, evaporation and plant transpiration leading to differences in crop growth and development. One can test the mitigating effects of crop variety, sowing, and harvest dates. Currently, no model is able to simulate the effects of slope and aspect on crop yield, because typically meteorological data are not available on slopes. Some models (e.g. SUCROS; van Ittersum et al. 2003) account for direct and diffuse components of radiation, and it needs to be generalized for sloping areas. It is important to simulate the different crop growth dynamics in a typical catchment of variable slope and aspect, and to evaluate using specific risk indicators. Therefore, we have developed a model system as a part of the STAMINA project (Stability and mitigation of arable systems in hilly terrain; Richter et al. 2003) with the specific objective of simulating crop growth in small catchments (100–1000 ha). The modelling system developed in the STAMINA project integrates a distributed micrometeorological, a hydrological and a crop growth model (CGM). It also computes a set of indicators, which allow a synthetic overview over the results and comparison of different management options.

The objectives of this paper are to (a) briefly describe the components of the model system and (b) evaluate the model system for a sample site. A specific objective is to estimate the meteorological variables in various points in the catchment in comparison with data from a reference location. We assess the interactive effects of the environment (terrain, soil, weather) on yield and discuss how the model could be used to develop management strategies in response to risk indicators for agriculture in hilly areas.

Materials and methods

The STAMINA modelling system

The STAMINA model system simulates agro-meteorological variables, soil water balance (SWB), crop growth and development, and finally derives a set of agro-ecological (risk) indicators. Three components integrate (1) a micrometeorological model (MM) simulating the energy balance for single point of the terrain characterized by elevation, slope, aspect and crop state; (2) a CGM which uses the outputs of the MM model (radiation and temperature), and (3) a SWB model. The model outputs are aggregated as risk indicators, which are related to crop performance and management.

The time step of the MM model is hourly; the CGM works with the same time step for radiation absorption and photosynthesis. Crop development and the SWB model is run at a daily time step. The catchment is divided in to squared cells; the size of the cell is defined by the user. The seasonal indicators are calculated, for each cell, by accumulating hourly values, e.g. temperature in excess of a threshold or as daily values, e.g. moisture deficit. The daily outputs are generated for a single cell. The model system is developed using object-oriented programming (OOP) to ensure modularity and reusability of components. The programming language adopted for all model components is Visual Basic 6.0.

The micrometeorological model

The MM model simulates the energy balance (latent and sensible heat fluxes, net radiation and soil heat flux) and agrometeorological variables (air temperature, global radiation, surface temperature and wind speed) for the centre of each cell. The input variables, measured at a reference weather station in the catchment, are hourly solar radiation, air temperature and humidity, wind speed and direction, and rainfall. The MM model also uses outputs from the CGM and SWB model, such as crop height, leaf area index (LAI) from the CGM, and/or soil cover, rooting depth and soil water content from the SWB. Each cell of the catchment is described with its slope and aspect (azimuth with respect to North), shape of the hill, its length and its maximum height, usually supplied by a Digital Terrain Model or a Geographic Information System. Specific conversion routines are implemented to derive hourly data when only daily data exist.

The crop growth model

The CGM was developed on the basis of SUCROS2 (Goudriaan and van Laar 1994). For the purpose of interfacing CGM and MM for the simulation of processes in hilly terrain, we introduced several modifications to the original SUCROS algorithms. First, we changed the temporal scale of radiation absorption, using hourly outputs from the MM model. The developmental stage (DVS) was simulated according to Enz and Dachler (1997). Second, we added alternative algorithms for the calculation of (a) germination and emergence, (b) root water uptake, and (c) actual evapotranspiration. The new module for the simulation of crop germination and emergence (Bechini et al. 2004) follows a more mechanistic approach, which is an alternative to the thermal time used by many models. While root front advancement is taken from SUCROS root water uptake is simulated following the approach from CERES (Ritchie and, Otter 1985).

The soil water balance model

The soil water dynamic is simulated with a modified tipping bucket model, which allows the soil water content to reach values higher than field capacity. Water moves from one layer to the next by calculating a travel time for the soil layers exceeding field capacity using the SWAT approach (Neitsch et al. 2002). Potential crop evapotranspiration is simulated through calculation of the latent heat flux in the MM energy balance, and actual ET is calculated iteratively from the actual water uptake in the profile.

The indicators module

In addition to yield and hydrological variables we calculate indicators summarizing the climate-related crop performance and production risks: a productivity index (PI) based on actual in reference to potential yield, a crop water stress (CWSI), a thermal stress index (TSI), and water use efficiency (WUE). Indicators, other than yield, will be presented in another paper.

Model calibration

The MM model was tested using data from four micrometeorological stations, running simultaneously on different slopes and aspects, situated in two different catchments, one in Southern Italy (41°46′ N 15°17′ E) and the other in the UK (52°00′ N 00° 25′ W), and not further modified.

Parameters for the CGM were originally taken from Boons-Prins et al. (1993) and calibrated for winter wheat and sugar beet on "plot 1" (non-N-limited treatment) of the Müncheberg Experimental Station (Germany, 52°52′N, 6°15′E) on a sandy loam soil (Kersebaum et al. 2007). We have chosen this treatment because our CGM does not simulate the effects of nitrogen availability on crop growth.

Hydrological parameters were either taken from the original data set (bulk density, field capacity and wilting point, soil water content at saturation). Saturated hydraulic conductivity was using HYPRES (Wosten et al. 1999) pedotransfer functions, on the basis of bulk density, soil particle densities and carbon contents.

Scenario simulations

After calibrating the integrated model for the plain, we simulated scenarios of varying yields on cells of different slopes and aspects using the same soil in every cell. The model was run using Müncheberg weather data for 1992–1997. Winter wheat was sown on day of the year (DOY) 289, and sugar beet on DOY 142. Simulations were carried out for the plain and the sloping land, with two aspects (North and South) and three different slopes (11.5°, 5.7° and 2.9°), located at a height of 50 m above the reference point situated on the plain.

Results and discussion

Table 1 shows some fitting indices for model calibration; for what concerning winter wheat the model was calibrated using Muncheberg data sets.

The simulated flowering date of winter wheat ranged between DOY 146 and 181 (late May to late June). Physiological maturity was simulated to occur between DOY 202 and 239 for wheat and between DOY 261 and 332 (mid-September – end of November) for sugar beet. These dates are close to those observed (Kersebaum et al. 2007).

Table 2 summarizes outputs of the integrated model for the multi-annual scenarios (winter wheat season) assuming different slopes and aspects. On average, the cumulative global radiation was 300–600 MJ m^{-2} higher on the south-facing slope than in the

Table 1 Fitting indices for winter wheat in Müncheberg, plot 1 in year 1993/94 (BBCH = coded phenological stages according to Enz and Dachler, 1997; ABG = aboveground biomass)

	MAE	RMSE	EF	CRM	Slope	Intercept	R^2
BBCH	1.17	0.69	0.998	0.02	1.01	−1.18	0.9988
AGB (kg ha^{-1})	1267.12	877.20	0.853	0.22	0.86	−470.68	0.9114
Yield (kg ha^{-1})	642.03	352.60	0.944	−0.09	1.16	−211.31	0.9975

Table 2 Mean values of selected model outputs for the multi-year simulations (1992/93 to 1996/97) for a plain location and on three different slopes for winter wheat season in Müncheberg

Slope (degrees)	Cumulative global Radiation (MJ m^{-2})		Cumulative actual ET (mm)		Grain yield (t DM ha^{-1})	
	North	South	North	South	North	South
11.5	2578 (±122)	3194 (±158)	726 (±308)	737 (±307)	7.26 (±1.20)	8.21 (±1.36)
5.7	2669 (±127)	3015 (±147)	728 (±308)	734 (±307)	7.47 (±1.25)	7.99 (±1.39)
2.9	2730 (±130)	2919 (±141)	728 (±308)	731 (±308)	7.53 (±1.23)	7.95 (±1.42)
0 (plain)	2601 (±173)		726 (±310)		8.22 (±0.65)	

plain whereas the three north-facing slopes showed relatively smaller differences compared to plain (see to Bennet et al. 1972). The simulated air temperature (not shown in the table) was on average 0.4°C lower on the slopes than in the plain, due to the vertical thermal gradient. The simulated actual evapotranspiration was very similar in all cells, with maximum difference of 1.1%, in spite of radiation being higher (up to +23%). This shows the compensating effect of lower temperatures for evapotranspiration on elevated locations.

The corresponding average wheat grain yields were smaller on the slopes than in the plain and smallest on the north-facing slopes. On average, wheat yield was simulated to be between 0.7 and 1.0 t DM ha^{-1} lower on north-facing slopes than in the plain, decreasing with increasing slope gradient. On southern aspects the lower mean temperature was almost fully compensated by higher radiation interception (maximum average difference of 0.3 t DM ha^{-1}). For a better understanding of these average results, we display simulation outputs for two contrasting growing seasons (dry vs. wet) for each region with respect to their variation in time.

In Fig. 1 the time course of simulated cumulative global solar radiation during winter wheat growth is shown for two contrasting seasons (Fig. 1a, wet 1993/94 and 1b, dry 1996/97). Cumulative global radiation is always higher on the south-facing slope than in the plain. It is not always smallest on the north-facing slope because growth stops about 2 weeks earlier in the warmer plain, while plants still intercept light and grow on north-facing slopes. The diffuse radiation is similar for all three aspects, differences are calculated as direct radiation, which will have an impact on the temperature regime of the different slope locations.

Total biomass production of winter wheat, during the wet season in 1993/94 (Fig. 2a) is higher in the plain than on southern and northern aspects. This is due to the higher mean air temperature in the plain which increases rate of growth and development. This allows the crop to escape the final period characterized by diminishing biomass (from DOY 211) because of air temperatures and the diminishing photosynthetic capacity cause a negative net photosynthesis, which is evident from the yield loss at the end of the wet season (Fig. 2c). In the dry season (1996/97), crop production is higher on the south-facing than on the north-facing aspect (Fig. 2b) due to higher cumulative global radiation. In the plain, higher air temperatures cause a more rapid rate of development and a shorter growth period in than on both elevated slopes. The differences in yields among locations (Fig. 2d) are smaller in the dry than in the wet year (1 compared to 2.3 t DM ha^{-1}).

For sugar beet, simulated, production during the dry-warm year (1997) showed a slight advantage of

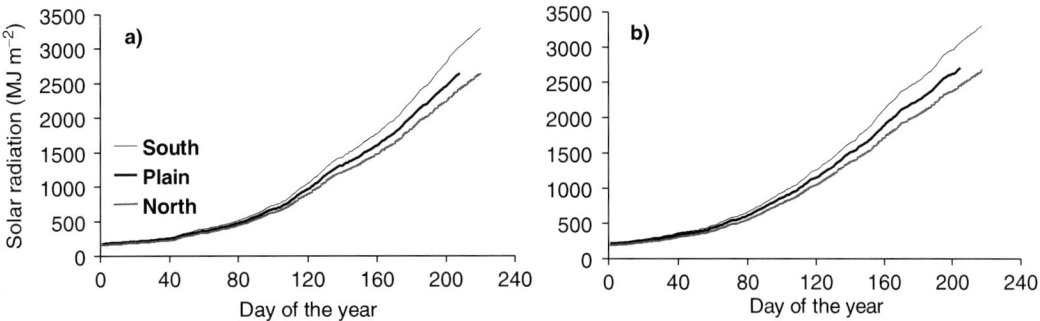

Fig. 1 Global radiation simulated during winter wheat growth in two contrasting seasons at three topographic sites (north - and south-facing with slope 11.5° and plain), in Müncheberg: (a) wet season (1993/94) and (b) dry season (1996/97)

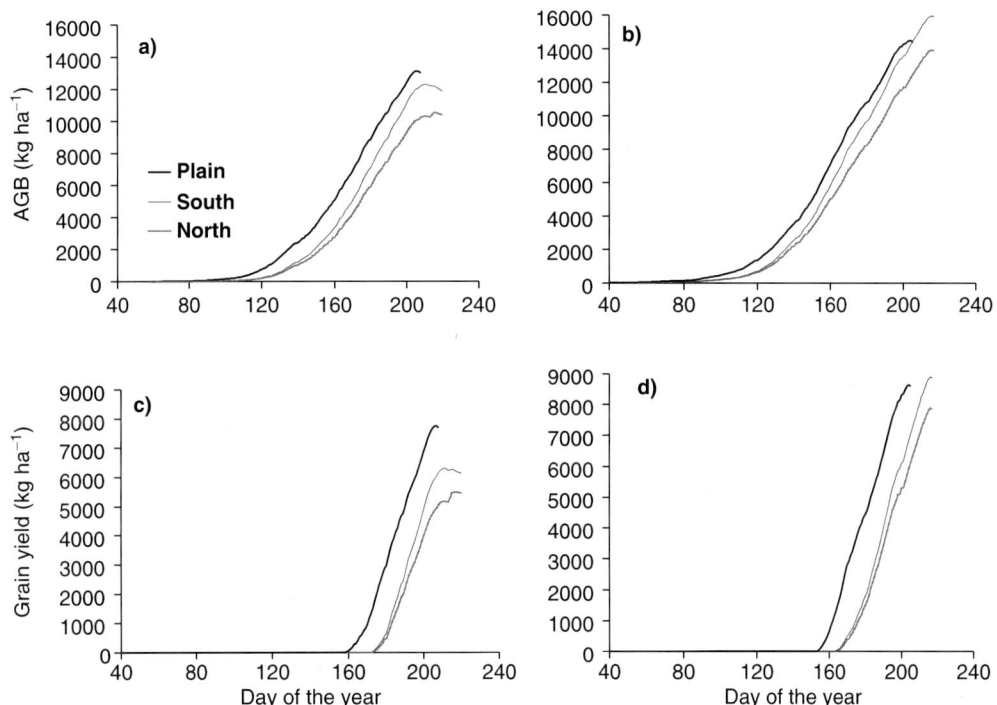

Fig. 2 Crop biomass accumulation simulated in Müncheberg for three topographic sites (north- and south-facing with slope 11.5° and plain); (a, b): above ground biomass (kg ha^{-1}) (c, d): biomass in the storage organ (kg ha^{-1}); (a) and (c): wet season (1993/94); (b) and (d) dry season (1996/97)

productivity (1 t DM ha^{-1}) for the south-facing aspect (Fig. 3a). There is little difference between yields in the plain and in the north-facing slope. As for wheat production in this location, the effect can be explained by lengths of growing period and greater radiation interception on south-facing slopes. Biomass accumulation is faster in the plain, but ceases earlier than on the north-facing aspect due to higher temperature. We found a marginally higher soil moisture deficit on north-facing slopes, which is due to a longer period of evaporation (Fig. 3b).

Overall, these results show the sensitivity of modelled yields to topographic position and weather patterns, a phenomenon which is considered in precision farming (Godwin and Miller 2003). The results show also that simulation is a useful tool to assess interactions

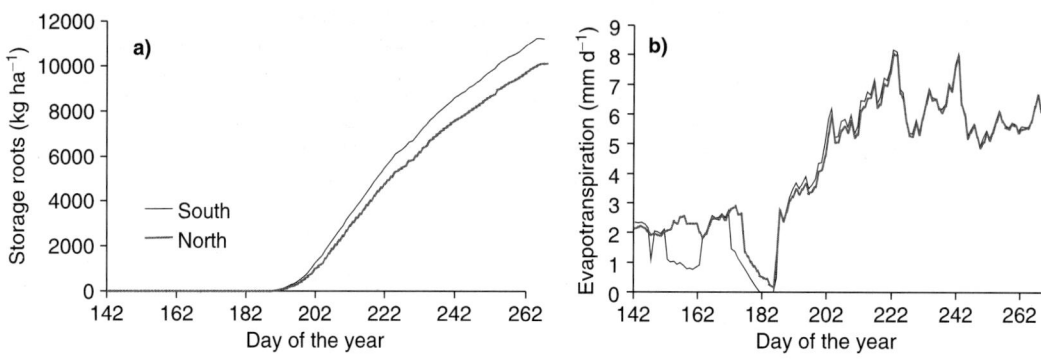

Fig. 3 (a) Biomass accumulation in the storage roots of sugar beet (kg DM ha^{-1}) and (b) actual evapotranspiration (mm day^{-1}), simulated in a dry year (1997) at two sites (north- and south-facing, with 11.5° slope), in Müncheberg

of environmental variables, which may give different results in the same way as annual yield maps differ from year to year: crop yields on slopes depend on the interaction between aspect and seasonal moisture deficit. Evapotranspiration never exhausted available soil water, and so longer runs for future climate scenarios are needed.

Conclusions

We developed a new model system which we subjected to limited testing and which needs more evaluation and development. We can visualize topographic impacts on radiation interception, microclimate and yield, differentiated according to region and annual weather pattern. In Germany, yield variations due to terrain can be as much as 1 t ha^{-1}, 12% in unfavourable years. In the continuation of the STAMINA project, the model will be used to simulate alternative management options for two areas in northern and southern Europe (UK and Italy). Different crops, varieties, sowing dates and other management options will be analysed by comparing specific risk indicators for arable crops in hilly terrain (e.g. crop establishment, water and heat stress) under contrasting climate scenarios (low vs. high emissions). This will enable us to assess current and future risks of crop production in hilly terrain and help to mitigate crop management. An extended version of this model might also be used to calculate differential fertilizer application rates according to landscape position.

Acknowledgements The STAMINA project was funded by the European Commission (QLK-5-CT-2002-01313). Rothamsted Research receives grant-aided support from the Biotechnology and Biological Sciences Research Council of the UK.

References

Bechini L, Laudato M, Trevisiol P, Richter GM, Rinaldi M, Acutis M (2004) A dynamic simulation model for seed germination, seedling elongation and emergence. Proceedings of the 8th European Society for Agronomy Congress, Copenhagen, Denmark, pp 217–218

Bennet OL, Mathias EL, Henderlong PR (1972) Effects of north and south facing slope on yield of Kentucky Bluegrass (Poa pratensis L.) wheat variable rate and time of nitrogen application. Agron J 64:630–635

Boons-Prins ER, de Koning GHJ, van Diepen CA, Penning de Vries FWT (1993) Crop specific simulation parameters for yield forecasting across the European community. Simulation Reports CABO-TT, No. 32, Wageningen, The Netherlands

Enz M, Dachler C (1997) Compendium of growth stage identification keys for mono- and dicotyledonous plants extended BBCH scale, 2nd edn., Novartis

Godwin RJ, Miller PCH (2003) A Review of the technologies for mapping within-field variability. Biosyst Eng 84(4): 393–407

Goudriaan J, van Laar H (1994) Modelling potential crop growth processes. Kluwer, Dordrecht, The Netherlands, pp 238

van Ittersum MK, Leffelaar PA, van Keulen H, Kropff MJ, Bastiaans L, Goudriaan J (2003) On approaches and applications of the Wageningen crop models. Eur J Agron 18(3–4):201–234

Kersebaum KC, Hecker J, Mirschel W, Wegehenkel M (2007) Modelling water and nutrient dynamics in soil-crop systems: a comparison of simulation models applied on common data sets. In: Kersebaum KC, Hecker J, Mirschel W, Wegehenkel M (eds) Modelling water and nutrient dynamics in soil-crop systems. Springer, Dordrecht, pp 1–17

Malhi SS, Johnston AM, Gill KS, Pennock DJ (2004) Landscape position effects on the recovery of 15N-labelled urea applied to wheat on two soils in Saskatchewan, Canada. Nutr Cycl Agroecosyst 68:85–93

Neitsch SL, Arnold JG, Kiniry JR, Williams JR, King KW (2002) Soil and water assessment tool theoretical documentation. Texas TWRI report TR-191. Texas Water Resources Institute, College Station, pp 509

Ritchie JT, Otter S (1985) Description and performance of CERES-Wheat: a user oriented wheat yield model. pp 159–175. In ARS Wheat Yield Project. ARS-38. National Technology Information Service, Springfield, VA

Richter GM, Acutis M, Mayr T, Rana G, Simota C (2003) How realistic is it to use model-based agro-ecological indicators for risk assessment? Contribution to the OECD-Expert Meeting, Land Conservation in Japan, May 2003

Whitman CE, Hatfield JL, Reginato RJ (1985) Effect of slope position on the microclimate, growth, and yield of barley. Agron J 77:663–669

Wosten JHM, Lilly A, Nemes A, Le Bas C (1999) Development and use of a database of hydraulic properties of European soils. Geoderma 90:169–185

CHAPTER SIX

Modelling soil–crop interactions with AGROSIM model family

Wilfried Mirschel* and Karl-Otto Wenkel

Abstract Models for winter cereals (winter wheat, winter barley, winter rye) and sugar beet of the agro-ecosystem model family AGROSIM are presented with their most important algorithms for describing plant (crop growth, development, water and nitrogen stresses, influence of atmospheric CO_2) and soil (temperature, water, nitrogen) processes. The AGROSIM models were tested with data sets obtained from crop rotations at agricultural field plots of Müncheberg (1992–1998) and Bad Lauchstädt (1999–2002) Experimental Stations, Germany. The test runs were realized with non-calibrated model versions. Additional for Bad Lauchstädt site the models were run with calibrated parameters. For analysing the model accuracy time courses, 1:1 diagrams and different statistical values were used like the coefficient of determination, the mean absolute deviation (MABS), the root mean square error (RMSE), the mean bias error (MBE), and a modelling efficiency index (index of agreement (IA) after Willmott (1982). On the basis of model-experiment comparisons their individual importance is shown and discussed.

Wilfried Mirschel and Karl-Otto Wenkel
Institute of Landscape Systems Analysis, Leibniz-Centre for Agricultural Landscape Research (ZALF) e. V. Müncheberg, Eberswalder Strasse 84, D-15374 Müncheberg, Germany
*Corresponding author:
tel.: +49 33432 82277; fax: +49 33432 82334
e-mail: wmirschel@zalf.de

K. Ch. Kersebaum et al. (eds.), Modelling Water and Nutrient Dynamics in Soil–Crop Systems, 59–73. © 2007 Springer.

For both sites, Müncheberg and Bad Lauchstädt, the ontogenesis was calculated with an accuracy of 5, 4 and 7 days for shoot initiation, flowering and maturity, respectively. The non-calibrated model results for Müncheberg site with AGROSIM show that yield and soil water in the rooting depth up to 90 cm are underestimated and biomass and N uptake are overestimated a little, but not significant. In the 0–30 cm soil layer there is the most turbulent soil water dynamic and also the greatest deviation between simulated and observed values.

For Bad Lauchstädt site the usage of parameter recalibrated AGROSIM models in comparison with non-recalibrated ones shows that the model recalibration focus in a significant better model accuracy. The mean absolute deviation (MABS) could be reduced, four times for biomass, three times for yield and by 30 % for N uptake in comparison with the non-calibrated simulation variant. The influence of parameter recalibration on the model accuracy for soil water was not significant and lead to a little underestimation by the AGROSIM models.

For the AGROSIM models in general the model accuracy for plant process values is higher than the model accuracy for soil process values for both tested crop rotations at Müncheberg and Bad Lauchstädt sites.

Keywords Agro-ecosystem model, Crop growth, Model accuracy, Model recalibration, Model test, Nitrogen uptake, Soil water, Statistical measures, Sugar beet, Winter cereals, Yield

Introduction

In mainly agricultural used landscapes the regional water, matter, and energy turnover and balance are characterized by agro-ecosystems with different types of production systems (intensive, organic, extensive, etc.). On the basis of actual knowledge it is necessary to evaluate the medium- and long-term consequences of economically or politically induced land use changes, changes in the agricultural production systems or expected climate changes. None of the influencing factors operates separately. To investigate the urgent questions with real experiments on regional level is excluded for practical reasons concerning time and space scale. Therefore, the evaluation of the complex influences requires another approach. Agro-ecosystem models or integrated models proved to be a promising attempt to become able to follow ramifications of acting chains and feedback circles in agro-ecosystems quite reliable. Agro-ecosystem models play an important role within the landscape modelling.

With the aid of agro-ecosystem models it is possible to describe the influences of agro-management (sowing date, fertilization, irrigation, etc.), soil characteristics, and climate (temperature, precipitation, radiation, atmospheric CO_2 content) on the most important ecosystem processes, e.g. soil water balance, soil nitrogen dynamics, ontogenesis of crops, biomass production, carbon balance, and yield.

Models for agricultural crops vary in complexity from simple ontogenesis and regression-based crop growth models, to detailed and mechanically based soil process and crop growth models, to very complex agro-ecosystem models which examine the influences of pests and management options on soils and plants. The mathematical formalism applied extends from statistical approaches, to algorithmic approaches, and to neuro-fuzzy models. Overviews of actually available agro-ecosystem models are given by Mirschel et al. (1997) and in the CAMASE register of agro-ecosystem models (CAMASE 2001). Only a few of the models described in the literature are really suitable for modelling landscapes. Most of the models are validated only for special sites.

The agro-ecosystem model family AGROSIM was developed especially for the agriculturally used moraine landscapes of northeast Germany. AGROSIM is an acronym for **AGRO**-ecosystem **SIM**ulation. The main goal for developing AGROSIM with focus on crop growth was to have a tool to investigate the consequences of different management strategies on farm level, modes of land use and climate change effects at regional level on biomass production and yield within crop rotations on the basis of computer-based simulation runs. Until now, there are models for winter wheat (AGROSIM-WW), winter barley (AGROSIM-WG), winter rye (AGROSIM-WR), sugar beet (AGROSIM-ZR), and different catch crops (AGROSIM-ZF), which are validated for weather and soil conditions at various locations in northeast Germany. The application of the agro-ecosystem model AGROSIM-WW for winter wheat was investigated for different European countries between the western part of France and Krasnodar region (Russia) in the west–east direction and Kaliningrad region (Russia) and south Italy in the north–south direction (Mirschel et al. 2004).

Coherent agro-ecosystem data sets consisting obtained data courses for soil, plant and weather processes are an important prerequisite for site-specific model parameterization and model applications to other sites. Comparative agro-ecosystem model studies applied to the same data set(s) for different locations were realized on a limited scale only (e.g. Diekkrüger et al. 1995, Jamieson et al. 1998).

This paper is a report about the agro-ecosystem model AGROSIM applied to data sets from the two different experimental stations Müncheberg and Bad Lauchstädt located in the eastern part of Germany. The two data sets, as well as the application results were presented at the international workshop "Modelling water and nutrient dynamics in soil crop systems" on the validation of agro-ecosystem models held at the Leibniz-Centre for Agricultural Landscape Research (ZALF e.V.) Müncheberg, Germany, in June 2004, organized by the Institute of Landscape Systems Analysis.

Material and methods

Test site and data

The AGROSIM model family was applied to data sets from a field experiment (crop rotation, 1999–2002) at the Bad Lauchstädt Experimental Station of the Centre for Environmental Research Leipzig-Halle (UFZ) located in Saxony-Anhalt, and from a field experiment (crop rotation with an intensive production system, 1992–1998) at Müncheberg Agricultural Experimental Station of the Leibniz Centre for Agricultural Landscape Research (ZALF) e.V. located in the eastern part of the Federal State of Brandenburg, Germany. The experimental sites, the experimental design, the management practises, as well as the data sets with soil, crop, and weather data are described in detail by Franko et al. (2007) for Bad Lauchstädt and by Mirschel et al. (2007) for Müncheberg.

Description of AGROSIM models

To fit the demands on agro-ecosystem models imposed by landscape modelling objectives, all crop models in the AGROSIM family describe the response of entire crop stands as affected by environmental factors, site characteristics, and development. In the models, homogeneous crop stands are assumed. All models require only meteorological standard values as driving forces and regionally available input parameters. The individual AGROSIM models are based on the same modelling philosophy, have a similar model structure on the basis of modules, use rate equations to describe process dynamics, operate on a minimum time step of one day, and are weather-, site- and management-sensitive. In all models, time step related interactions between the modules for climate, biomass increase, ontogenesis, and soil process description occur. Dependent on meteorological driving forces and agro-management in the models ontogenesis, yield, and growth are described between sowing and harvest. In the following the AGROSIM models for winter cereals (winter wheat, winter barley, winter rye) are described more in detail and the AGROSIM model for sugar beet is represented in a short form only because of the same model philosophy and the similar model structure.

Model for winter cereals

The general structure of the AGROSIM models for winter cereals including the couplings of soil and plant processes within the model are illustrated in Fig. 1.

Within AGROSIM for winter cereals, state variables and rates, such as actual and potential evapotranspiration (AET/PET); soil water content in different soil layers; soil nitrogen availability; mineralization; nitrogen and water uptake; percolation; soil temperature; ontogenesis; root, aboveground, green, dead and grain biomass; respiration; rate of grain filling, are obtained as outputs with daily time resolution.

Crop growth and development One of the most important subprocesses within AGROSIM is the process of ontogenesis which acts as a time-related control variable for other subprocesses. Other processes are initiated, stopped, accelerated or slowed down by ontogenesis. The ontogenesis is influenced by temperature, water and nitrogen stress, and is described according to the Feekes code (Feekes 1941) for crop development and transformed into the decimal code for crop development of Zadoks (Zadoks 1974) and into the BBCH code for crop development (Hack et al. 1992). A detailed model description of the ontogenesis subprocess is given by Mirschel et al. (1990) for winter wheat and Mirschel et al. (2005) for winter rye and winter barley.

Ontogenesis acts on the second important subprocess within AGROSIM models for cereals, namely carbon assimilation, which obtains daily increments via stand photosynthesis. The assimilation rates are based on a maximum photosynthetic rate per unit green biomass which is modified by environmental and management factors:

$$\Delta A = \text{PHO}_{max} \cdot \text{GBM} \cdot f_1(\text{QP}) \cdot f_2(\text{TP}) \cdot f_3(\text{BM}) \\ \cdot f_4(\text{WS}) \cdot f_5(\text{WL}) \cdot f_6(\text{NF}) \cdot f_7(\text{CO}_2) \\ \cdot (2 - \text{PE}) \quad (1)$$

where ΔA is the daily gross photosynthetic rate (g m^{-2}), PHO$_{max}$ is the maximum daily gain per unit green biomass, GBM is the current green biomass (g m^{-2}), f_1(QP) adjusts for the effects of daily photosynthetic active radiation input ($0 \leq f_1(\text{QP}) \leq 1$), f_2(TP) adjusts the daily temperature effects ($0 \leq f_2(\text{TP}) \leq 1$), f_3(BM) adjusts for the influences of vegetative biomass ($0 \leq f_3(\text{BM}) \leq 1$), f_4(WS) is a reduction factor for short-term water stress ($0 \leq f_4(\text{WS}) \leq 1$), f_5(WL) is a factor for long-term water stress ($1 \leq f_5(\text{WL}) \leq 1.2$), f_6(NF) is a reduction factor for nitrogen stress ($0 \leq f_6(\text{NF}) \leq 1$), f_7(CO$_2$) expresses the influence of atmospheric CO$_2$ concentration according to a reference concentration of 350 ppm and (2-PE) and PE are daytime and nighttime length, respectively. The 24-h period day length is scaled to the interval (0,2).

Reference to green biomass instead of leaf area index (LAI) is essential for cereals because not only the leaves, but also the steams and, especially, ears

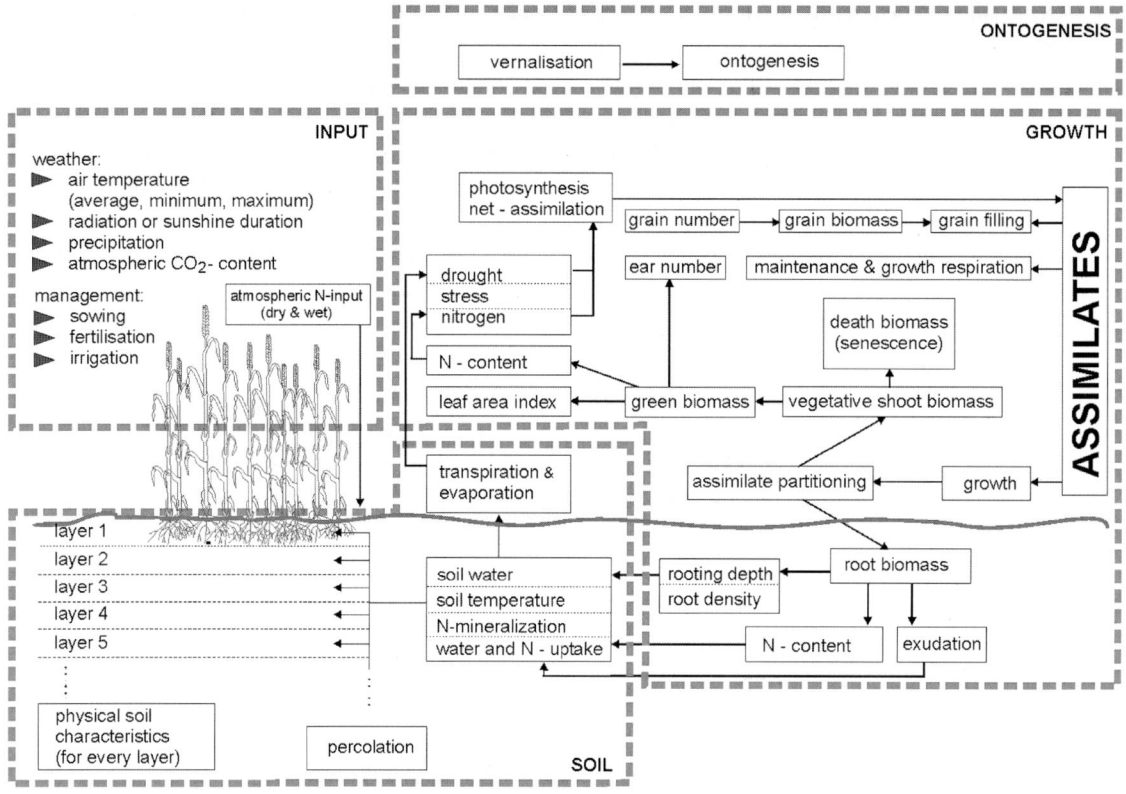

Fig. 1 Model structure of AGROSIM models for winter wheat, winter barley and winter rye as described in the text

contribute to assimilate production. For winter barley up to 30% of ΔA is achieved in this manner, as reported above (Reining 1995).

The radiation influence on carbon gain is described by the following empirically fit function:

$$f_1(QP) = \begin{cases} 0; & QP \leq Q_{min} \\ \left[\dfrac{\dfrac{(QP - Q_{min})^{0.9}}{(QP - Q_{min})^{0.9} + Q_S}}{\dfrac{(13,000 - Q_{min})^{0.9}}{(13,000 - Q_{min})^{0.9} + Q_S}}\right]^{1.13}; & Q_{min} \leq QP < 13,000 \\ 1; & QP \geq 13,000 \end{cases} \quad (2)$$

where QP is the daily photosynthetically active radiation [kJ m^{-2} day^{-1}], Q_{min} is the lower radiation threshold for assimilation and Q_s is a half-saturation constant.

The temperature effects on daily photosynthetic rate are taken into account using ontogenesis-dependent optimum function ($f_2(TP)$) for a temperature range between 0°C and 40°C (see Fig. 2).

Until flowering, a significant influence of accumulated vegetative biomass occurs which is described in the AGROSIM models for cereals by the following empirically fit function:

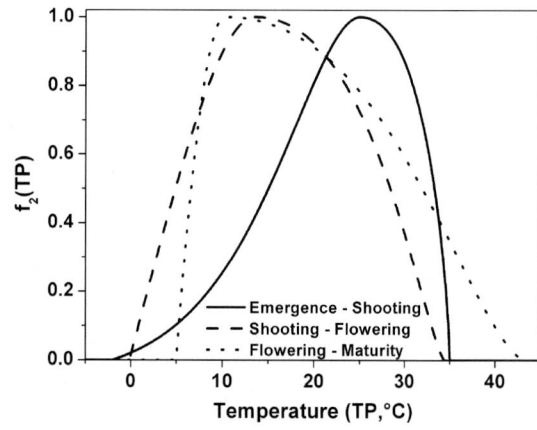

Fig. 2 Ontogenesis-dependent temperature effects ($f_2(TP)$) on daily photosynthetic rate parameterized for winter barley

$$f_3(\text{BM}) = 1 - \text{DENS} \cdot \text{BM} - \text{DELN} \cdot \ln(\text{BM}) \quad (3)$$

where BM is the total vegetative biomass (g m^{-2}) and DENS and DELN are linear and logarithmic density feedback parameters, respectively.

Daily photosynthetic gain is influenced in the model by water and nitrogen stress. The daily ratio of AET to PET is utilized as an indicator of whether water limitations on canopy gas exchange occur. The degree of influence on photosynthesis is established by comparing calculated actual AET/PET with a crop- and ontogenesis-dependent AET/PET threshold for an optimal plant water supply. If the actual AET/PET drops below this threshold, the influence of short-term water stress ($f_4(\text{WS})$) on photosynthesis is calculated as follows:

$$f_4(\text{WS}) = \begin{cases} 1; & \text{WG} \leq \dfrac{\text{AET}}{\text{PET}} \\ \left(\dfrac{\dfrac{\text{AET}}{\text{PET}} - \text{WTHR}}{\text{WG} - \text{WTHR}}\right)^{\text{WEXP}}; & \text{WTHR} \leq \dfrac{\text{AET}}{\text{PET}} < \text{WG} \\ 0; & \dfrac{\text{AET}}{\text{PET}} \leq \text{WTHR} \end{cases} \quad (4)$$

where AET is the actual evapotranspiration (mm), PET is the potential evapotranspiration (mm), WG is the crop- and ontogenesis-dependent AET/PET threshold, WTHR is the minimum AET/PET value allowing transpiration and WEXP is a curvature exponent.

Green biomass grown under water stress conditions is more effective in photosynthesis because of a higher stomata density. When a given duration of water stress and cumulative stress experience are exceeded, growth compensation effects are permitted. This positive long-term water stress effect linearly increases the factor $f_5(\text{WL})$ up to 1.2 over a period of 18 days and when linearly decreases $f_5(\text{WL})$ over a 40-day period until the initial level of the photosynthetic efficiency is reached.

The daily AET/PET values are calculated on the basis of the soil water model included in the AGROSIM models using daily plant values for stage of ontogenesis, rooting depth, and degree of coverage from the crop module.

The nitrogen stress influence on carbon gain ($f_6(\text{NF})$) reflects the nitrogen deficit of the aboveground biomass. The degree of influence on photosynthesis is established by comparing calculated actual nitrogen content in the aboveground biomass with a crop- and ontogenesis-dependent nitrogen content threshold for an optimal plant nitrogen supply. If the actual nitrogen content in the aboveground biomass drops below this threshold, the influence of nitrogen ($f_6(\text{NF})$) on photosynthesis is calculated as follows:

$$f_6(\text{NF}) = \begin{cases} 1; & \text{NG} \leq N_{\text{SP}} \\ \left(\dfrac{N_{\text{SP}} - \text{NTHR}}{\text{NG} - \text{NTHR}}\right)^{\text{NEXP}}; & \text{NTHR} < N_{\text{SP}} < \text{NG} \\ 0; & N_{\text{SP}} \leq \text{NTHR} \end{cases} \quad (5)$$

where N_{SP} is the nitrogen content of the aboveground biomass, NG is the crop- and ontogenesis-dependent nitrogen content threshold for an optimal plant nitrogen supply, NTHR is the minimum nitrogen content value in the aboveground biomass allowing crop growth and NEXP is a curvature exponent.

The daily values for plant-available soil nitrogen are calculated using the simple soil nitrogen balance model which is included in the AGROSIM models and which receive daily information on plant canopy status and crop development from the crop growth module. The algorithm for describing the influence of atmospheric CO_2 concentration on stand photosynthesis ($f_7(CO_2)$) basis on a Michaelis-Menten equation and is expressed according to a reference concentration of 350 ppm as follows (Mirschel and Wenkel 1998):

$$f_7(CO_2) = \dfrac{\dfrac{CO_2 - C_0}{k_1 + CO_2 - C_0}}{\dfrac{350 - C_0}{k_1 + 350 - C_0}}; \quad k_1 = 220 + 0.395 \cdot \text{QP};$$
$$C_0 = 80 - 0.09 \cdot \text{QP} \quad (6)$$

where CO_2 is the atmospheric CO_2 concentration (ppm), k_1 and C_0 are the half-saturation and the compensation CO_2 content, respectively. Both parameters proved to be dependent on the radiation are calculated according to Hoffmann (1993).

The produced assimilates initially enter an assimilate pool from which they are used sequentially for respiration, grain filling, and biomass growth (see Fig. 1). If there are not sufficient assimilates for respiration and grain filling, additional assimilates are supplied by translocation from green or root biomass. The vegetative biomass is partitioned into shoot and root biomass with fixed shoot to root ratio. Losses of assimilates allocated to roots occur by exudation and by root mortality. At the end of ontogenesis, senescence occurs, resulting in a loss of green biomass and a reduction in photosynthesis. Additional senescence is accelerated by water and nitrogen stress.

The grain filling rate depends on ontogenesis, grain number per ear, and the ear number per square meter and is calculated between flowering and maturity. The grain number per ear depends on water and nitrogen stress, while ear density depends on the dynamics of aboveground vegetative biomass growth between shooting and flowering.

Soil temperature The soil temperature within the AGROSIM models is calculated using the SOIL_TEM model according Suckow (1986). The mean air temperature as the only model input is the upper boundary condition and is damped in dependence on crop coverage. The soil temperature dynamic is described using a parabolic differential equation for the one-dimensional soil heat flux as follows:

$$c_T \rho_D \frac{\partial}{\partial t} T(z,t) = \frac{\partial}{\partial z}\left(\lambda_T(W(z,t)) \frac{\partial}{\partial z} T(z,t)\right) \quad (7)$$

where c_T is the specific soil heat capacity, ρ_D is the dry soil density, $T(\cdot,\cdot)$ is the time- and depth-dependent soil temperature, z is the soil depth, and $\lambda_T(\cdot)$ is the heat conductivity function.

Soil water The soil water component of AGROSIM models bases on the empirical soil water capacity models BOWA according to Koitzsch (1977) and Koitzsch et al. (1990) and BOWET, a further development of BOWA, according to Mirschel et al. (1995). The soil water model used in AGROSIM models takes into account the PET, the interception by crop stands, milting of snow, gravity-driven water percolation, and AET. Al calculations are realized in 1-day steps for 20 soil layers of 10 cm thickness each down to a depth of 200 cm on a semiempirical way. The model inputs are weather data (precipitation, mean air temperature, global radiation), soil and crop parameters. The soil water balance in each layer is assumed to be equal to the sum of precipitation plus the change in storage minus the losses due to AET and gravity-driven percolation.

$$\begin{aligned}W_{t,1} &= W_{t-1,1} + P_t - E_{t,1} - G_{t,1}\\ W_{t,k} &= W_{t-1,k} + G_{t,k-1} - E_{t,k} - G_{t,k};\\ &\quad k = 2,\ldots,20\end{aligned} \quad (8)$$

where t is the Julian day, k is the number of the layer, $W_{t,k}$ is the soil water content (mm), P_t is the precipitation (mm), $E_{t,k}$ is the contribution of layer k to the evapotranspiration (mm), and $G_{t,k}$ is the gravity-driven percolation from layer k (mm).

Soil nitrogen The soil nitrogen component of AGROSIM models bases on a simple balance model. The amount of soil mineral nitrogen NS_t (g N m^{-2}) in the rooted layer is calculated for each day t by the following balance:

$$NS_t = NS_{t-1} + NM_{t,0-60} + FE_t + NDD_t + NDW_t - NU_t - NL_t \quad (9)$$

where $NM_{t,0-60}$ is the mineralized nitrogen in the soil layer 0–60 cm (g N m^{-2} day^{-1}), FE_t is the nitrogen fertilization (g N m^{-2}), NDD_t and NDW_t are the dry and the wet atmospheric nitrogen deposition (g N m^{-2}), respectively, NU_t is the nitrogen uptake by the crop (g N m^{-2}), and NL_t is the nitrogen leaching out of the rooting depth (g N m^{-2}).

Mineralization depends on soil water and soil temperature, which are calculated in the soil water an soil temperature model components of AGROSIM models, respectively. $NM_{t,0-60}$ is calculated according to Stanford (1977) and Rausch et al. (1985).

A detailed description of existing AGROSIM models for winter cereals including a representation of validation results for different locations, years and cultivars, is given by Wenkel and Mirschel (1995) and Mirschel et al. (2001).

Model for sugar beet

The structure of the AGROSIM model for sugar beet including the couplings of soil and plant processes within the model are illustrated in Fig. 3.

Based on the daily potential transpiration, the atmospheric CO_2 concentration, and the LAI a daily potential assimilation rate is calculated. Taking into account the daily actual transpiration a daily water stress is defined, which is the sum of the water stress effect from the actual day and the effects of water stress events from 20 days before. The last ones are different in their effects on the plant reactions against water stress. Using the actual water stress value the daily potential assimilation rate is reduced to the actual assimilation rate. After assimilate partitioning the daily growth rates for leaf

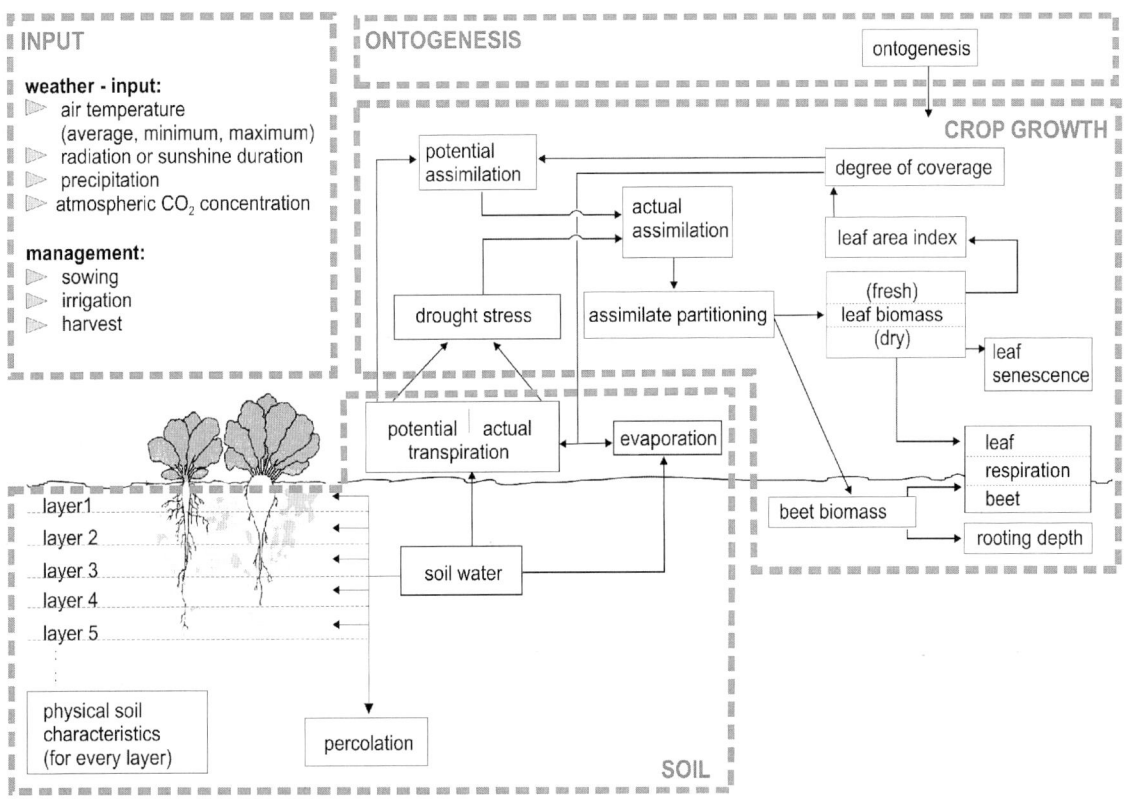

Fig. 3 Model structure of AGROSIM model for sugar beet

and beet can be obtained. The assimilate partitioning depends on ontogenesis, actual water stress and the development stage after emergence. The respiration rates are influenced by the accumulated biomass and the potential transpiration and are different for leaf and for beet. On the basis of the accumulated assimilates the dry leaf biomass, as well as the beet biomass can be calculated using separate ontogenesis-dependent assimilate – dry biomass – ratio functions for beet and for leaf. From the 70th day after emergence onwards a senescence process for leaf biomass is taken into account.

Because the LAI is an important input value for the calculation of potential assimilation and LAI highly is correlated with the fresh biomass of the leaves it is necessary to model the fresh leaf biomass first. This is realized on the basis of dry leaf biomass using the actual water supply status of the leaves taking into account a short-term water stress factor for the last decade on the one hand, and a long-term water stress factor cumulative over the whole time period between emergence and the actual day on the other hand.

All soil water-related state variables (actual transpiration, potential transpiration, etc.) are calculated using the same soil water model like in the AGROSIM models for winter cereals described above.

In the following an overview of the qualitative functional connections/interactions most important for the growth processes is given:

Ontogenesis (ON) $= f(TM)$
Potential transpiration (PT) $= f(TM, GR)$
Actual transpiration (AT) $= f(SWC, BG, RD)$
Potential assimilation (PA) $= f(PT, LAI, CO2, GR)$
Actual assimilation (AA) $= f(PA, BG, PT, AT)$
Assimilate portitioning (AP) $= f(ON, BG, PT, AT, SD)$
Dry beet biomass (DBB) $= f(AA, AP, BR, ON)$
Dry leaf biomass (DLB) $= f(AA, AP, LR, ON)$
Fresh leaf biomass (FLB) $= f(DLB, PT, AT, ON)$
Leaf area index (LAI) $= f(FLB)$
Beet respiration (BR) $= f(DBB, PT, CO2)$
Leaf respiration (LR) $= f(DLB, PT, CO2)$
Rooting depth (RD) $= f(ON)$
Degree of coverage (BG) $= f(LAI)$

where TM is the daily mean air temperature, GR is the global radiation, SWC is the soil water content in the rooting depth, SD is the site-specific optimal sowing date, and CO_2 is the atmospheric CO_2 concentration.

Within AGROSIM model for sugar beet, state variables and rates, such as actual and potential transpiration; evaporation; soil water content in different soil layers; water uptake; percolation; ontogenesis; beet, leaf (fresh and dry) and dead biomass; respiration; LAI; degree of coverage are obtained as outputs with a daily time resolution.

A detailed description of the AGROSIM model for sugar beet including the representation of validation results for different locations, years and cultivars, is given by Wenkel and Mirschel (1995) and Mirschel et al. (2002).

Model simulation runs

The simulation runs with the AGROSIM models for winter wheat, winter barley and winter rye were realized using the SONCHES (simulation of nonlinear complex hierarchical ecological system) simulation system (Wenzel et al. 1986) with the help of which the models have been implemented. The simulation runs with the AGROSIM model for sugar beet basis on the software solutions AGROSIM-ZR for the growth and development processes and BOWET for the soil water processes, both outside of SONCHES.

The model simulations were run for two data sets, first for a crop rotation with an intensive management (plot 1) from Müncheberg experimental site on a loamy sandy soil from 1992 until 1998 (Mirschel et al. 2007), and second for a crop rotation from Bad Lauchstädt experimental site on Chernozem soil from 1999 until 2002 (Franko et al. 2007). For the Müncheberg site all crops within the crop rotation were simulated without any model recalibration, while for the Bad Lauchstädt site (Franko et al. 2007) only sugar beet (1999) and winter wheat (2001/2002) were simulated both, with and without cultivar parameters adaptation. The adaptation was necessary because Bad Lauchstädt is located in central Germany, outside of the AGROSIM model validation region.

All crop models in the simulation started anew at sowing dates both for Müncheberg site and for Bad Lauchstädt site. The initial conditions to start the models were derived either from observed values or from mean assumptions like in the case of the starting value for green and root biomass which are calculated on the basis of sowing density (kg seed grains m^{-2}). For winter wheat as starting values for green and root biomass at the beginning of growth, i.e. emergence, 8 and 2 g m^{-2} were assumed, respectively. For calculation of nitrogen mineralization in the layer 0–60 cm an available mineralization potential of 550 kg N ha^{-1} for the Müncheberg site and of 850 kg N ha^{-1} for the Bad Lauchstädt site was assumed.

On the basis of atmospheric deposition measurements it is known that 40–50 kg N ha^{-1} are from the atmosphere as nitrogen input every year for Müncheberg Experimental Station and 47.3 kg N ha^{-1} for Bad Lauchstädt Experimental Station (Franko et al. 2007). For the model calculations carried out with the Müncheberg and Bad Lauchstädt data sets 25 kg N ha^{-1} were assumed for wet and 25 kg N ha^{-1} for dry nitrogen deposition, i.e., 0.0068 g N m^{-2} day^{-1} for dry and 0.005 g N m^{-2} (mm precipitation)$^{-1}$ for wet deposition, respectively.

Statistical methods

For the evaluation of the degree of simulation quality and the model-experiment comparison of different model approaches a lot of statistical methods can be used. In this paper the following methods were selected for a result evaluation:

– coefficient of determination (R^2; Rasch 1987)

$$R^2 = 1 - \frac{\sum_{i=1}^{n}(O_i - S_i)^2}{\sum_{i=1}^{n}(O_i - \overline{O})^2} \quad (10)$$

– the mean absolute deviation (MABS; Rasch 1987)

$$MABS = \frac{1}{n}\sum_{i=1}^{n}|O_i - S_i| \quad (11)$$

– the root mean square error (RMSE; Fox 1981)

$$RMSE = \sqrt{\frac{\sum_{i=1}^{n}(S_i - O_i)^2}{n}} \quad (12)$$

– the mean bias error (MBE; Addiscott and Whitmore 1987)

$$MBE = \sum_{i=1}^{n}\frac{S_i - O_i}{n} \quad (13)$$

– the "index of agreement" (IA; Willmott 1982)

$$IA = 1 - \frac{\sum_{i=1}^{n}(S_i - O_i)^2}{\sum_{i=1}^{n}\left(|S_i - \overline{O}| + |O_i - \overline{O}|\right)^2} \quad (14)$$

where n is the number of samples, S_i and O_i are the simulated and the observed values, and $\overline{O}$ is the mean of the observed data.

MABS, RSME and MBE are used for describing the average deviation between simulated and observed values. The lower the RSME the more accurate the simulation is. While the MABS can take only positive values, the MBE can take positive or negative values and the result gives an indication of the bias error. The IA as an additional method for evaluation of modelling efficiency results in a range between 0 and 1. The closer it is to 1 the better the simulation quality, similar like for R^2.

Results and discussion

Accuracy in describing ontogenesis as the biological timescale for plant-intern controlling of crop growth processes is very important for achieving accurate predictions of all process rates and variables subsequently calculated in agro-ecosystem models. Considering all winter cereals from Müncheberg experimental site (plot 1, crop rotation 1992–1998 (intensive management)) and from Bad Lauchstädt experimental site (crop rotation 1999–2002) the ontogenesis stages for shoot initiation, flowering, and maturity were calculated with an MABS of 4.5, 4.2, and 6.6 days, respectively. Between the accuracy results from Müncheberg and from Bad Lauchstädt there are not significant differences. The ontogenesis time courses are shown in the upper parts of Fig. 4 for the Müncheberg site and Fig. 5 for the Bad Lauchstädt site.

For Müncheberg different statistical values (Eqs. 10–14) for assessment of accuracy in describing biomass (without root), yield, crop nitrogen uptake, as well as soil water in the 0–90 cm depth soil layer using the AGROSIM models without any parameter calibration are listed in Table 1. The assessment basis on 24, 16, 13, and 21 measurements distributed over the growing season, respectively.

The MBE shows that the yield and the soil water in the rooting zone up to 90 cm depth is underestimated and the biomass (without root) and the N uptake are overestimated by the AGROSIM models a little. Yield and biomass (without root) were calculated with a MABS of 0.77 t ha^{-1} and 1.04 t ha^{-1}, i.e. with a mean relative deviation of 14.08% and 14.84%, respectively. Nitrogen uptake by plant is calculated with an MABS of 3.42 g m^{-2}. The MABS of soil water in the 0–90 cm layer is equivalent to 19.3% of the layer-specific field capacity. The 0–30 cm soil layer has the most turbulent soil water dynamic and also the greatest deviation between simulated and observed values. A comparison with AGROSIM simulation results for the Müncheberg experimental site realized with parameter calibrated AGROSIM models (Mirschel et al. 2001) shows that the statistical values for the parameter uncalibrated AGROSIM models used on plot 1 reflect a not so good model accuracy.

For Bad Lauchstädt the different statistical values for assessment of accuracy in describing biomass (without root), yield, crop nitrogen uptake, as well as soil water in 30–60 cm and 60–90 cm depths using the AGROSIM models are listed in Table 2, on the one hand for conditions without any parameter calibration

Table 1 Statistical values for assessment of accuracy in describing biomass (without root), yield, crop nitrogen uptake, as well as soil water in 0–90 cm depth using AGROSIM models without any parameter recalibration for the Müncheberg site (plot 1)

Statistical measure	Biomass (without root)	Yield	N uptake	Soil water (0–90 cm depth)
R^2	0.924	0.940	0.811	0.721
MABS	104.16 g m^{-2}	77.24 g m^{-2}	3.42 g m^{-2}	34.66 mm
RMSE	171.79 g m^{-2}	105.34 g m^{-2}	4.23 g m^{-2}	40.32 mm
MBE	49.32 g m^{-2}	−1.43 g m^{-2}	2.90 g m^{-2}	−30.63 mm
IA	0.978	0.984	0.906	0.738
N	24	16	13	21

Table 2 Statistical values for assessment of accuracy in describing biomass (without root), yield, crop nitrogen uptake, as well as soil water in 30–60 cm and 60–90 cm depths using AGROSIM models without any and with crop growth parameter recalibration for the Bad Lauchstädt site

Statistical measure	Biomass (without root)	Yield	N uptake	Soil water 30–60 cm	Soil water 60–90 cm
Without crop growth parameter recalibration					
R^2	0.929	0.999	0.393	0.739	0.279
MABS	199.16 g m^{-2}	77.93 g m^{-2}	4.80 g m^{-2}	9.65 mm	11.638 mm
RMSE	236.25 g m^{-2}	113.66 g m^{-2}	5.83 g m^{-2}	11.36 mm	18.839 mm
MBE	74.29 g m^{-2}	−32.58 g m^{-2}	4.06 g m^{-2}	2.87 mm	2.081 mm
IA	0.977	0.993	0.382	0.871	0.702
With crop growth parameter recalibration					
R^2	0.994	0.998	0.443	0.738	0.276
MABS	48.88 g m^{-2}	28.54 g m^{-2}	3.40 g m^{-2}	9.65 mm	11.78 mm
RMSE	66.61 g m^{-2}	37.50 g m^{-2}	4.39 g m^{-2}	11.35 mm	18.91 mm
MBE	5.78 g m^{-2}	5.96 g m^{-2}	3.40 g m^{-2}	−1.77 mm	−1.85 mm
IA	0.998	0.999	0.454	0.871	0.703
N	7	4	4	539	539

and on the other hand with a parameter recalibration for crop growth processes. For assessment of modelled soil water accuracy the soil water in 45 cm depth measured by Frequency Domain Reflectometry (FDR) (Franko et al. 2007) was compared with the simulated soil water from the 30–60 cm soil layer and the measured soil water in 90 cm depth by FDR was compared with the simulated soil water from the 60–90 cm soil layer.

Because the Bad Lauchstädt site is located in central Germany, outside of the AGROSIM model validation region, for this location the model simulations were realized both, without and with a parameter adaptation. The comparison of both simulation variants shows that the statistical values for assessment of accuracy in describing crop growth variables (biomass, yield, N uptake) of the parameter recalibrated variant are quite better and underline a better model accuracy (Table 2). The MABS's could be reduced, four times for biomass, three times for yield and by 30% for N uptake in comparison with the non-calibrated simulation variant. The influence of parameter recalibration on model accuracy of soil water is not so noticeable and lead to a little underestimation by the model. While the statistical values R^2, MABS, RMSE, and IA are in the same range for both the recalibrated and the non-calibrated simulation variants in the case of soil water, the MBE values show a change from model overestimation to model underestimation. The RMSE values show that after parameter recalibration simulation outliers not exist.

In the recalibrated simulation variant yield, biomass (without root) and N uptake were calculated with a MABS of 0.28 t ha^{-1}, 0.49 t ha^{-1}, and 3.4 g m^{-2}, i.e. with a mean relative deviation of 2.8%, 7.1% and 28.4%, respectively. In the non-calibrated simulation variant the mean relative deviations are much higher, i.e. 7.2%, 22.8%, and 41.1% for yield, biomass and N uptake, respectively. The MBE values show a little overestimation for biomass, yield and N uptake and a little underestimation for soil water. The R^2 and IA values of soil water in the 30–60 cm and 60–90 cm layers show that the accuracy in the upper layer is higher in comparison with the lower layer.

The accuracy of the AGROSIM models in describing agro-ecosystem plant (yield, biomass, N uptake) and soil water values also can be shown using the model-experiment comparison first as time courses and second as time-independent 2D visualization.

For plot 1 at the Müncheberg site in Fig. 4 the model-experiment comparisons for ontogenesis, biomass, yield, N uptake and soil water in 0–60 cm depth are shown as time courses for the whole crop rotation (sugar beet–winter wheat–winter barley–winter rye–sugar beet–winter wheat) in 1993–1998. Ontogenesis in decimal code stages only is calculated for winter cereals, as well as N uptake by plant. Fig. 4 also shows that the simulation runs were started anew with sowing date every year. The statistical shown overestimation (Table 1) in describing the plant N uptake by the AGROSIM models clearly is visualized in Fig. 4. In Fig. 4 there also is seen the discrepancy between modelled and observed biomasses at harvest. In many cases the biomass observations at harvest time are smaller than the biomass observations at sampling time before harvest. The reason is the dry biomass losses by breaking down between sampling times. In the AGROSIM models the biomass losses are not modelled, all biomass is accumulated.

Exemplary for the crop rotation at the Bad Lauchstädt site in Fig. 5 the model-experiment comparisons for ontogenesis, biomass, yield, N uptake and soil water in 45 and 90 cm depths are shown as time courses for winter wheat grown in 2001/2002. In Fig. 5 the simulation results from the recalibrated and non-calibrated simulation variants are compared. The influence of parameter calibration on calculated plant and soil values is seen very clear for biomass, yield and N uptake. The influence on soil water is not noticeable, but exists. An adaptation of original ontogenesis parameters was not necessary for Bad Lauchstädt. These visualized model-experiment comparison results underline the statements based on statistical values (Table 2) given above.

For both Müncheberg and Bad Lauchstädt sites Fig. 6 gives a model-experiment comparison for biomass, yield and nitrogen uptake in a 1:1 diagram form. In addition the linear regression between simulations and observations is given. The slopes of the regression line with 0.945 ... 0.965 is near the slope of the 1:1 line. The coefficient of determination is 0.920, 0.953 and 0.746 for biomass, yield and nitrogen uptake, respectively. Here for nitrogen uptake the same overestimation is detected like by the statistical values and the time courses already was seen.

Figure 7 shows the model-experiment comparison for soil water in the layers 30–60 cm and 60–90 cm depths for both sites, Müncheberg and Bad Lauchstädt. Because for Bad Lauchstädt there are not

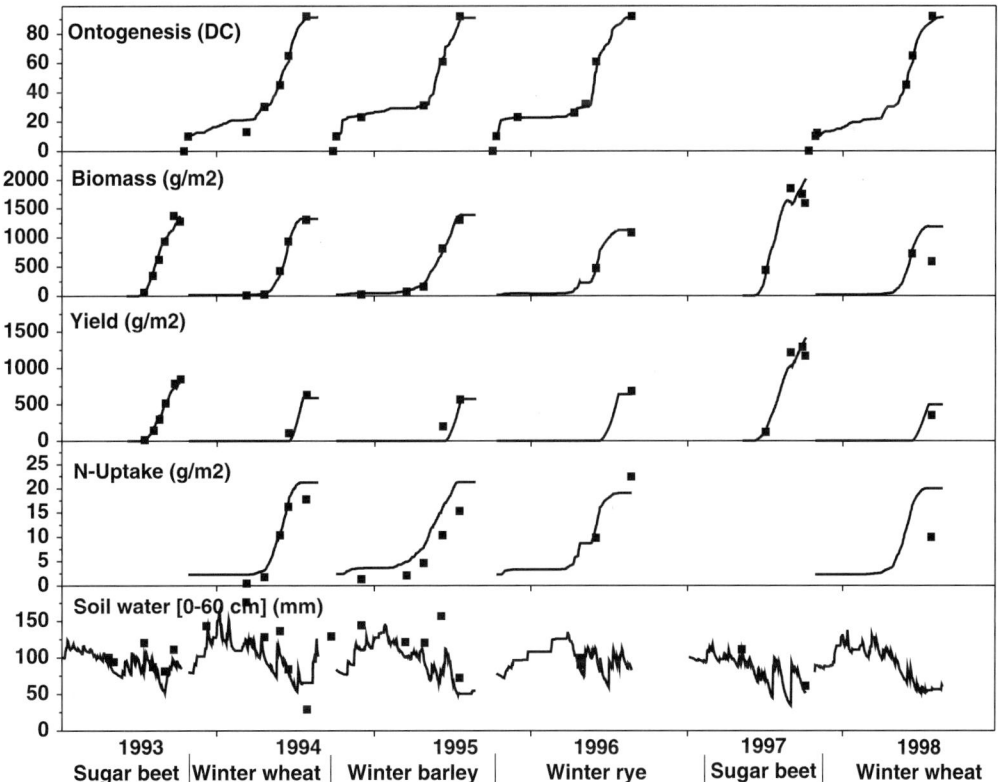

Fig. 4 Model-experiment comparisons for ontogenesis, biomass, yield, N uptake and soil water (0–60 cm depth) as time courses for the whole crop rotation (1993–1998) at the Müncheberg site (plot 1) (lines – simulation with AGROSIM models; squares – observation)

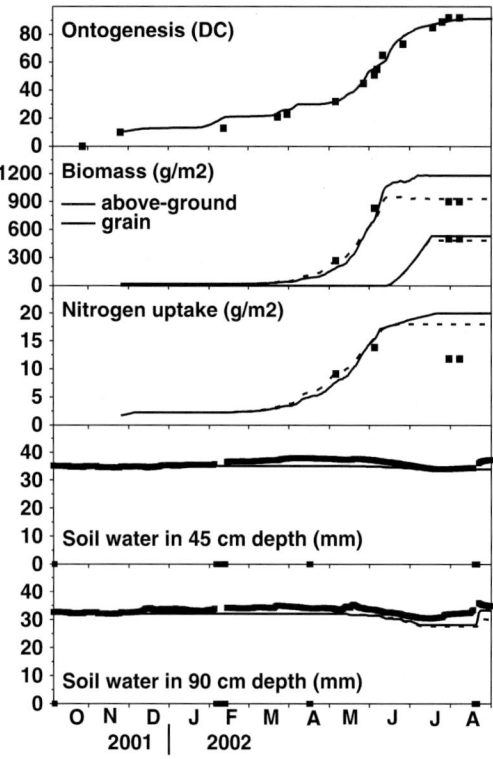

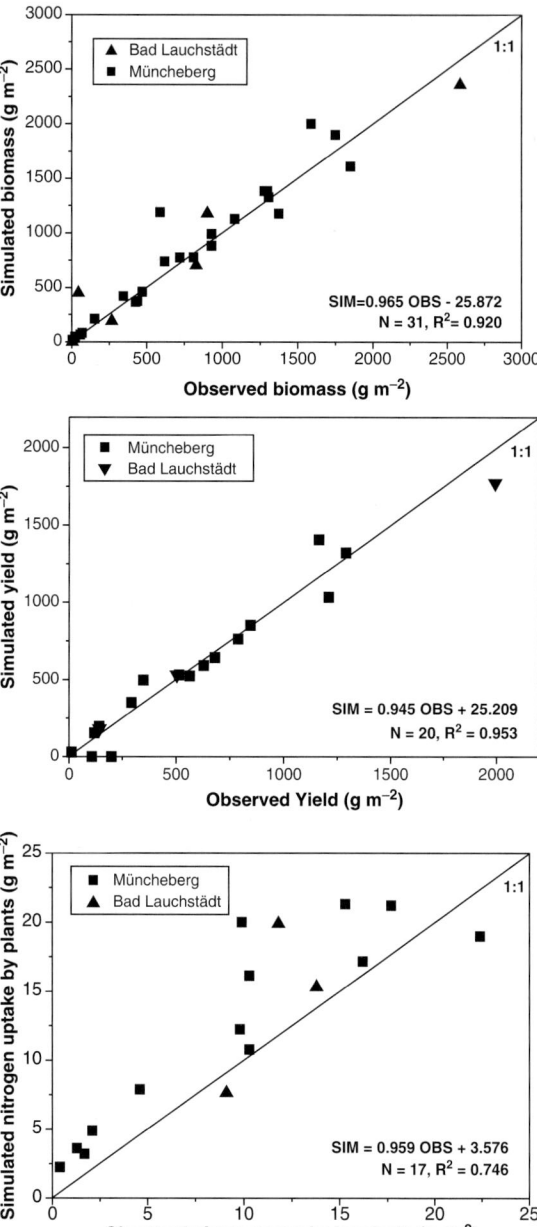

Fig. 5 Model-experiment comparisons for ontogenesis, biomass, yield, N uptake and soil water in 45 and 90 cm depths as time courses for winter wheat grown in 2001/2002 at the Bad Lauchstädt site (plot 1) (full lines – simulation with non-calibrated AGROSIM-WW model; dotted line – simulation with calibrated AGROSIM-WW model; squares – observation)

exist soil water observations for the 0–30 cm neither using auger nor using FDR method a joint comparison for the 0–30 cm depth is not possible.

The difference in soil water level in all depths between the Müncheberg and the Bad Lauchstädt sits reflects the difference in soil type. While the Müncheberg site is characterized by sandy soils with a low field capacity (about 54 mm for 0–30 cm depth) the Bad Lauchstädt site is characterized by a loess soil with a high field capacity (about 126 mm for 0–30 cm depth). In the depth of 30–60 cm the observed soil water values can be described with a sufficient accuracy for both sites ($R^2 = 0.776$) regardless of different soil water measurement methods (auger and FDR methods). In the soil layer of 60–90 cm depth the model accuracy also is sufficient with a little underestimation

Fig. 6 Model-experiment comparison for yield (top), biomass (middle), and nitrogen uptake (below) using the AGROSIM models as 1:1 diagram for both sites, Müncheberg and Bad Lauchstädt

only with the exception of the year 1999 at the Bad Lauchstädt site there sugar beet were grown. In that year an overestimation by the model was simulated. Starting with an agreement between modelled and observed values in spring 1999 the measured water uptake up to October 1999 from the upper soil layer is

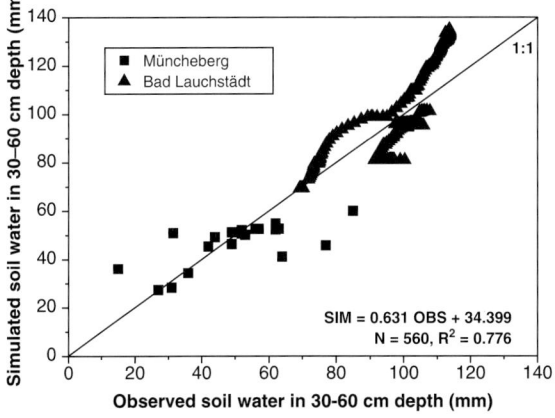

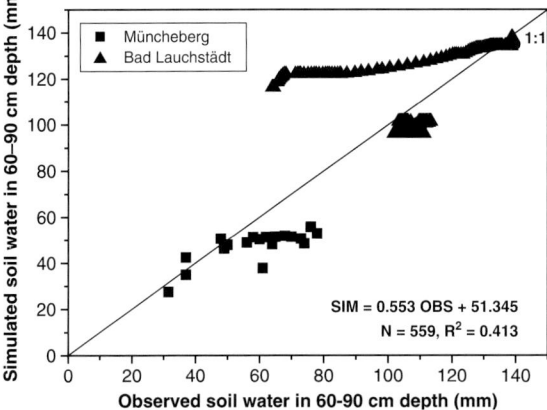

Fig. 7 Model-experiment comparison for soil water in the layers 30–60 cm (top) and 60–90 cm (below) depths using the AGROSIM models as 1:1 diagram for both sites, Müncheberg and Bad Lauchstädt

not described satisfactory by the model. The reason may be the not correct estimation of active rooting depth for this dry loess soil in the AGROSIM model for sugar beet. From the site of measurement using the FDR method there could be exist some calibration problems which are responsible for moving the whole soil water level. Similar difficulties in 1999 also are obtained in simulation runs with the OPUS and THESEUS models discussed by Wegehenkel et al. (2007).

The discrepancies in 1999 partly also can be caused by the comparison method. Here, the FDR measurements were realized in a depth of 90 cm only. For a comparison with modelled values for the 60–90 cm soil layer the measured values at the upper layer boundary in 90 cm depth were transferred to the soil layer of 30 cm thickness located above.

Conclusions

In general, the AGROSIM models are sensitive to climate inputs, site characteristics, and management practices. Since the data requirements of AGROSIM are small, the models can be used over a wide geographical range for systems analysis, for assessment of sustainability of different types of land use, to assess climate change effects, and as parts of landscape models.

The AGROSIM models can describe the crop rotation at plot 1 of the Müncheberg experimental station without parameter recalibration with a sufficient accuracy for the plant and soil state variables expected the nitrogen uptake by plants in some cases and expected winter wheat biomass in 1998. In the result of a non-excellent experiment management in 1998 there were a relative high weed population and relative high pest and diseases levels. Algorithms for such influences are not incorporated into the AGROSIM models as separate modules and so the models can not describe competition interactions between weeds and agricultural crops for nutrients and water or between the plant and diseases or/and pests.

Because the AGROSIM models only exist for winter wheat, winter barley and winter rye, sugar beet and some catch crops it was not possible to realize calculations for the whole crop rotation at the Bad Lauchstädt experimental station. The AGROSIM models for sugar beet and winter wheat can describe the real situation at the Bad Lauchstädt experimental station in 1999 and 2001/02 with a sufficient accuracy after a parameter recalibration only.

The workshop results for Müncheberg and Bad Lauchstädt sites show that for a model transfer to other geographical and sits conditions model parameters representing crop, site and other properties must be re-estimated or newly derived. In general a model transfer without any adaptation is not useful. The better considered the influence of site, weather, agromanagement and cultivar properties the more accurate the simulation results and the greater the possibilities to transfer a model from one geographical site to another. The chances of a broad model application increase if model adaptation could be limited to weather and soil information and only a few clearly defined crop parameters. For this coherent data sets are needed.

Because the AGROSIM models are agro-ecosystem models with the focus on crop growth processes it is clear that the accuracy for describing crop growth process values is higher than the accuracy for describing of soil process values like is shown in the tables above with statistical measures for Müncheberg and Bad Lauchstädt sites.

Disadvantages of the AGROSIM models are first that the models not exist for all important agricultural crops and second that the calculations for a whole crop rotation only are possible with interruptions between harvest of the first crop and sowing of the next crop, i.e. calculations only are possible between sowing and harvest, starting with sowing anew.

A well-founded assessment of model accuracy in describing agro-ecosystem values for plant and soil processes must base on different statistical values because each statistical value has their own specific in the result interpretation. The MABS gives a direct measure on the amount of deviation between modelled and observed values without information about underestimation or overestimation. The RMSE as a direct measure underlines big deviations and informs about possible existing deviation outliers. The MBE takes into account the direction of deviation and gives a direct measure for the mean degree of model overestimation or model underestimation. The coefficient of determination (R^2) and the IA are statistical measures normalized to (0,1) and are relative measures, i.e. the closer to 1 the better the model accuracy. In comparison with the last one R^2 gives a stronger assessment of accuracy. All stastistical measures mentioned above take into account only the modelled and observed values not relating to time. For a better process comparison between calculations and measurements and a better assessment of model accuracy a comparison of time courses in a visualization form and model-experiment comparisons as a 1:1 diagram are necessary.

Acknowledgements This contribution was supported by the German Federal Ministry of Consumer Protection, Food and Agriculture and the Ministry of Agriculture, Environmental Protection and Regional Planning of the Federal State of Brandenburg (Germany). A special thanks is given to the data providers from Bad Lauchstädt and Müncheberg Experimental Stations and also to the organizers of the modelling workshop.

References

Addiscott TM, Whitmore AP (1987) Computer simulation of changes in soil mineral nitrogen and crop nitrogen during autumn, winter and spring. J Agric Sci Cambridge 109:141–157

CAMASE (2001) The CAMASE register of quantitative methods and models for research on agricultural systems and the environment. Online in Internet: URL: http://www.bib.wau.nl/camase/ [last update: October 2001]

Diekkrüger B, Söndgerath D, Kersebaum KC, McVoy CW (1995) Validity of agroecosystem models – a comparison of results of different models applied to the same data set. Ecol Model 81:3–29

Fox DG (1981) Judging air quality model performance: a summary of the AMS Workshop on dispersion model performance. Bull Am Meteorol Soc 62:599–609

Franko U, Puhlmann M, Kuka K, Böhme F, Merbach I (2007) Dynamics of water, carbon and nitrogen in an agricultural used Chernozem soil in Central Germany. In: Kersebaum KC, Hecker J-M, Mirschel W, Wegehenkel M (eds) Modelling water and nutrient dynamics in soil-crop systems. Springer, Dordrecht, pp 245–258

Hack H, Bleiholder H, Buhr L, Meier U, Schnock-Fricke U, Weber E, Witzenberger A (1992) Einheitliche Codierung der phänologischen Entwicklungsstadien mono-und dikotyler Pflanzen – Erweiterte BBCH-Skala, Allgemein. Nachrichtenbl. Deut Pflanzenschutzd 44:265–270

Hoffmann F (1993) Die CERES-Modelle – Übersicht, Weiterentwicklung und Erfahrungen. Schriftenreihe Agrarinformatik 24:139–150

Jamieson PD, Porter JR, Goudriaan J, Ritchie JT, Keulen H van, Stol W (1998) A comparison of the models ARCFCWHEAT, CERES-WHEAT, SIRIUS, SUCROS2 and SWHEAT with measurements from wheat under drought. Field Crops Res 55:23–44

Koitzsch R (1977) Zur Schätzung der Bodenfeuchte aus meteorologischen Daten, Boden-und Pflanzenparametern mit einem Mehrschichtenmodell. Z Meteorologie 27:302–306

Koitzsch R, Günther R (1990) Modell zur ganzjährigen Simulation der Verdunstung und der Bodenfeuchte landwirtschaftlicher Nutzflächen mit und ohne Bewuchs. Archiv Acker-Pflanzenbau Bodenkd 34:803–810

Mirschel W, Förkel H, Franko U (2002) Modulares dynamisches Wachstumsmodell für Zuckerrüben als integrativer Bestandteil von komplexen agrarökologischen Simulationsmodellen. In: Gnauck, A. (ed) Systemtheorie und Modellierung von Ökosystemen. UmweltWissenschaften, Physika-Verlag, Heidelberg, pp 136–156

Mirschel W, Kretschmer H, Matthäus E (1990) Dynamisches Modell zur Abschätzung der Ontogenese von Winterweizen unter Berücksichtigung des Wasser-und Stickstoffversorgungszustandes. Arch Acker-u Pflanzenbau u Bodenkd 34:691–699

Mirschel W, Schultz A, Wenkel K-O (1997) Agroökosystemmodelle als Bestandteile von Landschaftsmodellen. Archiv für Naturschutz und Landschaftsforschung 35:209–225

Mirschel W, Schultz A, Wenkel K-O (2001) Assessing the impact of land use intensity and climate change on ontogenesis, biomass production, and yield of northeast German agro-landscapes. In: Tenhunen JD, Lenz R, Hantschel (eds) Ecosystem approaches to landscape management in Central Europe. Ecological studies, 147. Springer-Verlag, Berlin, Heidelberg, New York, pp 299–313

Mirschel W, Schultz A, Wenkel K-O, Wielend R, Poluektov RA (2004) Crop growth modelling on different spatial scales – a wide spectrum of approaches. Arch Agron Soil Sci 50:329–343

Mirschel W, Wenkel K-O (1998) Estimation of consequences of climate changes using the agroecosystem model family AGROSIM. In: Dalezios NR (ed.) COST 77,79,711-International Symposium on Applied Agrometeorology and Agroclimatology (Volos, Greece, 24 to 26 April 1996), Proceedings, Office for Official Publication of the European Commission (Luxembourg), 1998 (EUR 18328 EN), pp 67–72

Mirschel W, Wenkel K-O, Koitzsch R (1995) Simulation of soil water and evapotranspiration using the model BOWET and data sets from Krummbach and Eisenbach, two research catchments in North Germany. Ecol Model 81:53–69

Mirschel W, Wenkel K-O, Schultz A, Pommerening J, Verch G (2005) Dynamic ontogenesis model for winter rye and winter barley. Eur J Agron 23(2):123–135

Mirschel W, Wenkel K-O, Wegehenkel M, Kersebaum K-C, Schindler U, Hecker J-M (2007) Müncheberg field trial data set for agro-ecosystem model validation. In: Kersebaum KC, Hecker J-M, Mirschel W, Wegehenkel M (eds) Modelling water and nutrient dynamics in soil-crop systems. Springer, Dordrecht, pp 219–243

Rasch D (1987) Biometrisches Wörterbuch. 3rd edn. VEB Deutscher Landwirtschaftsverlag Berlin, 764 pp

Reining F (1995) Klimakammerexperimente zur CO_2-Gasaustauschmessung bei Winterrogen und Wintergerste. In: Wenkel K-O, Mirschel W (eds) Agroökosystemmodellierung, Grundlage für die Abschätzung von Auswirkungen möglicher Landnutzungs- und Klimaänderungen. ZALF-Bericht 24, Müncheberg, pp 27–55

Suckow F (1986) Ein Modell zur Berechnung der Bodentemperatur unter Brache und unter Pflanzenbestand, Akademie der Landwirtschafts-Wissenschaften der DDR, Dissertation A, pp 121

Wegehenkel M, Mirschel W (2007) An application of the models THESEUS and OPUS with the experimental data sets from Bad Lauchstädt and Berlin-Dahlem. In: Kersebaum KC, Hecker J-M, Mirschel W, Wegehenkel M (eds) Modelling water and nutrient dynamics in soil-crop systems. Springer, Dordrecht, pp 37–49

Wenkel K-O, Mirschel W (1995) Agroökosystemmodellierung. Grundlagen für die Abschätzung von Auswirkungen möglicher Landnutzungs-und Klimaänderungen. ZALF-Bericht 24, Müncheberg, pp 187

Wenzel V, Matthäus E, Flechsig M (1986) The simulation system SONCHES. In: Ebert W (ed.) Computer-aided modelling and simulation of the winter wheat agro-ecosystem (AGROSIM-W) for integrated pest management. Tagungsb Akad Wiss DDR 242, pp 29–41

Willmot CJ (1982) Some comments on the evaluation of model performance. Bull Am Meteorol Soc 64:1309–1313

Zadoks JC, Chang TT, Konzak CF (1974) Decimal code for the growth stages of cereals. Weed Res 14:415–431

CHAPTER SEVEN

Crop simulation model of the second and the third productivity levels

Ratmir Aleksandrovich Poluektov* and Vitaly Viktorovich Terleev

Abstract The structure and functionality of the model of the second and the third productivity level is described. Special attention was paid to simulation of real plant transpiration and soil evaporation, as well as to calculation of shoot to root distribution in crops. Shoot to root ratio is determined according to C:N balance in plant organs. Information support system intended to calculation of soil hydraulic parameters is presented. Comparison of simulation results with experimental data obtained in various sites of East Germany is discussed.

Keywords Comparison of simulation and experiment results, Crop dynamic model, Plant transpiration, Shoot to root relation, Soil evaporation

Introduction

During the last decades several tens crop models have been published in the world literature. Some of them are generally recognized. It is possible to note the following models, which are widely used in west countries: European models WOFOST (Diepen et al. 1988) and AGROSIM (Wenkel and Mirschel 1995), American models CERES (Hanks and Ritchie 1991) and EPIC (Williams et al. 1983) and some others. In Agrophysical Research Institute St. Petersburg, Russia, the model family AGROTOOL was developed (Poluektov et al. 2002). It was calibrated on the data collected in northwest Russia region. All these models are semiempirical, have similar structure and distinguished by detailed description of separate processes. Progress in model development consists in partly change the regression relations by the including units, which have strong physical or biological basis. It concerns, for example, the description of photosynthesis, plant transpiration, or carbon/nitrogen interaction in plants. It is necessary to note that all the existing models were developed and identified according to concrete soil and climatic conditions. It means that only such limiting environment factors are included into models, which are typical for concrete site of plant cultivation. So the application of models adapted to one geographic zone to other region is rather problematic one (Poluektov et al. 2000). In this paper we try to use AGROTOOL model family to predict crop dynamics of some cultures cultivated in East Germany.

Model structure and functionality

Model family AGROTOOL is described by Poluektov et al. (2002) in details. So we present here only short model structure. Model consists of following

---

Ratmir Aleksandrovich Poluektov and Vitaly Viktorovich Terleev
Agrophysical Research Institute, Laboratory of agroecosystem simulation Grazhdansky pr. 14, 195220 St.-Petersburg, Russia
*Corresponding author: tel.: +812 5344640; fax: +812 5341900, e-mail: ratmir@mail.wplus.net

units: weather and turbulence in crop, radiation regime and photosynthesis, soil water dynamics, plant transpiration and soil evaporation, soil temperature regime, plant growth and development, nitrogen regime in soil and plants. General time step of the model is set to 24 h with two exceptions. Photosynthesis unit uses hourly time step. It means that the rate of assimilates is being calculated during each hour. Afterwards the accumulated assimilates are summarized over the day and this amount is being used in other model units. Time step is also less than 1 day in the unit describing soil water dynamics. It is caused by necessity to realize the computation algorithm for the solution of differential equation of water movement in the soil. In order to run the model in real time the programme converting the Julian calendar date into day number and vice versa is being used. It permits to carry out computer experiments from real or generated sowing (planting) date up to harvesting. Model shell permits to execute model proper in an interactive regime. Strictly speaking it provides the possibility to organize and carry out computer experiments with the model (case study) and to solve some practical problems. The model shell has a universal structure and provides the work with several models, a set of cultures, locations and types of management. Taking into account the purpose of this paper directed to compare the simulation and experiment results we describe below three questions in more detail: (i) new method of simulation of plant transpiration and soil evaporation, (ii) new method of root to shoot ratio calculation according to N:C interaction in plants, and (iii) information support of calculation of soil hydraulic parameters.

Simulation of plant transpiration and soil evaporation

New method of plant transpiration and soil evaporation is realized in the model. It is based on the approach suggested primarily by Penman (1948) and modified by Monteith (1981), but differs from it in principal way. The method of Penman-Monteith is widely used by west researches for estimating the potential evapotranspiration only. It was modified to calculate real evaporation rates, which characterize both values of soil water loss, transpiration and soil evaporation. As in the original method daily heat balance equation of crop is used. But in distinction from Penman-Monteith method only the short-wave radiation absorbed by plants and accordingly only heat exchange between plants and their environment is described by this equation. So, the following relation is accepted:

$$R_p - H_p - \lambda E_p = 0, \tag{1}$$

where

- R_p : short-wave radiation absorbed by crop [J m^{-2} day^{-1}]
- H_p : rate of heat exchange between crop and air [J m^{-2} day^{-1}]
- E_p : plant transpiration [mm]
- λ : latent heat of evaporation [J kg^{-1}]

Following Penman-Monteith concept and using the equations of water and heat transfer from plant canopy to atmosphere it is possible to obtain the following relation (Poluektov et al. 1997).

$$E_p = \frac{r_\Sigma^T b R_p}{c_p r_\Sigma^q + b\lambda r_\Sigma^T} + \frac{\rho_a c_p (q_l(T_a) - q_a) k_{pL}}{c_p r_\Sigma^q + b\lambda r_\Sigma^T} \cdot 86,400, \tag{2}$$

where

- c_p : specific heat capacity [J g^{-1} °C^{-1}]
- r_Σ^T : summary resistance for heat flow [s cm^{-1}]
- r_Σ^q : summary resistance for water vapour flow [c cm^{-1}]
- b : coefficient of linearization in Magnus equation [°C^{-1}]
- T_a : average daily air temperature
- q_a : specific air humidity [g g^{-1}]
- $q_{sat}(T_a)$: specific air humidity by saturating point at the temperature T_a [g g^{-1}]
- k_{pt} : fraction of crop cover [–]
- 86,400 : number of seconds per day

Resistances r_Σ^T and r_Σ^q included in Eq. 2 depend on wind speed. The latter of them contains stomata resistance, r_{st} as an item, which in its own turn depends on leaf water potential ψ_L.

$$r_\Sigma^q = r_\Sigma^T + r_{st}(\psi_L), \tag{3}$$

where

- ψ_L : Leaf water potential [MPa]

Therefore, in accordance with Eqs. 2–3, plant transpiration is a function of daily weather conditions and leaf water potential:

$$E_p = F(R_p, T_a, q_a, r_{st}(\psi_L)). \tag{4}$$

On the other hand water uptake by roots can be expressed by

$$V_r = \xi \sum_{i=1}^{N_R} \omega_{si}(\psi_{si} - \psi_R), \tag{5}$$

where

- V_r : water uptake by roots [mm day^{-1}]
- ω_{si} : root area index [m^2 m^{-2}] in layer i
- ψ_{si} : soil water potential in calculation layer i [MPa]
- ξ : root conductivity [mm MPa^{-1}]
- N_R : number of calculation layers [–]
- ψ_R : root water potential [MPa]

Ignoring the loss of water potential across xylem and assuming that total water uptake for the whole day is equal to total transpiration it can be written as:

$$E_p \approx V_r, \qquad \psi_L \approx \psi_R. \tag{6}$$

Latter equations are used for the excluding of unknown variable $\psi_L \approx \psi_R$ from Eqs. 4 and 5 and for obtaining the final relation connecting daily sum of real transpiration with current data for soil water potential and meteorological parameters of the simulation day. The iterative algorithm was proposed in (Poluektov et al. 1997) for the solution of this problem. The situation is illustrated in Fig. 1 where E_p and V_r are functions on $\psi_L = \psi_R$ and ψ^* is their steady-state value. Analogues method for simulation of real soil water evaporation was described in Poluektov et al. (1992). It is based on the use of heat balance equation on the soil surface.

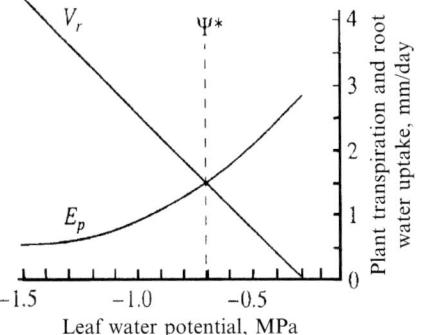

Fig. 1 Dependence of plant transpiration and root water uptake on leaf water potential

Calculation of root to shoot ratio

Deterministic distribution keys were proposed for the description of dry matter allocation for many annual crops such as wheat, maize, potatoes, rice and others. But it is not evident how to apply this approach to such cultures like alfalfa, clover and other perennial forage crops. Really, after each cut for harvest a sufficient root dry matter remains in the soil and the accumulated amount of new assimilates limits plant growth. The root dry matter can even decrease due to respiration while aboveground biomass rises. After some period when leaf and stem biomass increases and daily photosynthesis exceeds plant demand, an extra amount of carbohydrates must be transported into the roots causing their growing renewal. So, the whole situation could be under control of C and N content in plant organs, CO_2 assimilation by green plant parts, and N uptake by roots. Our aim was to describe an alternative mechanism of dry matter partitioning between root and shoot, which reflects adaptive crop reactions on ambient conditions and especially on assimilation of CO_2 by leaves and nitrogen uptake by roots.

Let us describe this situation formally. The method described lower is a universal one. It can be used to annual as well as to perennial cultures. We suppose in the first place that soil nitrogen does not limit plant productivity. In such a case the rate of N absorption is proportional to the integral root surface and the absorption rate of root unit is constant and equal to saturation value V_{max}. The amount of assimilates accumulated for a day k, ΔW must be distributed over root and shoot according to the following equation:

$$\begin{aligned}\Delta W_r(k) &= c_{rs}\Phi(k)\\ \Delta W_s(k) &= (1-c_{rs})\Phi(k),\end{aligned} \tag{7}$$

where

- $\Phi(k)$: primary assimilates accumulated by crop at the day k, g m^{-2}
- $\Delta W_r(k)$ and $\Delta W_s(k)$: increments of root and shoot dry mass, g m^{-2}
- c_{rs} : coefficient of root to shoot distribution

Our aim is the determination of the last value. Following procedure was proposed for the realization of this aim. The increments in nitrogen demands by shoot ΔN_{sd} and root ΔN_{rd} can be described by following notations:

$$\Delta N_{sd}(k) = \Delta W_s(k) S_c N_{shoot},$$
$$\Delta N_{rd}(k) = \Delta W_r(k) R_c N_{root}, \qquad (8)$$

where

N_{shoot} and N_{root} : nitrogen concentration in structural mass of shoot and root, respectively

S_c and R_c : coefficients of the conversion carbohydrates in structural mass of shoot and root, respectively, kg kg^{-1}

Let $\Omega(k-1)$ denote root area for a previous day, $k-1$. The increase in this value at the day k can be calculated by the formula:

$$\Delta\Omega(k) = s_r \Delta W_r(k), \qquad (9)$$

where s_r is specific root area, m^2 ha^{-1}. The nitrogen balance equation can be written as:

$$\int_{(k-1)T}^{kT} N_L(t) w_L(t) dt + \int_{(k-1)T}^{kT} N_R(t) w_R(t) dt$$
$$= V_{max} \int_{(k-1)T}^{kT} (\Omega(k-1) + \omega(t)) dt$$

where w_L and w_R are the rates of biomass growth, N_L and N_R are nitrogen concentrations in shoot and root, $\omega(t)$ is current increment in root area, $T = 24$ h. It is supposed that all nitrogen is utilized completely and no reserve is stored. For $N_L(t)$ and $N_R(t)$ change slowly during the vegetation the last equation can be rewritten in the form:

$$N_L \Delta W_L(k) + N_R \Delta W_R(k) = V_{max}(\Omega(k-1) + \xi \Delta\Omega(k)), \qquad (10)$$

where

$$\xi\Delta\Omega(k) = \int_{(k-1)T}^{kT} \omega(t) dt. \qquad (11)$$

It is clear that the coefficient ξ ranges in the limits $0 < \xi < 1$. It can be approximately assumed to be equal to 0.5.

Assimilate distribution which is controlled by partitioning coefficient c_{rs} in Eq. 7 is determined firstly by the relationship between available forms of carbon and nitrogen, and secondly by demand of plant organs in these nutrient elements. If nitrogen absorbed by the roots [kg ha^{-1} day^{-1}]

$$\Delta N_R = V_{max} \cdot \Omega(k) \qquad (12)$$

exceeds a demand of aboveground crop part, $N_L \Delta W$ all assimilates remain in the leaves, and $c_{rs} = 0$. In opposite case when

$$V_{max} \cdot \Omega(k) \leq N_L \Delta W$$

all assimilates are allocated into the roots and $c_{rs}=1$. Intermediate case leads to the equality, which follows from Eqs. 8–10:

$$c_{rs} = \frac{N_L \Delta W(k) - V_{max} \Omega}{\Delta W(k)(\xi s_r V_{max} + N_L - N_r)}. \qquad (13)$$

Actually, however, a small part of assimilates is always transported into the roots or remains in the leaves. So c_{rs} is greater than 0 and less than 1. We can assume that $\alpha \leq c_{rs} \leq (1-\alpha)$, where α is a small positive value.

The whole situation is clarified in Fig. 2. Line 1 corresponds to N demand of a total crop for the day k. Indeed, as it is known, N concentration in roots is lower than in shoot during the whole vegetation. So, the dependence of a crop N demand decreases linearly in accordance with Eqs. 7 and 8. On the other hand, N availability (line 2) increases with the increase of c_{rs}. The crossing of these two lines gives the c_{rs} value providing the balance in nitrogen availability and demand.

The described method was generalized in the model of the third productivity level, which means the limiting of plant production by nitrogen. Change the maximum rate of nitrogen absorption by roots V_{max} in Eq. 12 by the real rate depended on nitrogen content in soil was done for this purpose. Figure 3 demonstrates the dependence of shoot to root ratio on

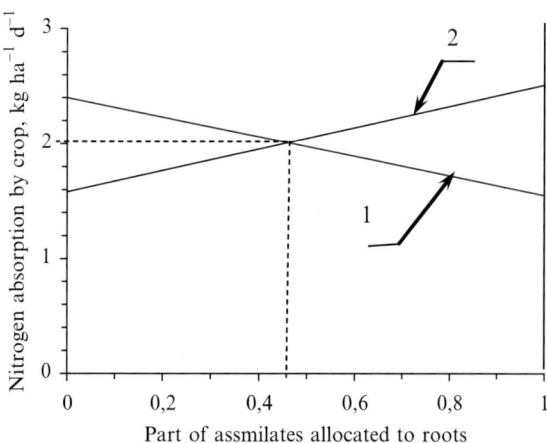

Fig. 2 Determination of part of assimilates allocated to roots

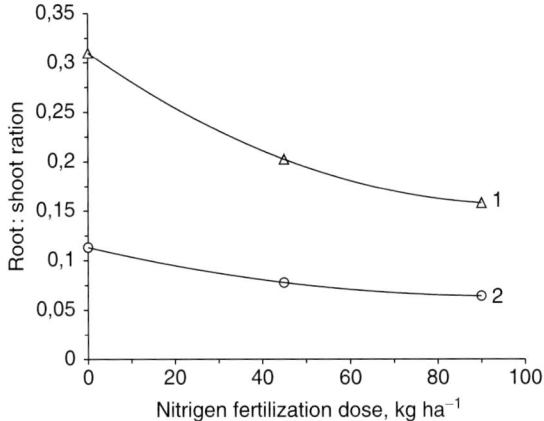

Fig. 3 Dependence of root to shoot relation on N-fertilization dose

N-fertilization dose.

Information support of calculation of soil hydraulic parameters

Numeric integration of Richard's equation was used for simulation of soil water dynamics in the model. In its own turn, model includes three soil layers with different hydrological characteristic: 0–20 cm (or 0–30 cm), 20–50 cm (or 30–60 cm) and 50–100 cm (or 60–100 cm). Separate pedotransfer functions (PTFs) were used for each soil layer. They have the same form but different parameters in correspondence with soil properties of each layer. The following PTFs were used in the model:

$$\theta = MH + \frac{SP - MH}{1 + b(-p)^a} \qquad (14)$$

$$k = k_f (-p)^{-c}, \qquad (15)$$

where

θ	: volumetric water content, cm^3 cm^{-3}
p	: matrix potential, cm
MH	: soil hygroscopy, cm^3 cm^{-3}
SP	: saturated point, cm^3 cm^{-3}
k	: hydraulic conductivity, cm day^{-1}
k_f	: saturated hydraulic conductivity, cm day^{-1}
a, b, c	: empirical parameters

Equation 14 includes four parameters: MH, SP, a and b. So it is necessary to have minimum four points on pF curve for their estimation. Usually we use the following data. Two points MH and SP are used directly. Two other points field capacity (FC) corresponding to $p = -330$ cm and wilting point (WP) corresponding to $p = -15{,}000$ cm are used additionally for estimation the values a and b. Mualem's method was employed for determination of parameter c in Eq. 15.

In many cases all data necessary for estimation of PTF parameters are not available. But there are other soil hydraulic constants, which permit to calculate the lacking constants using some approximation formulae. A special software, called AGROHYDROLOGY, (Fig. 4) was developed for this purpose (Terleev 2000). It includes four basic variants and four additional ones, which are distinguished by a set of input data that can be used for calculation of the total parameter set included in model 14. These variants are shown in Table 1.

The following notations are included in Table 1: ρ – soil bulk density, ρS – solid phase density, LC – lower capillary moisture, UC – upper capillary moisture. Additional variants (1'–4') are distinguished from the basic ones by using the measured value ρS instead of SP.

Comparison of simulated and experimental results

Model family AGROTOOL was calibrated by the experimental data from Russian sites: Saratov region (Middle Volga) and Leningrad region (north west of Russia). Our problem was to prove how this model may reproduce the field experiments fulfilled in other Europe regions, in particular, in East Germany. Crop simulation system AGROTOOL contains the models of the cultures: spring and winter wheat, spring barley, winter rye, oats, potatoes and perennial grasses. So the following variants were chosen for model test: winter wheat for the years 1993/94, 1997/98 and winter ray for the years 1995/96 for the Müncheberg site, spring barley for the year 2000, potatoes for the year 2001 and winter wheat for the years 2001/02 for the Bad Lauchstaedt site and lysimeters 1, 2, 5, 6, 11, 12 for the years 1996, 1997 and 1998 for the Berlin site. It is necessary to determine the following data for model run: hydrological soil constants, daily weather parameters and initial model conditions. Soil water model contains three soil layers with possibly various characteristics, such as maximum hygroscopy, wilting point, field capacity and saturation point for the three layers. All these constants were estimated by the pre-

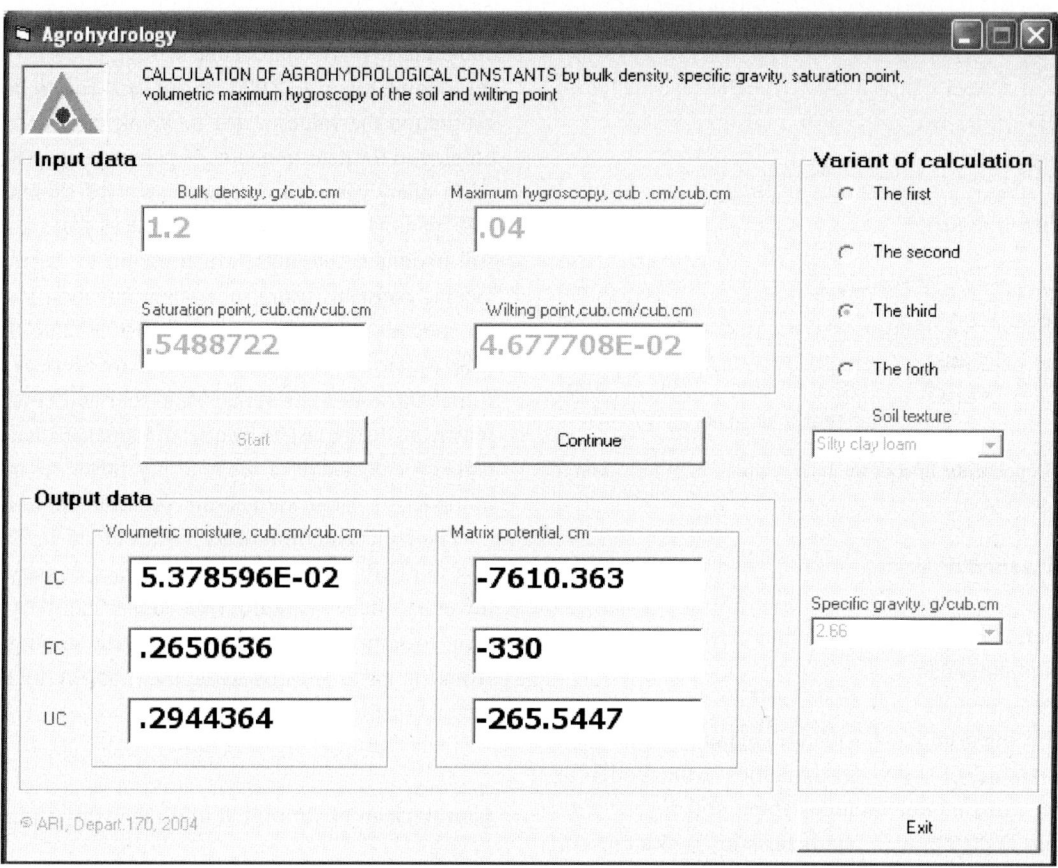

Fig. 4 Interface of the AGROHYDROLOGY software for estimation of soil hydrological parameters

Table 1 Variants of calculation of hydrological parameters

Variants	Input data	Calculated data
1	ρ, MH, WP, SP	LC, FC, UC
2	ρ, WP, SP, soil texture	MH, LC, FC, UC
3	ρ, MH, SP soil texture	LC, WP, FC, UC
4	ρ, SP	MH, LC, WP, FC, UC
1'	ρ, ρS, MH, WP	LC, FC, UC, SP
2'	ρ, ρS, WP, soil texture	MH, LC, FC, UC, SP
3'	ρ, ρS, MH, soil texture	LC, WP, FC, UC, SP
4'	ρ, ρS	MH, LC, WP, FC, UC, SP

sented soil data. Six daily parameters must be used as weather data: maximum and minimum air temperature, air humidity, precipitation, wind speed and coefficient of solar radiation extinction. Five of them were included in the weather file directly. For estimation of the last parameter a special programme using measured data and extraterrestrial radiation calculated for appropriate geographic attitude was used. Besides of that there were no data for maximum and minimum air temperature in Berlin weather data. These parameters were rewritten from Muencheberg weather file. Initial model data are the date of sowing or planting and soil water content in one meter layer for this date. The dates of sowing or planting are used directly. Soil water content was estimated according to measured data nearest to initial date. All calculations were made without model recalibration.

Comparison results for the Muencheberg site

Pedotransfer functions

Experimental data sets presented by Mirschel et al. (2007) for Muencheberg were used. The information about soil texture and bulk density of the layers 0–30 cm, 30–60 cm, 60–90 cm for plots 1–3 were employed for estimation of PTF parameters. Such parameters as WP and SP were estimated firstly using the method published in "Estimating generalized soil water characteristics from texture" (Saxton et al. 1986). Then all parameters included in Eq. 14 were determined according to the Program AGROHYDROLOGY, variant 2. The calculation results are shown in Tables 2–4.

Yield and dry mass

The simulation and measurement results for the grain yield and harvesting date are presented in Tables 5 and 6. Some calculation results of the formation of plant organs are shown in Fig. 5, where dynamics of winter wheat above-ground dry mass in 1993/94 years on plots 1a and 3b are presented.

The results show a sufficient qualitative conformity between simulation and experiment results with exception of grain yield for winter rye 1995/96 years and winter wheat 1997/98 years on plot 2 and harvesting date for winter wheat 1997/98 years on plots 1 and 2.

Water status

Experimental data for volumetric water content in 1 m layer have been calculated according to formula:

Table 2 Parameters of water retention curve (plot 1)

Depth, cm	SP, $cm^3\ cm^{-3}$	MH, $cm^3\ cm^{-3}$	FC, $cm^3\ cm^{-3}$	WP, $cm^3\ cm^{-3}$	a	b
0–30	0.26	0.03	0.14	0.06	0.474	0.07
30–60	0.26	0.03	0.15	0.06	0.52	0.045
60–90	0.26	0.04	0.16	0.08	0.442	0.064

Table 3 Parameters of water retention curve (plot 2)

Depth, cm	SP, $cm^3\ cm^{-3}$	MH, $cm^3\ cm^{-3}$	FC, $cm^3\ cm^{-3}$	WP, $cm^3\ cm^{-3}$	a	b
0–30	0.52	0.03	0.21	0.06	0.573	0.062
30–60	0.51	0.03	0.2	0.06	0.55	0.078
60–90	0.49	0.03	0.18	0.06	0.507	0.11

Table 4 Parameters of water retention curve (plot 3)

Depth, cm	SP, $cm^3\ cm^{-3}$	MH, $cm^3\ cm^{-3}$	FC, $cm^3\ cm^{-3}$	WP, $cm^3\ cm^{-3}$	a	b
0–30	0.26	0.04	0.14	0.08	0.346	0.161
30–60	0.26	0.06	0.14	0.09	0.348	0.199
60–90	0.26	0.05	0.14	0.1	0.229	0.353

Table 5 Simulated and experimental results for grain yield, t ha^{-1}

Culture, vegetation season	Plot 1		Plot 2		Plot 3	
	Simulated	Experimental	Simulated	Experimental	Simulated	Experimental
Winter wheat, 1993/94	3.58	4.52	3.60	3.22	3.37	4.80
Winter rye, 1995/96	5.63	6.81	5.62	2.45	5.24	5.13
Winter Wheat, 1997/98	4.85	3.49	4.50	1.43	4.66	4.37

Table 6 Simulated and experimental results for harvesting date

Culture, vegetation season	Plot 1		Plot 2		Plot 3	
	Simulated	Experimental	Simulated	Experimental	Simulated	Experimental
Winter wheat, 1993/94	08. Aug	29. Jul	10. Aug	29. Jul	07. Aug	29. Jul
Winter rye, 1995/96	11. Aug	21. Aug	13. Aug	21. Aug	12. Aug	21. Aug
Winter Wheat, 1997/98	05. Sep	27. Jul	22. Aug	27. Jul	02. Aug	27. Jul

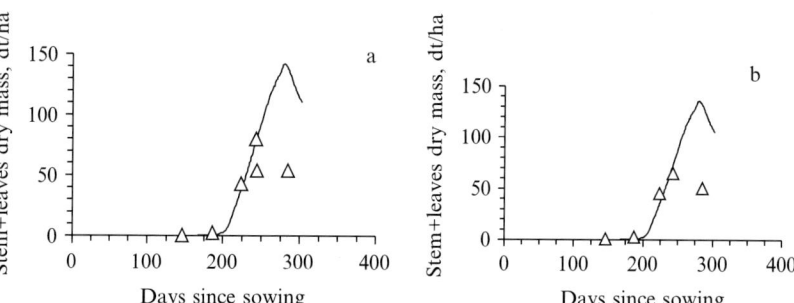

Fig. 5 Dynamics of stem + leaves dry biomass of winter wheat during the vegetation period 1993/94 years on plot 1a and 3b: Δ – experimental data

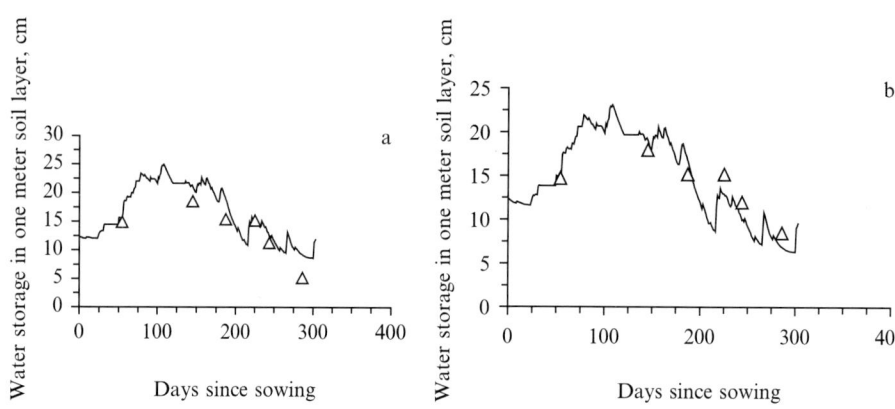

Fig. 6 Dynamics of soil water under winter wheat crop, 1993/94 years

WC (0 – 100) = 0.3 * WC (0 – 30) + 0.3 * WC (30 – 60) + 0.4 * WC (60 – 90)

where
WC $(x-y)$ is water content in the layer $[x, y]$. Examples of results of water content dynamics for plots 1 and 2 are presented in Fig. 6.

Statistical data

The statistical data for the difference between simulated and measured results were calculated according to formulae:

$$\overline{X} = \sum_{i=1}^{n} \frac{X\text{sim}_i - X\text{exp}_i}{n}$$

$$\text{MSE} = \left(\frac{1}{n} \sum_{i=1}^{n} \left(X\text{sim}_i - X\text{exp}_i \right)^2 - \overline{X}^2 \right)^{1/2},$$

where

$X\text{sim}_i$: simulated values
$X\text{exp}_i$: experimental values
$\overline{X}$: mean difference (error) between simulated and experimental data
MSE : mean square error of simulation

The following statistical data where obtained:

$\overline{X} = 0.739$ t ha^{-1},   MSE = 1.96 t ha^{-1}
for the dry mass

$\overline{X} = 0.19$ cm,   MSE = 1.97 cm
for water content in 0–100 cm soil layer

$\overline{X} = -0.201$ t ha^{-1},   MSE = 0.935 t ha^{-1}
for grain yield

Summary results for dry mass, grain yield and water content are shown in Figs. 7–9.

Comparison results for the Bad Lauchstaedt site

Pedotransfer functions

Experimental data sets presented by Franko et al. (2007) for Bad Lauchstaedt were used. The information about bulk density, hygroscopy and saturated point of the layers 20–24 cm, 45–49 cm, 115–119 cm for V 521 was employed for estimation of PTF parameters. Parameters included in Eq. 15 were determined according to the software AGROHYDROLOGY, variant 3. The calculation results are shown in Table 7. For example, the comparison of simulated pF curve with experimental data for layer 20–24 cm is presented in Fig. 10.

Yield and dry mass

The simulation and measurement results for the grain yield and harvesting date are presented in Table 8.

Simulation results for dynamics of plant organs dry mass of spring barley, winter wheat and potatoes are presented in Figs. 11–13.

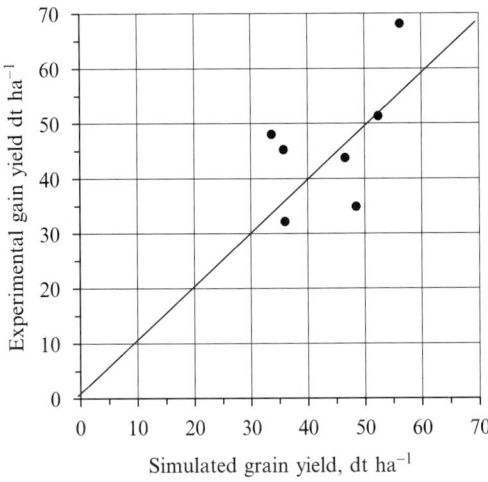

Fig. 7 Comparison of simulated and experimental data for grain yield: plots 1–3 for 1993/94 years, plots 1, 3 for 1995/96 years, plots 1, 3 for 1997/98 years

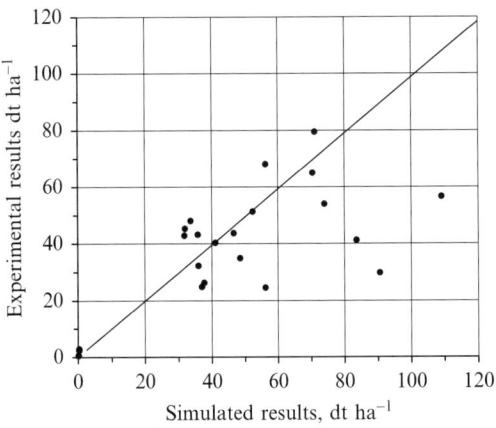

Fig. 8 Comparison of simulated and experimental results for dry mass: plots 1–3 for 1993/94 and 19965/96 years and plots 1, 3 for 1997/98 years

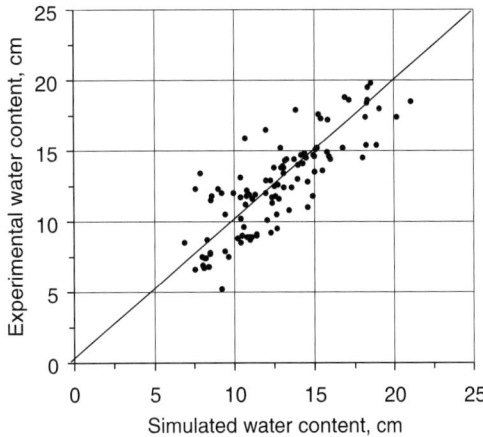

Fig. 9 Comparison of simulated and experimental water content in 1 m soil layer, plots 1–3

Table 7 Parameters of water retention curve (V 521)

Depth, cm	SP, cm³ cm⁻³	MH, cm³ cm⁻³	FC, cm³ cm⁻³	WP, cm³ cm⁻³	a	b
20–24	0.5083	0.0725	0.270	0.098	0.6787	0.02356
45–49	0.4864	0.0779	0.262	0.104	0.6515	0.02787
115–119	0.4303	0.0441	0.206	0.059	0.75692	0.01722

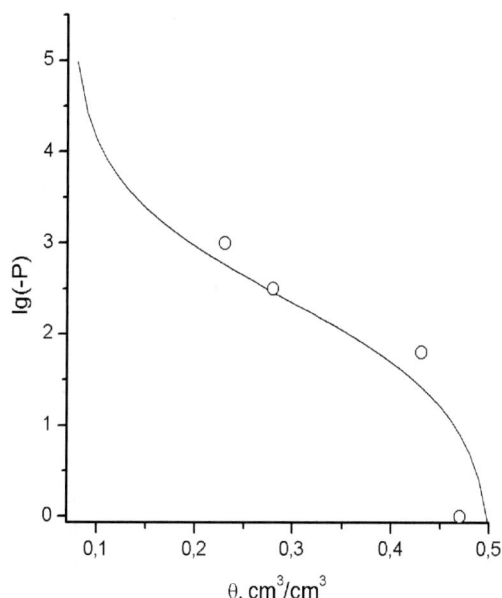

Fig. 10 Comparison between simulated water retention curve and experimental data (circles) for layer 20–24 cm (V 521)

Water status

With the help of simulation system AGROTOOL water storage in the layer 0–100 cm under spring barley and potatoes crops was calculated. For comparing the experimental and simulation results the experimental water content in the same layer was estimated.

The data of FDR measurement with approximately 10 days time step were used. We assume that the measured data of FDR-05 present the water content in the soil layer 0–10 cm, the data of FDR-45 present the water content in the soil layer 10–80 cm, and the data of FDR-90 present the water content in the soil layer 80–100 cm. So the following formula was used for estimation of soil water storage in the 0–100 cm layer:

WS exp = 0.1 * (FDR–05) + 0.7 * (FDR–45)
 + 0.2 * (FDR–90).

Soil water dynamics for spring barley and potatoes are presented in Fig. 14.

Statistical data

The results of statistical treatment for water status are shown in Table 9.

The error was calculated according to formula:

$$E_w = \frac{1}{n}\sum Ew_i = \frac{1}{n}\sum \text{WSsim}_i - \text{WS exp}_i$$

where WSsim$_i$ is simulated soil water storage corresponding to ith measurement, n is total number of measurements. MSE is the square root from the variance of the errors Ew_i.

The results of statistical treatment for dry mass dynamics are shown in Table 10.

Relative error was calculated using the following formula:

$$\text{RE} = \frac{1}{n}\sum \frac{B\text{sim}_i - B\exp_i}{B\text{sim}_i},$$

where $B\text{sim}_i$ is simulated value of dry mass, $B\exp_i$ corresponding experimental value, n – total number of measurements. MSE was calculated as square root from variance of RE$_i$.

Summary results for water content are presented in Figs. 15 and 16.

Comparison results for the Berlin site

Experimental data sets presented by Diestel et al. (2007) for Berlin were used. It is not provided the

Table 8 Yield and harvesting date

Culture/year	Yield of crop, dt ha⁻¹		Harvesting date	
	Simulated	Experimental	Simulated	Experimental
Spring barley/2000	35.9	34.6	29. Jun	31. Jul
Potatoes/2001	540.3	549.2	02. Sep	14. Sep
Winter wheat/2001/02	51.4	56.4	31. Jul	31. Jul

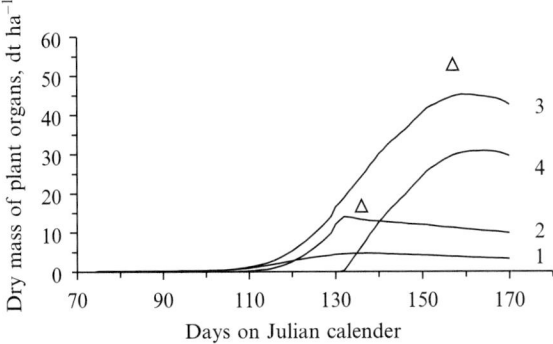

Fig. 11 Dynamics of dry biomass of spring barley plant organs, Bad Lauchstaedt, 2000: 1 – leaves, 2 – stems, 3 – above-ground biomass, 4 – ears, Δ – experimental data for above-ground biomass

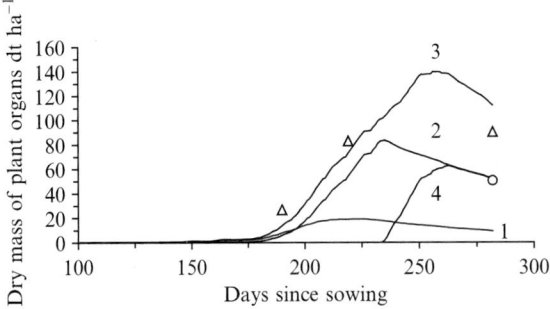

Fig. 12 Dynamics of dry biomass of winter wheat plant organs, Bad Lauchstaedt, 2001: 1 – leaves, 2 – stems, 3 – above-ground, 4 – ears, Δ – experimental data for above-ground

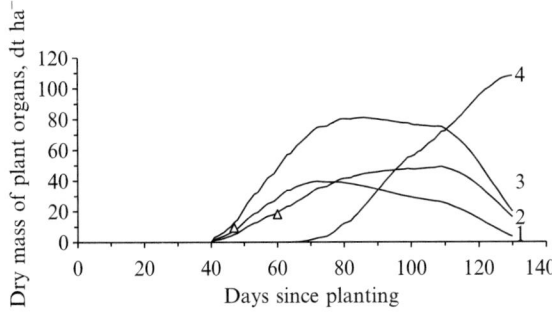

Fig. 13 Dynamics of dry biomass of potatoes plant organs, Bad Lauchstaedt, 2001: 1 – leaves, 2 – stems, 3 – leaves + stems, 4 – tubers, Δ – experimental data for leaves

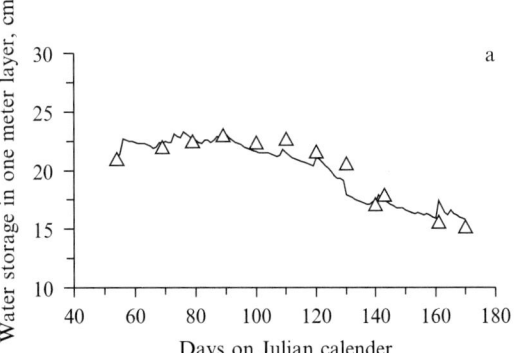

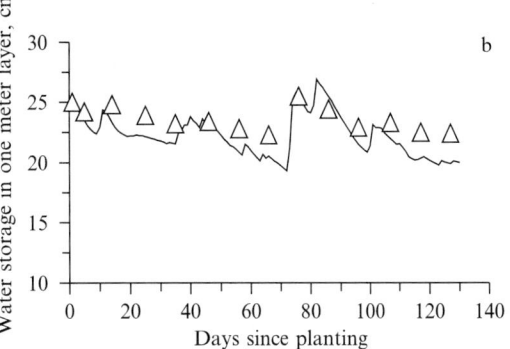

Fig. 14 Soil water dynamics under spring barley (a) and potatoes (b) crops, Bad Lauchstaedt, Δ – measurement data

Table 9 Results of statistical treatment for water content

Culture/year	Mean value, cm	Error, cm	MSE, cm
Spring barley/2000	19.0	−0.208	1.04
Potatoes/2001	22.4	−1.16	0.95

Table 10 Results of statistical treatment for dry biomass

Culture/year	Relative error	MSE
Spring barley/2000	0.065	0.24
Potatoes/2001	0.061	0.30
Winter wheat/2001/02	−0.21	0.44

description of soil water dynamics with high level of water table in the system AGROTOOL. So the following variants were chosen for calculation in the Berlin site: Lysimters 1–2, 5–8 and 11–12 with water table placed on 2.1 m. The years of vegetation were 1996, 1997 and 1998. Total number of variants was 12. It is necessary to determine the following data for model run: hydrological soil constants, daily weather parameters and initial model conditions. Hydrological soil constants determine PTFs. Correspondence of the soil monoliths is depicted in Table 11.

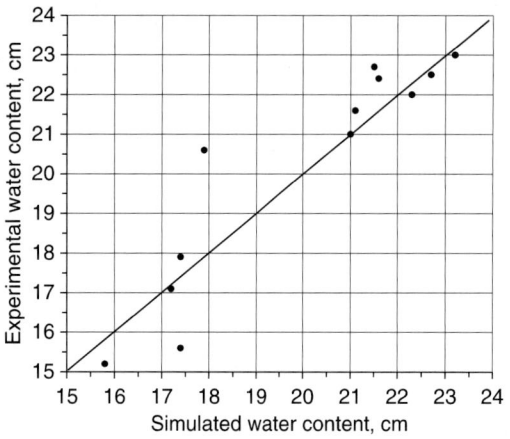

Fig. 15 Comparison of simulated and experimental data for water content in 1 m soil layer, spring barley, 2000

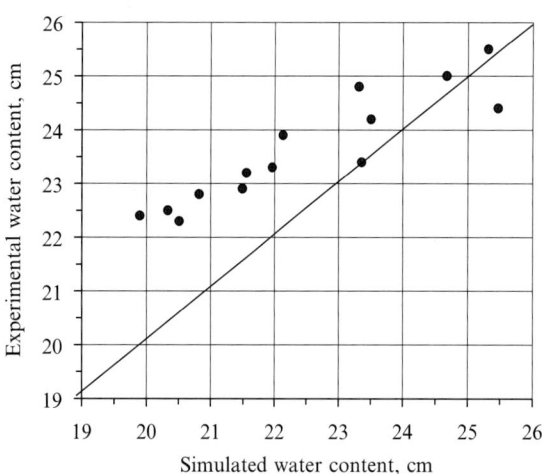

Fig. 16 Comparison of simulated and experimental data for water content in 1 m soil layer, potatoes, 2001

Table 11 Correspondence of the soil monoliths with lysimeters

Lysimter no	Origin of the soil monoliths	Soil type
1–2	Wildeshausen/ Nieders.	Podsol
5–8	Parlow-Glambeck/ Brdbg.	Parabraunerde Pseudogley
11–12	Weckesheim/ Hessen	Braunerde

Six daily parameters must be used as weather data: maximum and minimum air temperature, air humidity, precipitation, wind speed and coefficient of solar radiation extinction. Five of them were included in the weather file directly. For estimation of the last parameter a special programme was applied using measured data for incoming solar radiation and extraterrestrial radiation calculated for appropriate geographic attitude. Because there were no data for maximum and minimum air temperature in Berlin weather data, these parameters were rewritten from Muencheberg weather file. There were chosen standard data for initial model conditions: April 5, the date of vegetation renewal and water content in 1 m layer close to field capacity.

Pedotransfer functions

The information about soil texture and field capacity of the layers 0–40 cm, 40–60 cm, 60–150 cm for Wildeshausen (Podsol monolith), the layers 0–20 cm, 20–60 cm, 60–150 cm for Weckesheim (Braunerdesoil monolith), the layers 0–40 cm, 40–90 cm, 90–150 cm for Parlow-Glambeck (Parabraunerde Pseudogley soil monolith) was employed for estimation of PTF parameters. Such parameters as ρ, WP and SP were estimated firstly using the method published in "Estimating generalized soil water characteristics from texture" (Saxton et al. 1986). Then all parameters included in Eq. 14 were determined according to the software AGROHYDROLOGY, variant 2. The calculation results are shown in Tables 12–14. For example, the comparison of simulated pF curve with experimental data for layer 0–40 cm for Wildeshausen (Podsol monolith) is presented in Fig. 17.

Evapotranspiration

For simulation of evapotranspiration the model of perennial grasses included in simulation system AGROTOOL was used. An example of simulated and experimental results for lysimters 1 and 2 in 1996 is presented in Fig. 18. The results of statistical treatment for all investigated variants are shown in Table 15 and in Fig. 19.

Acknowledgement Investigation was fulfilled in the framework of Russian–German programme of collaboration in the field of agriculture (theme 128) and under financial support of RFBR (Project No. 04-05-64980).

Table 12 Parameters of water retention curve for Wildeshausen, Podsol monolith

Depth, cm	SP, $cm^3\,cm^{-3}$	MH, $cm^3\,cm^{-3}$	FC, $cm^3\,cm^{-3}$	WP, $cm^3\,cm^{-3}$	a	b
0–40	0.35	0.020	0.151	0.03	0.67042	0.0295
40–60	0.36	0.0129	0.156	0.03	0.68129	0.02741
60–150	0.39	0.0134	0.174	0.04	0.40851	0.11467

Table 13 Parameters of water retention curve for Weckesheim, Braunerdesoil monolith

Depth, cm	SP, $cm^3\,cm^{-3}$	MH, $cm^3\,cm^{-3}$	FC, $cm^3\,cm^{-3}$	WP, $cm^3\,cm^{-3}$	a	b
0–20	0.52	0.14	0.32	0.16	0.70193	0.02108
20–60	0.51	0.13	0.31	0.15	0.72969	0.01614
60–150	0.49	0.10	0.28	0.12	0.72408	0.01751

Table 14 Parameters of wter retention curve for Parlow-Glambeck, Parabraunerde Pseudogley soil monolith

Depth, cm	SP, $cm^3\,cm^{-3}$	MH, $cm^3\,cm^{-3}$	FC, $cm^3\,cm^{-3}$	WP, $cm^3\,cm^{-3}$	a	b
0–40	0.40	0.04	0.20	0.08	0.48636	0.07447
40–90	0.46	0.06	0.25	0.12	0.42825	0.09224
90–150	0.44	0.05	0.23	0.10	0.46186	0.08012

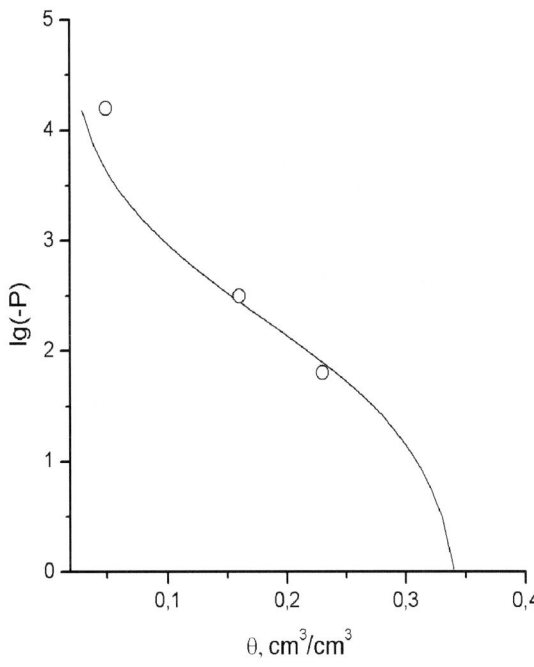

Fig. 17 Comparison between simulated water retention curve and experimental data (circles) for layer 0–40 cm for Wildeshausen (Podsol monolith)

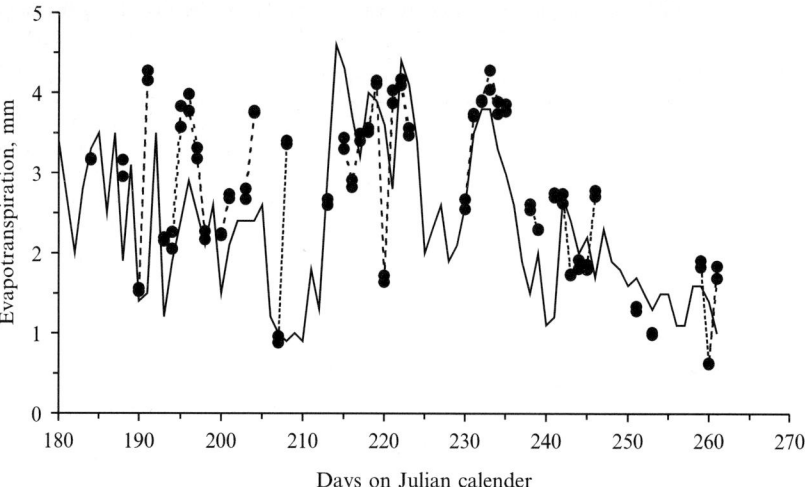

Fig. 18 Comparison of annual course of simulated and experimental data, lysimeter 1 and 2 (1996)

Table 15 The results of statistical evapotranspiration treatment

Year of vegetation	Mean error, mm	MSE, mm
1996	−0.57	0.99
1997	−0.40	1.03
1998	−0.22	0.92
General	−0.40	0.98

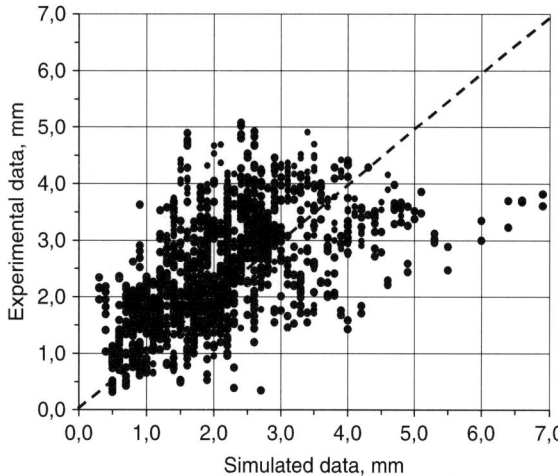

Fig. 19 Comparison of simulated and experimental results for evapotranspiration, lysimeters 1–2, 5–8, 11, 12 (1996–1998)

References

Diepen CA van, Rappold C, Wolf J, Keulen H van (1988) Crop growth simulation model WOFOST. Documentation version 4.1. Centre for World Food Studies, Wageningen, The Netherlands, pp 1–299

Diestel H, Zenkel T, Schwartengraeber R, Schmidt M (2007) The lysimeter station at Berlin-Dahlem. In: Kersebaum KC, Hecker J-M, Mirschel W, Wegehenkel M (eds) Modelling water and nutrient dynamics in soil-crop systems. Springer, Dordrecht, pp 259–266

Franko U, Puhlmann M, Kuka K, Böhme F, Merbach I (2007) Dynamics of water, carbon and nitrogen in an agricultural used Chernozem soil in Central Germany. In: Kersebaum KC, Hecker J-M, Mirschel W, Wegehenkel M (eds) Modelling water and nutrient dynamics in soil-crop systems. Springer, Dordrecht, pp 245–258

Hanks J, Ritchie JT (1991) Modelling plant and soil systems. Agronomy (A Series of Monographs). SSSAI Publishers, Madison, WI, 1–544

Monteith JL (1981) Evaporation and surface temperature. Q J R Meteorol Soc 101:1–27

Mirschel W, Wenkel K-O, Wegehenkel M, Kersebaum KC, Schindler U, Hecker J-M (2007) Müncheberg field trial data set for agro-ecosystem model validation. In: Kersebaum KC, Hecker J-M, Mirschel W, Wegehenkel M (eds) Modelling water and nutrient dynamics in soil-crop systems. Springer, Dordrecht, pp 219–243

Penning de Vries FWT, Jansen DM, Bergen HFM ten, Bakema A (1989) Simulation of ecophysiological processes of growth in several annual crops. Pudoc, Wageningen, The Netherlands, pp 1–271

Penman HL (1948) Natural evaporation from open water, bare soil, and grass. Proc R Soc, London A193, pp 120–146

Poluektov RA, Vasilenko GV (1992) Calculation of transpiration and physical evaporation in applied agroecosystem models (In Russian: Raschet transpiracii i fizicheskogo isparenija v prikladnyh modeljah agroekosistem). In: Pochva o rastenije – processy i sistemy, SPb, Proc. of ARI 58–66

Poluektov RA, Kumakov VA, Vasilenko GV (1997) Simulation of transpiration in crop stands. Russian plant physiol 44 (1):57–61

Poluektov RA, Oparina IV, Topaj AG, Fintushal SM, Mirschel W (2000) Adaptivity of agroecosystem dynamic models to varies soil and climatic conditions (In Russian: Adaptirujemostj dinamicheskih modelej agroecosistem krazlichnym pochvenno-klimaticheskim uslovijam). Matematicheskoje modelirovanije 12 (11):3–16

Poluektov RA, Fintushal SM, Oparina IV, Shatskikh DV, Terleev VV, Zakharova ET (2002) Agrotool – a system for crop simulation. Arch Acker-Pfl Boden 48:609–635

Saxton KE et al. (1986) Estimating generalized soil water characteristics from texture. Soil Sci Soc Am J 50(4): 1031–1036

Terleev VV (2000) Modelling of water retention curve as a capillary-porous body. (In Russian: Modelirovanije vodoudergivajushtej sposobnosti pochv kak kapiljarno-poristyh tel.) Chemistry Research Institute of St. Petersburg State University, pp 1–72

Wenkel K-O, Mirschel W (1995) Agrooerosystemmodellierung. Grundlage für die Abschätzung von Auswirkungen möglicher Landnutzungs- und Klimaänderungen. ZALF-Bericht, Müncheberg 24:1–187

Williams JR, Dyke PT, Jones CA (1983) EPIC -a model for assessing the effects of erosion on soil productivity. In: Laueroth WK, Skogerboe CV, Flug M (eds). Analysis of ecological systems. State-of-the art in ecological modeling, May 24–38. Colorado State University, Fort Collins, CO, pp 553–572

CHAPTER EIGHT

The NDICEA model, a tool to improve nitrogen use efficiency in cropping systems

Geert Jan H. M. van der Burgt[1,*], Gerard J. M. Oomen[1,2], A. S. J. Habets[1] and Walter A. H. Rossing[2]

[1]Louis Bolk Institute, Driebergen, The Netherlands; [2]Biological Farming Systems Group, Wageningen University, Wageningen, The Netherlands; *Author for correspondence (e-mail: g.vanderburgt@louisbolk.nl)

Received 25 August 2005; accepted in revised form 23 January 2006

Key words: Decision support, Leaching, Modelling, NDICEA, Nitrogen dynamics, Organic farming

Abstract

The effective management of nitrogen dynamics is essential for cropping systems which have the double objective of achieving acceptable yields and minimizing environmental impact. The decisions to be made are both particularly complex and of great urgency to farmers, including all organic farmers, who rely on organic sources of nitrogen. Models can be useful means of providing a better understanding of the nitrogen dynamics and of supporting decision-making at tactical and strategic levels. This paper presents a model that aims at providing support in the decision-making process based on a target-oriented description of nitrogen dynamics in a cropping system. The NDICEA model describes soil water dynamics, nitrogen mineralization and inorganic nitrogen dynamics in relation to weather and crop demand. Crop yields are put into the model, resulting in a target-oriented modelling approach which is distinctive from most other models. Parameter calibration is an inherent component of the modelling philosophy and is geared to establishing plot-specific factors. Using both quantitative and visual performance indicators, and different ratios of calibration to validation data, we evaluate the performance of NDICEA based on three treatments obtained from the Müncheberg dataset. Based on a maximum of 3 years of data for calibration, the root mean square error (RMSE) was found to vary between 14 kg N ha^{-1} and 37 kg N ha^{-1}, and in the majority of cases absolute prediction error was less than 20 kg N ha^{-1}. We introduce a user-friendly version of the model that is aimed at farmers and extension workers.

Introduction

The management of nitrogen is a critical element in cropping systems that aim at combining positive economic returns with minimizing environmental impact. Management issues are particularly complex and of great urgency to farmers, including all organic farmers, who rely on organic sources of nitrogen. In these systems, the management of nitrogen is not only related to chemical soil fertility, but interlinked with the management of physical and biological soil fertility, with the objective of providing crops with water and nutrients and suppressing soil-borne pathogens (Mäder et al. 2002).

The mineralization of organic nitrogen from manure starts after excretion from the animal source, and that from residues after harvest. Once started, operational management can scarcely control mineralization. Where residues and manure are important sources of nitrogen for crops, plant production systems have to fit into these patterns of inherent release of nitrogen. This release should not limit crop production, nor evoke pests and diseases or cause pollution of the environment. Tactical – i.e. within-season – decisions on timing and amount of organic nitrogen to apply are complex due to the prolonged release of inorganic nitrogen, which is influenced by uncontrollable factors such as rainfall and temperature and which is difficult to monitor.

From a strategic point of view, nitrogen in current organic farming systems in The Netherlands is an area of some concern. Currently, manure from conventional farming systems is a major source of nitrogen (Hendriks and Oomen 2000; Hofstad and Schröder 2002), which is not in line with EU regulations that aim at, ultimately, 100% organic origin of inputs. Oomen (1995) calculated that for The Netherlands self-reliance in organic agriculture implies that less than 95 kg manure-N is available per hectare. In such a situation, arable cropping systems have to be designed in such a way that this limited amount of manure is sufficient for the attainment of acceptable production levels.

Thus, both from a tactical and a strategic viewpoint there is a need for farmers to adjust nitrogen demand and supply in arable cropping systems. While trial and error will ultimately provide the necessary information, it is ultimately very costly in both time and expenditure: the measurements are expensive to carry out, and the economic stakes associated with crop failure are high. Models based on a system representative of key processes of nitrogen dynamics and location-specific inputs would enable an assessment of the current cropping system and allow an evaluation of alternative systems, and may thus accelerate learning among farmers. Of the models reported in the literature, those that aim at improving scientific knowledge of system functioning are not necessarily fit for application in practice (Koopmans and Bokhorst 2002). To contribute to an informed decision-making process, scientific rigor needs to be combined with an application-oriented philosophy in model design (Smith et al. 1996; Meynard et al. 2002; Jones et al. 2003; Keating et al. 2003) as is the case of the NDICEA model presented in this paper.

The NDICEA model was developed as a tool for representing the dynamics of water, organic matter and inorganic nitrogen in organic cropping systems to enable an assessment of organic fertilization strategies using relatively easily obtainable information on initial states, parameters and driving variables. Evaluations of the model with data from The Netherlands have been published elsewhere (Koopmans and Bokhorst 2002; van der Burgt 2004). The objective of this paper was to present the structure of NDICEA within the framework of the application-oriented goal for which it was developed and to assess model performance based on experimental data obtained from the Müncheberg cropping system experiment (Mirschel et al. 2007).

The NDICEA model

Model description

Modelling concept

NDICEA is an acronym for Nitrogen Dynamics In Crop rotations in Ecological Agriculture. It is a process-based simulation model which calculates the dynamics of the state variables soil water (W_i; cubic metres per hectare), soil carbon (C_i; kilograms per hectare), soil organic matter (OM_i; kilograms per hectare), apparent initial age of a source of organic matter (A_i; years) and soil organic nitrogen (NO_i) and inorganic nitrogen (N_i; both kilograms per hectare) for each soil layer (i) over the course of a crop rotation with a 1-week time interval. The model applies to homogenous layers in well-drained, mineral soils. The purpose of NDICEA is to enhance experiential learning by farmers and extension agents by reconstructing the dynamics of water and nitrogen in experiments on cropping systems or in farmers' fields. This objective resulted in a target-oriented approach: the target of crop production, expressed in terms of biomass, water use and nitrogen accumulation, is input for the model, and the model 'reconstructs' the dynamics of the state variables. When alternative scenarios are explored, targets may be infeasible because of a lack of nitrogen. There is no feedback of such resource constraint on the target in NDICEA. The user is required to ascertain adequate resource availability.

The soil is divided into two horizontal layers: the topsoil (indicated by suffix $i = 1$), where tillage and the application of organic matter take place, and the subsoil (indicated by suffix $i = 2$), into which no input of new organic matter occurs. The subsoil extends to the depth from which crops can take up water and nutrients. It is assumed that a complete mixing of soil water and dissolved nutrients takes place in both layers within each time interval.

The model consists of three components, which will be described in more detail in the following sections. In the first component, the dynamics of soil water are calculated taking into account rainfall, irrigation, evapotranspiration and capillary rise or its opposite, percolation. In the second component, the decomposition of organic matter and mineralization of organic nitrogen from the initial soil organic matter stock and from successive additions of crop residues and organic manure are calculated by a modified one-parameter carbon dissimilation model based on Janssen (1984) and a nitrogen mineralization model based on Verveda (1983; referenced in Janssen 1984), taking into account soil temperature, soil moisture content, soil pH and organic matter protection capacity of soil. In the third model component, soil inorganic nitrogen dynamics are calculated based on nitrogen input by mineralization, atmospheric deposition, irrigation, fertilizers, capillary rise, and biological fixation, and on nitrogen loss by crop uptake, denitrification and leaching.

The model comprises 46 parameters and input functions, nine of which cannot be determined independently from NDICEA. Although default values are available for these parameters, the model was developed based on the premise these parameters would be calibrated using location-specific data. A calibration algorithm that searches for optimum parameter values within plausible ranges (Price 1979; Klepper and Rouse 1991) is therefore described as an integral component of the approach.

In the following description, rates are indicated with starting letter 'd', followed by N for (organic or inorganic) nitrogen, W for water, C for carbon, A for apparent initial age and OM for organic matter, and specified by further capitals and suffices. Parameters and forcing functions are indicated in lowercase italics.

Soil water dynamics

Precipitation and irrigation constitute inflows to the topsoil. Capillary rise may increase the water content of each layer, the rate depending on the suction properties (pF) of the soil and the depth of the groundwater table. Plants can take up water from both layers, with the amount of each fraction taken up depending on the crop rooting depth and moisture content in each layer. Water uptake by the crop is determined by the developmental stage of the crop, potential evapotranspiration and soil pF. Water in the topsoil above field capacity is assumed to instantaneously percolate to the subsoil and that from the subsoil to deeper layers. Water drained from the subsoil is lost from the system.

Capillary rise (dWC_i) in each layer i (millimetres per week) is driven by a matrix suction gradient from a soil layer to the groundwater table. Matrix suction in layer i, ψ_i, is calculated according to Driessen (1988) as

$$\psi_i = \exp(\text{sqrt}\,(\text{gam}^{-1} * \ln(\text{smo}/\theta_i))) \qquad (1)$$

where smo is saturated soil moisture content (in meters per cubic meter of soil) and gam is a texture-specific coefficient describing the shape of the soil pF curve. Values for gam and smo for different soil types can be found in Driessen (1988). Volumetric soil moisture content θ_i [cubic metres (water) per cubic metre (soil)] is derived from the state variable W_i and the thickness of soil layer i, thl_i.

The actual capillary rise is equal to the maximum capillary rise as determined by soil type, corrected for the dryness of the adjacent layers. The maximum capillary rise for layer i ($dWC_{\max, i}$) is an empirical function of the distance from the centre of a layer to the groundwater table.

$$dWC_{\max,1} = \max(7 * \exp((cr_{c,1} - thl_1/2 \\ - \psi_{FC,2})/cr_{x,1}), 21) \qquad (2a)$$

and

$$dWC_{\max,2} = \max(7 * \exp((cr_{c,2} - \psi_{GWT,2})/cr_{x,2}), 21) \qquad (2b)$$

where, $\psi_{GWT,i}$ = the equilibrium matrix suction in layer i, i.e. the distance from the centre of layer i to the water table (in centimetres), $\psi_{FC,i}$ = the matrix suction in soil layer i at field capacity (in centimetres) and thl_i = the thickness of the soil layer i (in centimetres).

The coefficients of the function, cr_c and cr_x, are based on literature data from Anonymous (1989) after Wösten et al. (1987). As in these data capillary rise never exceeded 3 mm day^{-1}, maximum capillary rise was limited to 21 mm week^{-1}. Based on the same dataset we postulate that maximum capillary rise to the topsoil is reached when the difference between pF in the topsoil and pF in the subsoil exceeds 0.4. For the subsoil, we assume that maximum capillary rise is reached when the difference between the actual pF and the pF in equilibrium with the ground water table exceeds 0.4. If in any layer this difference is less, capillary rise is reduced proportionately. The resulting capillary rise in the topsoil (dWC_1) and subsoil (dWC_2) is now:

$$dWC_1 = dWC_{max,1} * \max(\min((-\log(\psi_1) + \log(\psi_2))/0.4, 1), 0) \quad (3a)$$

and

$$dWC_2 = dWC_{max,2} * \max(\min((-\log(\psi_2) + \log(\psi_{GWT,2}))/0.4, 1), 0) \quad (3b)$$

The depth of the groundwater table is introduced as a forcing function, following an annual pattern derived by interpolation between a highest and a lowest level around the start and end of the growing season, respectively. Default values are week 12 and week 38, respectively. The maximum and minimum depths of the water table are assumed to be greater than the depth of the subsoil.

Actual water uptake by a crop, dWU_{act}, is calculated for each soil layer as minimum of the potential water uptake and the amount available for the plant. Potential water uptake (dWU_p; in millimetres per week) is calculated by multiplying reference evapotranspiration (ET_0; in millimetres per week) calculated according to Makkink (Makkink 1957; de Bruin 1987) using correction factors to account for crop-related evapotranspiration or evaporation from bare soil. Bare soil is assumed to have a constant evaporation factor, $et_{bare} = 0.25$ (Huinink 1998). The crop evapotranspiration factor is calculated by multiplying crop cover (cc) and a crop-specific maximum crop evapotranspiration factor (et_{max}). Crop cover (cc) is a function of the stage of crop development. Four developmental stages are distinguished: sowing, full cover, ripening and harvest. The crop cover factor (cc) increases linearly from 0 at sowing to 1 at full cover, then remains constant until ripening, subsequently decreasing linearly to 0.5 at harvest. Crop cover and the maximum crop evapotranspiration factor were derived for a range of crops based on Huinink (1998).

$$dWU_p = \max(cc * et_{max}, et_{bare}) * ET_0 \quad (4)$$

where, dWU_p = the potential water uptake by the crop (in millimetres per week), cc = the crop cover factor (–, no units), et_{max} = the maximum crop evapotranspiration factor (–), et_{bare} = the evaporation factor of bare soil (–) and ET_0 = the reference crop evapotranspiration (in millimetres per week).

When soil pF exceeds 2.7, actual water uptake from a layer is reduced, reaching zero at a pF of 4.2. This proportional decrease is summarized in the reduction factor wuf_i (–).

The fraction of root biomass rf in topsoil and subsoil determines the partitioning of water uptake between the soil layers. The fraction rf in a layer is calculated from the rooting depth, rd, which is assumed to increase linearly from zero at sowing to a maximum value at full cover, after which it remains constant. A linear decrease in root biomass with depth is assumed. The fraction of roots in the topsoil (rf_1) is calculated from actual rooting depth:

$$\begin{array}{lll} rf_1 = 1 & \text{for} & rd < thl_1 \\ rf_1 = 1 - ((rd - thl_1)/rd)^2 & \text{for} & rd \geq thl_1 \\ rf_2 = 1 - rf_1 & & \end{array} \quad (5)$$

where rd = rooting depth (in centimetres), thl_i = the thickness of the soil layer, i (in centimetres) and rf_i = the fraction of roots in soil layer i (–).

After calculating the uptake from the topsoil, the remaining amount of required water is assumed to be taken up from the subsoil, but only if the rooting depth (rd) exceeds the thickness of the topsoil (i). When the pF of the subsoil exceeds 2.7, relatively more water is taken up from the topsoil, independent of rooting depth. For bare soil, the rooting depth is zero and evaporation takes place from the topsoil only. In summary:

$$d\mathrm{WU}_{a,1} = \mathrm{rf}_1 * \mathrm{wuf}_1 * d\mathrm{WU}_p \quad \text{for pF in subsoil} < 2.7$$
$$d\mathrm{WU}_{a,1} = \mathrm{wuf}_1 * d\mathrm{WU}_p \quad \text{otherwise} \quad (6a)$$

$$d\mathrm{WU}_{a,2} = 0 \quad \text{for } \mathrm{rf}_1 = 1$$
$$d\mathrm{WU}_{a,2} = \mathrm{wuf}_2 * (d\mathrm{WU}_p - d\mathrm{WU}_{a,1}) \quad \text{otherwise} \quad (6b)$$

where $d\mathrm{WU}_{a,i}$ is the actual water uptake (in millimetres per week) in layer i. Actual water uptake ($d\mathrm{WU}_a$; in millimetres per week) is calculated as the sum of actual uptake from each layer.

Outflow or drainage from each layer ($d\mathrm{WO}_i$; in millimetres per week), which only occurs in the absence of capillary rise, is calculated as the balancing item in the weekly water balance.

Decomposition of organic matter and mineralization of organic nitrogen

The decomposition of organic matter and mineralization of organic nitrogen is described as a function of the totality of the sources of organic matter, soil moisture content, temperature, soil pH, organic matter protection capacity of the soil and composition of the food web in soil. The approach used by Janssen (1984) to calculate the annual decomposition of applied organic matter is modified to enable the calculation of weekly decomposition rates. The mineralization of nitrogen during decomposition is calculated using a modification of the approach of Verdeda (1983, referenced in Janssen 1984).

Calculation of the decomposition of organic matter. Janssen (1984, 1996) proposed describing the time course of carbon remaining after the application of C_0 (in kilograms per hectare) in organic carbon at time $t = 0$ by:

$$C_y = C_0 * \exp(4.7 * (A_y^{-0.6} - A_0^{-0.6})) \quad (7)$$

where A_y is the apparent age of the organic substrate at time y (in years), and A_0 the apparent age at the time of application (in years). Janssen's equation was derived from annual measurements of organic matter on bare and permanently moist soils in The Netherlands. We adapted this approach to describe decomposition on a cropped soil during a time interval of 1 week. A correction factor f was introduced, which expresses the effects of temperature, moisture, organic matter protection capacity and pH. Although interaction, for example between temperature and moisture (Cassman and Munns 1980), is possible, it is assumed that all effects act independently on decomposition:

$$f = f_T * f_\theta * f_\mathrm{prot} * f_\mathrm{pH} \quad (8)$$

where f = the overall correction factor = rate of aging (–), f_T = the temperature correction factor (–), f_θ = the moisture correction factor (–), f_prot = the protection correction factor (–) and f_pH = the pH correction factor (–).

The apparent age of the substrate at time t (weeks), A_t, is described by Woli (2000) as

$$A_t = 1/52 * f * t + A_0 \quad (9)$$

We assume the carbon (C; in kilograms) and organic matter (OM, in kilograms) in soil organic matter to be related by $C = \mathrm{com} \times \mathrm{OM}$, where com is a function which describes the carbon content of organic matter as a dependent of its apparent age. Based on data of Goudriaan and van Laar (1994) and Allison (1965) we assumed organic matter of an apparent age up to 1 year to have a com value of 0.45, increasing linearly to 0.58 at an apparent age of 3 years and higher. The decomposition rate of organic matter ($d\mathrm{OM}$; in kilograms of OM per week) is found by combining the derivative of Eq. 7 with 9:

$$d\mathrm{OM} = (-2.82 * A_t^{-1.6} * \mathrm{OM} * f)/52 \quad (10)$$

Decomposition of the organic matter pool already present in the soil, referred to as 'initial organic matter', is described using the same equation.

The effect of temperature on decomposition is described by an empirical modification of the Q9-equation by Yang (1996), which was based on annual average temperature, to take into account the influence of weekly average temperature T_av on decomposition rate:

$$\begin{aligned} f_\mathrm{Yang} &= 0 & T_a &< = -1 \\ f_\mathrm{Yang} &= 0.09 * (T_\mathrm{av} + 1) & -1 &< T_\mathrm{av} < = 9 \\ f_\mathrm{Yang} &= 0.88 * 2 \wedge (T_\mathrm{av} - 9)/9) & 9 &< T_\mathrm{av} < = 27 \\ f_\mathrm{Yang} &= 3.5 & T_\mathrm{av} &> 27 \end{aligned} \quad (11)$$

The effect of a lack of moisture on mineralization of organic matter is described as a linear decrease to zero between pF = 2.7 and pF = 4.2 according to Rijtema (1980).

The decomposition rate of organic matter is affected by the organic matter protection capacity, which depends on soil texture, soil structure and soil organic matter content (Hassink et al. 1993). Soil texture and soil organic matter content can be considered to be constant during a year, but soil structure fluctuates due to soil tillage, soil compaction, activity of soil fauna and fluctuations of moisture content and temperature (freezing). A single correction factor f_{prot} was introduced for all three effects. The value of f_{prot} is found by calibration.

Under acid conditions the decomposition of organic matter can be reduced to practically zero. Baht et al. (1980) assumed that above pH = 4.5 decomposition proceeds unhampered and below pH = 3.5 it comes to a complete standstill; Rijtema and Kroes (1991) assume a range from pH 6 to pH 4 for the same process. In normal agricultural practice, a pH correction of OM decomposition is hardly of interest, as soil pH is usually maintained above 5. However, for completeness sake, a pH correction factor has been included, which is about 1 at a pH of 7 and decreases to 0.5 at a pH of 4:

$$f_{pH} = 1/(1 + \exp(-1.5 * (pH - 4))) \quad (12)$$

Calculation of mineralization of organic nitrogen. Organic matter is decomposed via a chain of transformations involving various organisms in the soil. Fungi and bacteria form the base of the pyramid; these are consumed by organisms such as nematodes and protozoa, which in their turn are eaten by other organisms (Brussaard 1998). Verveda (1983) proposed a procedure to calculate the amount of nitrogen released through decomposition by micro-organisms. We applied the procedure to the whole of the chain of transformations and established process parameters by calibrating these to location-specific data instead of deriving them from experiments with single organisms.

Carbon in an organic matter application is used by micro-organisms as a source of energy and building material. Hence, as mineralization proceeds, the remaining substrate will contain less carbon in the original form and more in the form of biomass (bacteria, fungi, etc), and decay and conversion products. Different rates may be distinguished: the rate of consumption of organic carbon, also referred to as the turnover of carbon (dCT; in kilograms per hectare per week); the assimilation rate of carbon (dCA; in kilograms per hectare per week) describing the amount of carbon that is re-used as organic building material, with or without modification; the rate of dissimilation of organic carbon (dCD; in kilograms per hectare per week), which causes the production of CO_2. The dissimilation rate of organic carbon is described by Eq. 10. Assuming the ratio (ad_{micro}) of assimilation rate dCA (in kilograms per hectare per week) and dissimilation rate dCD to be known, dCT can be calculated using the mass balance $dCT = dCD + dCA$:

$$dCT = (1 + ad_{micro}) * dCD \quad (13)$$

The assimilation rate of nitrogen (dNA; in kilograms per hectare per week) follows from the carbon assimilation rate, dCA, and the carbon-to-nitrogen ratio of the micro-organisms (cn_{micro}), which we assume to be constant:

$$dNA = 1/cn_{micro} * dCA \quad (14)$$

The nitrogen turnover rate (dTN; in kilograms per hectare per week) is related to carbon turnover rate (dCT; in kilograms per hectare per week) and the current carbon (C)-to-nitrogen (NO; kilogram per hectare) ratio of the substrate:

$$dTN = NO/C * dCT \quad (15)$$

The difference between turnover and assimilation of nitrogen equals net mineralization of nitrogen (dNM; in kilograms per hectare per week):

$$dNM = dTN - dNA = ((1 + ad_{micro})/(C/NO) - ad_{micro}/cn_{micro}) * dCD \quad (16)$$

Assuming $C = com \times OM$, Eq. 16 can be rewritten to express net nitrogen mineralization as a function of the decomposition of organic matter (dOM; kilogram per hectare per week):

$$dNM = ((1 + ad_{micro})/(OM / NO) - ad_{micro}/(cn_{micro}/com)) * dOM \quad (17)$$

Total nitrogen mineralization from organic matter sources is calculated as the sum of dNM

values for individual sources, distinguished by soil layer.

Soil inorganic nitrogen dynamics

The amount of inorganic nitrogen increases as a result of deposition, irrigation, capillary rise and the mineralization of organic compounds and decreases due to plant uptake, denitrification and drainage, as described per layer by:

$$d\mathrm{N}_i = d\mathrm{NI}_i + d\mathrm{NM}_i - d\mathrm{ND}_i - d\mathrm{NU}_i - d\mathrm{NO}_i \quad (18)$$

where, $d\mathrm{N}_i$ = the net rate of change of inorganic nitrogen in layer i (in kilograms per hectare per week), $d\mathrm{NI}_i$ = the inflow of inorganic nitrogen into layer i (in kilograms per hectare per week), $d\mathrm{NM}_i$ = the nitrogen mineralization in the soil layer (i; in kilograms per hectare per week), $d\mathrm{ND}_i$ = the denitrification in the soil layer i (in kilograms per hectare per week), $d\mathrm{NU}_i$ = the modelled nitrogen uptake by the crop from soil layer i (in kilograms per hectare per week), $d\mathrm{NO}_i$ = the outflow of inorganic nitrogen out of layer i by drainage (in kilograms per hectare per week).

Inorganic nitrogen input to topsoil. Annual deposition of nitrogen from the atmosphere, a location-specific quantity, is assumed to be distributed homogeneously over the year. The input of inorganic nitrogen with irrigation water or fertilizers is calculated from dose and inorganic nitrogen content in the fertilizers or manure. Losses by volatilization are not calculated in the model and have to be taken into account before entering information into the model.

$$d\mathrm{NI}_i = d\mathrm{N}_{\mathrm{dep}} + d\mathrm{N}_{\mathrm{irr}} + d\mathrm{N}_{\mathrm{fert}} \quad (19)$$

where $d\mathrm{N}_{\mathrm{dep}}$ = the input of inorganic nitrogen by atmospheric deposition (in kilograms per hectare per week), $d\mathrm{N}_{\mathrm{irr}}$ = the input of inorganic nitrogen by irrigation (in kilograms per hectare per week), $d\mathrm{N}_{\mathrm{fert}}$ = the input of inorganic nitrogen by fertilizers (in kilograms per hectare per week).

Denitrification. Denitrification is calculated based on a modification of the procedure of Bradbury et al. (1993). Denitrification in the soil occurs in so-called 'hot spots' located close to decomposing organic matter and is correlated to the number and size of these hot spots and to the diffusion of oxygen and, therefore, to water content, soil texture and soil structure. In these hot spots, the concentration of nitrate is by definition near zero, and the denitrification rate is mainly determined by the rate of diffusion of nitrate to these sinks. In well-drained agricultural soils nitrate is the main compound of inorganic nitrogen. Total nitrate diffusion per unit area is related to soil water content, the number, size and locations of these hot spots and the nitrate concentration gradient around the hot spots. The effect of the distribution of nitrate and hot spots through the soil is assumed to be accounted for by a denitrification factor, dnf. This factor is estimated from location specific data by calibration. Denitrification is considered to be proportional to the rate of decomposition of the organic matter, $d\mathrm{OM}$, the amount of inorganic nitrogen, N_i, as well as the scaled moisture content above the wilting point:

$$d\mathrm{ND}_i = d\mathrm{nf}_i/1000 * d\mathrm{OM}_i * \mathrm{N}_i \\ * (\theta_i - \theta_{\mathrm{WP},i})/(\theta_{\mathrm{FC},i} - \theta_{\mathrm{WP},i}) \quad (20)$$

where $d\mathrm{nf}_i$ = the denitrification factor of the soil layer i (per 1000 kilograms OM), N_i = the amount of inorganic nitrogen (kilograms N per hectare), $\theta_{\mathrm{FC},i}$ = the volumetric moisture content of the soil layer (i) at field capacity (in cubic metres per cubic metre) and $\theta_{\mathrm{WP},i}$ = the volumetric moisture content of the soil layer (i) at wilting point (in cubic metre per cubic metre)

Nitrogen uptake by the crop. The rate of nitrogen uptake per week is calculated in a target-oriented fashion – i.e., as a fraction of the total nitrogen uptake by the crop, the target. Total nitrogen uptake may be equal to the amount measured in an experiment being analyzed, or to the amount expected when assessing alternative crop sequences or fertilizer application regimes. Total nitrogen uptake includes nitrogen in harvested and unharvested, and above and belowground parts of the crop. Nitrogen fixed by symbiotic bacteria is subtracted from the total nitrogen uptake to arrive at the amount to be supplied from the soil.

$$d\mathrm{NU}_{\mathrm{tar}} = \mathrm{N}_{\mathrm{crop,tar}} * \mathrm{I}_{\mathrm{N}}/\mathrm{SI}_{\mathrm{N}} - d\mathrm{NF}_{\mathrm{act}} \quad (21)$$

where $d\mathrm{NU}_{\mathrm{tar}}$ = the targeted rate of nitrogen uptake (kilograms of N per hectare per week), $\mathrm{N}_{\mathrm{crop,tar}}$ = the total nitrogen uptake by the crop (kilograms of N per hectare), $d\mathrm{NF}_{act}$ = the actual crop nitrogen fixation rate (kilograms of N per hectare per week), I_N = the indicator for the rate of nitrogen uptake, based on diffusion-corrected actual water uptake (–), SI_N = I_N summed over the growing season (–).

The indicator I_N is calculated as a function of actual weekly water uptake. At high soil moisture levels, nitrogen taken up by the roots will originate from both mass flow and diffusion. Because diffusion is small at low levels of soil moisture, nitrogen uptake may deviate from water uptake. The threshold thr_{diff} accounts for this contribution of nitrogen diffusion and is found by calibration:

$$I_N = (WU_{a,1} + WU_{a,2})/(WU_p * \mathrm{thr}_{diff}) * cc * et_{max} * ET_0 \quad (22)$$
for $(WU_{a,1} + WU_{a,2})/WU_p < \mathrm{thr}_{diff}$

$$I_N = cc * et_{max} * ET_0$$
for $(WU_{a,1} + WU_{a,2})/WU_p >= \mathrm{thr}_{diff}$

The targeted rate of nitrogen uptake is now distributed to the soil layers ($d\mathrm{NU}_{tar,i}$) by assuming nitrogen uptake to be proportional to actual water uptake ($dWU_{a,i}$) and inorganic nitrogen concentration (N_i/W_i) in each layer:

$$d\mathrm{NU}_{tar,i} = p * dWU_{a,i} * N_i/W_i \quad (23)$$

The proportionality factor p may vary in time but is assumed to be equal for both layers. It is eliminated from the equation by combining Eqs. 23 and 24:

$$d\mathrm{NU}_{tar} = d\mathrm{NU}_{tar,1} + d\mathrm{NU}_{tar,2} \quad (24)$$

After re-arranging, the actual rates of nitrogen uptake per layer are found:

$$d\mathrm{NU}_{tar,1} = d\mathrm{NU}_{tar}/(1 + (dWU_{a,2}/dWU_{a,1}) * (N_2/N_1) * (W_1/W_2)) \quad (25a)$$

and

$$d\mathrm{NU}_{tar,2} = d\mathrm{NU}_{tar} - d\mathrm{NU}_{tar,1} \quad (25b)$$

The rate of nitrogen uptake per layer cannot exceed the amount of nitrogen available, so that the actual rate of N uptake per layer ($d\mathrm{NU}_{act,i}$) is

$$d\mathrm{NU}_{act,i} = \min(d\mathrm{NU}_{tar,i}, N_i/\mathrm{DELT}) \quad (26)$$

where DELT is the time interval of integration in the model, i.e. 1 week. Total simulated whole crop nitrogen uptake $N_{crop,act}$ (kilograms of N per hectare) is found by summing $(d\mathrm{NU}_{act,2} + d\mathrm{NU}_{act,2})$ over time.

Nitrogen fixation by crops. Nitrogen fixation by crops is calculated from the potential nitrogen fixation rate $d\mathrm{NF}_{pot}$ (in kilograms per hectare per week), and a threshold inorganic nitrogen amount in the topsoil thr_{fix} (kilograms of N per hectare), above which fixation is negatively affected (van Mil 1981). Above this threshold, potential fixation is reduced in proportion to the inorganic nitrogen amount in the topsoil N_1 up to a level of twice the value of thr_{fix}.

$$d\mathrm{NF}_{act} = d\mathrm{NF}_{pot} * \max(\min(2 - N_1/\mathrm{thr}_{fix}, 1), 0) \quad (27)$$

The threshold value thr_{fix} is related to soil moisture content (Ennik 1960) and possibly also to soil structure and root health and is found by calibration.

The distribution of fixation over the growing period is calculated from the total potential amount of nitrogen fixed $N_{crop,fix}$ (kilograms per hectare) in a fashion similar to that for weekly nitrogen uptake:

$$d\mathrm{NF}_{pot} = N_{crop,fix} * I_N/SI_N \quad (28)$$

Nitrogen transport by water and leaching. Water transport down the soil profile is not homogeneous. A portion of the water follows larger pores and cracks (Bouma and Dekker 1978; Bronswijk et al. 1990) and does not transport inorganic nitrogen, especially nitrogen mineralized from soil organic matter that is present in the smaller pores. Reduced nitrification in winter and the adsorption of ammonium further reduce the amount of inorganic nitrogen available for mass flow. In NDICEA, a complete mixing of water and nitrogen is assumed at each time interval. Nitrogen flow out of a layer dNO_i (in kilograms per hectare per week) is proportional to water flow and inorganic nitrogen concentration in the layer. To account for preferential water flow and adsorption, a nitrogen-leaching factor, nlf_i, is introduced for each layer. Values for nlf_i are found by calibration.

$$dNO_i = dWO_i * N_i/W_i * nlf_i \quad (29)$$

In case of capillary rise, nitrogen import into a layer takes place proportional to the nitrogen concentration in the subsoil and the groundwater in the topsoil and subsoil, respectively.

Model calibration

In this paper, the view is adopted that model structure, measured inputs and parameter values derived from literature are valid. Parameters, for which the value cannot be determined independently from NDICEA, have to be adjusted by calibration to bring model output into acceptable agreement with the output of the real systems.

To calibrate parameters in NDICEA, we have adopted an approach that uses an adaptive random search algorithm (Price 1979). The algorithm starts with plausible ranges for each parameter and searches for the subset of each range, which results in acceptable model output (Klepper and Rouse 1991). In the algorithm, the initial random search is gradually intensified in the most promising parts of the parameter space.

NDICEA produces various outputs that may all be compared to the real world system. The degree to which a set of model output values associated with an input parameter set, P, is acceptable when compared with the real system is expressed as an objective function C_p:

$$C_p = \min_p(C'_1/w_1 + \cdots + C'_m/w_m) \quad (30)$$

where w_j is a weighting factor, calculated as the average measured value for output variable j, and the partial objective function C'_j measures the acceptability of output j:

$$C'_j = \frac{1}{n}\sum_{i=1}^{n}|e_{i,j}|, \text{ with } e_{i,j} = m_{i,j} - y_{i,j} \quad (31)$$

where $e_{i,j}$ = the residual at measurement time i ($i = 1 \ldots n$) for output variable j ($j = 1 \ldots m$); $m_{i,j}$ = the model output for variable j at time i; $y_{i,j}$ = the measured value for variable j at time i.

For discussion on the choice of the objective function, see Klepper and Rouse (1991).

The Price algorithm resembles random methods such as simulated annealing (Press et al. 1986) and genetic algorithms (see review by Mayer 2002) by not focussing on subsequent improvement of the best parameter combinations but on generating parameter sets that are better than the worst one so far, and by iterative removal of these worst sets. The procedure involves storing Q parameter sets of q parameters ($Q >> q$). In the first step, the Q parameter sets are generated at random, keeping each parameter value within the selected plausible range. For every parameter set, NDICEA is executed, and the associated objective function value C is calculated. A new parameter set P' is generated in two steps. First, a new parameter set G is generated by randomly selecting $q + 1$ sets of parameters from the Q stored ones and calculating each element of G as the average of the corresponding elements in the first q sets. Next, the $q + 1$st parameter set is reflected element by element in the centroid G to find P'. A corresponding C-value is the calculated and compared to the Q stored ones. If it is better than the previously worst one, this parameter set replaces the worst set, and so on. The iteration is continued until a stop criterion is reached.

Model evaluation

Data for the calibration and independent evaluation of NDICEA were derived from a nitrogen application experiment at the Leibniz Centre for Agricultural Landscape and Land Use Research in Müncheberg, Germany (Mirschel et al. 2007). In the experiment, which ran from 1993 to 1998, crops in a 4-year rotation were grown at different, unreplicated nitrogen application regimes. Weather, crop yields and nutrient dynamics were monitored on each field, but no information was collected on soil organic matter (Mirschel et al. 2007). In this paper, we use results from the so-called plots 1, 2 and 3. On plot 1, nitrogen was applied as ammonium urea and ammonium nitrate lime. The crops on plot 2 received nitrogen in the form of farm yard manure and liquid manure; these plots were managed in an organic-like fashion, but were not certified as such. A combination of inorganic and organic fertilizer was applied to plot 3. In 1998, wheat yield on plot 2 was extremely low due to severe infestation by weeds. The nitrogen content of the weeds was not measured; consequently, no information is available

Table 1. Summary of main agronomic data from the Müncheberg experiment (Mirschel et al. 2007) used for calibrating and evaluating NDICEA.

Year	Crop	Parameter	Plot 1	Plot 2	Plot 3
1993	Sugarbeet	Crop yield[a] (kg ha^{-1})	8455	9467	15427
		N-yield crop(kg N ha^{-1})	143	153	179
		Fertilizer application[b] 1	MF	FYM	FYM
		Application date[c]	18	Previous autumn	Previous autumn
		N amount (kg N ha^{-1})	80	142	142
		Fertilizer application 2			MF
		Application date			18
		N amount (kg N ha^{-1})			80
1994	Winter-wheat	Crop yield (kg ha^{-1})	7690	4544	7568
		N-yield crop(kg N ha^{-1})	233	110	206
		Fertilizer application 1	MF	FYM	FYM
		Application date	15,19,22	Previous autumn	Previous autumn
		N amount (kg N ha^{-1})	170	66	66
		Fertilizer application 2		LM	MF
		Application date		18,19	15,19
		N amount (kg N ha^{-1})		60	60
1995	Winter-barley	Crop yield (kg ha^{-1})	7321	4733	8953
		N-yield crop(kg N ha^{-1})	210	126	248
		Fertilizer application	MF	LM	MF
		Application date	12,20,21,22	11	12,2
		N amount (kg N ha^{-1})	145	41	95
1996	Winter-rye	Crop yield (kg ha^{-1})	7500	3700	7000
		N-yield crop(kg N ha^{-1})	215	101	201
		Fertilizer application	MF	LM	MF
		Application date	17, 21, 23	18	17,21
		N amount (kg N ha^{-1})	100	64	75
1996	Green manure	Crop yield (kg ha^{-1})	2200	1000	1500
		N-yield crop(kg N ha^{-1})	55	33	34
		Fertilizer application	MF	FYM	FYM
		Application date	36	36	36
		N amount (kg N ha^{-1})	60	198	198
1997	Sugarbeet	Crop yield (kg ha^{-1})	11,639	11,762	12,419
		N-yield crop(kg N ha^{-1})	215	146	292
		Fertilizer application	MF		
		Application date	15,18		
		N amount (kg N ha^{-1})	130		
1998	Winter-wheat	Crop yield (kg ha^{-1})	6000	1126	7000
		N-yield crop(kg N ha^{-1})	182	43	191
		Fertilizer application 1	MF	FYM	MF
		Application date	12	Previous autumn	12
		N amount (kg N ha^{-1})	50	66	35
		Fertilizer application 2		LM	
		Application date		19	
		N amount (kg N ha^{-1})		44	

[a]Yield, Dry weight.
[b]Application date of fertilizers represents week number, unless otherwise specified.
[c]MF, Mineral Fertilizer; FYM, farmyard manure; LM, liquid manure.

on total nitrogen uptake. A summary of the agronomic information is given in Table 1.

Preliminary inspection of the data revealed two outliers, which were removed from further analysis by us as well as by other dataset users (personal communication, K.C. Kersebaum, ZALF, Germany). The first data point concerned a measurement of 201 kg ha^{-1} of inorganic nitrogen in the top soil (0- to 30-cm depth) of plot 1 on 3 May 1993. This result was deemed to be highly unlikely in view of the measurement on 21 April of 19 kg ha^{-1} inorganic nitrogen, and the application, on 1 May, of 80 kg N ha^{-1}. The measurement of 125 kg N ha^{-1} in week 27 (28 June 1993) in the top soil of plot 3 was also discarded: in week 25, 47 kg N ha^{-1} had been measured, in week 29 14 kg N ha^{-1} was found, while nitrogen had been applied in week 18. Measurements in the subsoil (30– to 90–cm depth) were also removed from the analysis.

Data on the dry matter yield of crops, nitrogen contents of crop components and fertilizers, dates of seeding and harvest of the crops and dates of application of fertilizer were used as input for the model without adaptation. Yield data based on harvesting by hand were used as these were considered to be more accurate than combined harvest data, which were consistently lower. Weekly averages of temperature and precipitation were calculated from the daily weather data. Weekly average soil temperature at a depth of 20 cm was used in the NDICEA module on organic matter decomposition. Weekly evapotranspiration was calculated from daily values of global radiation and average temperature according to Makkink (1957). Based on the information on the soil, we classified it as loamy fine sand for the calculation of the water retention curve. We assumed a deep groundwater table without capillary rise to the root zone and a rate of nitrogen deposition from the atmosphere of 20 kg ha^{-1} year^{-1} (personal communication, K.C. Kersebaum, ZALF, 2004). On these well-drained sandy soils, denitrification in the subsoil is assumed to be absent.

For calibration, the objective function C_p was formulated as the sum of three partial objective functions with weights $w_i = 1$:

$$C'_1 = \frac{\sum_{i=1}^{n} |N^i_{1,sim} - N^i_{1,obs}|}{\sum_{i=1}^{n} N^i_{1,obs}}$$

$$C'_2 = \frac{\sum_{i=1}^{n} |N^i_{2,sim} - N^i_{2,obs}|}{\sum_{i=1}^{n} N^i_{2,obs}}$$

$$C'_3 = \frac{|N_{crop,act} - N_{crop,tar}|}{N_{crop,tar}}$$

where $N^i_{j,obs}$ and $N^i_{j,sim}$ represent i-th observed and simulated inorganic nitrogen amounts in topsoil ($j=1$) and subsoil ($j=2$), respectively, and $N_{crop,\ tar}$ and $N_{crop,\ act}$ designate observed and simulated crop nitrogen uptake, respectively. The calibration procedure involved 100 iterations per parameter set and resulted in parameter sets with C_p values within a decreasing range as the calibration progressed. The procedure was terminated when the range of C_p values was 0.005 or less. We used the averages of the 80 final parameter sets as the outcome of the calibration procedure.

Parameters included in calibration and their plausible ranges are summarized in Table 2.

The performance of the model was evaluated by visual inspection of the agreement between the simulation results and the observations using two statistics – index of agreement (Wilmott 1982) and root mean square error (RMSE) (see Wallach and Goffinet 1989) – and the envelope method proposed by Mitchell and Sheehy (1996). Visual inspection focussed on the closeness of the observations to the simulated dynamics and, therefore, on trends and representation of peaks and troughs.

Table 2. Parameters[a] in NDICEA, their default values and plausible ranges used during calibration.

Parameter	Description	Default value	Range
thr$_{diff}$	Threshold for nitrogen diffusion	0.75	0.6–1
cn$_{micro}$	C/N ratio soil micro-organisms	8.3	5–9
ad$_{micro}$	Assimilation/dissimilation ratio of soil micro-organisms	0.45	0.2–0.5
dnf$_1$	Denitrification factor in top soil	0.1	0–0.6
f_{prot}	Protection factor in top soil	1	0.8–2
nlf$_1$	Nitrogen-leaching factor topsoil	1	0.8–1.2
nlf$_2$	Nitrogen-leaching factor subsoil	1	0.8–1.2
A_0	Apparent initial age of soil organic matter	24	14–28

[a]All parameters have unit 1, except A_0 (year).

Willmott (1982) proposed the index of agreement IA, defined as

$$IA = 1 - \frac{\sum_{i=1}^{n}(\text{sim} - \text{obs})^2}{\sum_{i=1}^{n}(|\text{sim} - \overline{\text{sim}}| + |\text{obs} - \overline{\text{obs}}|)^2}$$

The statistic evaluates the sum of squared deviations between simulated and observed data, relative to the sum of squared absolute deviations from the mean for both simulated and observed data. Thus, it scales the simulation error on variation in observations and simulation results. A more transparent statistic is the RMSE (see Wallach and Goffinet 1989), which results in the average prediction error in the same unit as that of the observed variable:

$$\text{RMSE} = \sqrt{\frac{\sum_{i=1}^{n}(\text{sim} - \text{obs})^2}{n}}$$

Finally, we calculated the percentage of absolute prediction errors, i.e. |sim−obs|, which fell within a range of 20 kg ha^{-1} for inorganic nitrogen in the top soil. This statistic is based on a concept of 'sufficiently good' predictions for practical purposes, which are within the range and predictions of insufficient accuracy outside the range.

Model performance was assessed for different degrees of calibration of the parameters in NDICEA (Table 2). In the first step, only default values were used (Table 2, second column) to establish the performance of the uncalibrated model. Next, all observations on inorganic nitrogen content in the topsoil and subsoil as well as total nitrogen uptake by the crop were used to calibrate the parameters in Table 2. A comparison of the observations and model predictions demonstrates the extent to which the model can be tuned to the data, although without providing an independent assessment of model performance. Finally, the model was iteratively calibrated on observations covering an increasing period of time, starting with observations over the first 42 weeks up to the harvest of sugar beet, then using the first 83 weeks up to the harvest of the first winter wheat crop, and finally using the first 138 weeks up to the harvest of the winter barley crop. In all three cases, the performance of NDICEA was assessed using the last 3 years of the experiment, which consisted of five observations in each soil layer. The evaluation in this step represents an independent assessment of model performance and provides an indication of the need to use locally collected data.

Results

Figures 1 and 2 demonstrate the performance of NDICEA, when it was run with default values of the calibration parameters and calibrated on all observations for the topsoil and the subsoil, respectively.

Visual inspection showed that without calibration, NDICEA underestimated the amount of inorganic nitrogen in the topsoil of plots 1 and 3 and, to some extent, of plot 2, resulting in several episodes with a lack of simulated nitrogen, while the measurements indicated a sufficient availability of inorganic nitrogen. The modelled pattern of inorganic nitrogen coincided well with observations, but absolute values in peaks were represented less well, in particular for the subsoil. The runs with default values showed larger differences between simulated and observed data for the subsoil (Figure 2) than for the topsoil (Figure 1). Calibration on the full dataset visually improved the predictions for the first 3 years of the experiment, but errors remained larger for the subsoil than for the topsoil, and patterns were represented less well.

The simulated peak in inorganic nitrogen in topsoil of plot 2 during the final year deviated substantially from the measurements. This result was most likely caused by the fact that nitrogen uptake by the large weed population was ignored.

Results for the quantitative statistics show that calibration improved both the index of agreement (Table 3) and RMSE (Table 4) in all cases except for soil 0–90 cm depth of plot 3. The index of agreement was higher for the topsoil than for the subsoil, but RMSE for the subsoil was lower than that for the topsoil in two of the three plots. In all cases, differences between the results for the topsoil and subsoil were small.

The percentage of observations for which predictions differed by 20 kg (N) ha^{-1} or less from the observed values increased with calibration when

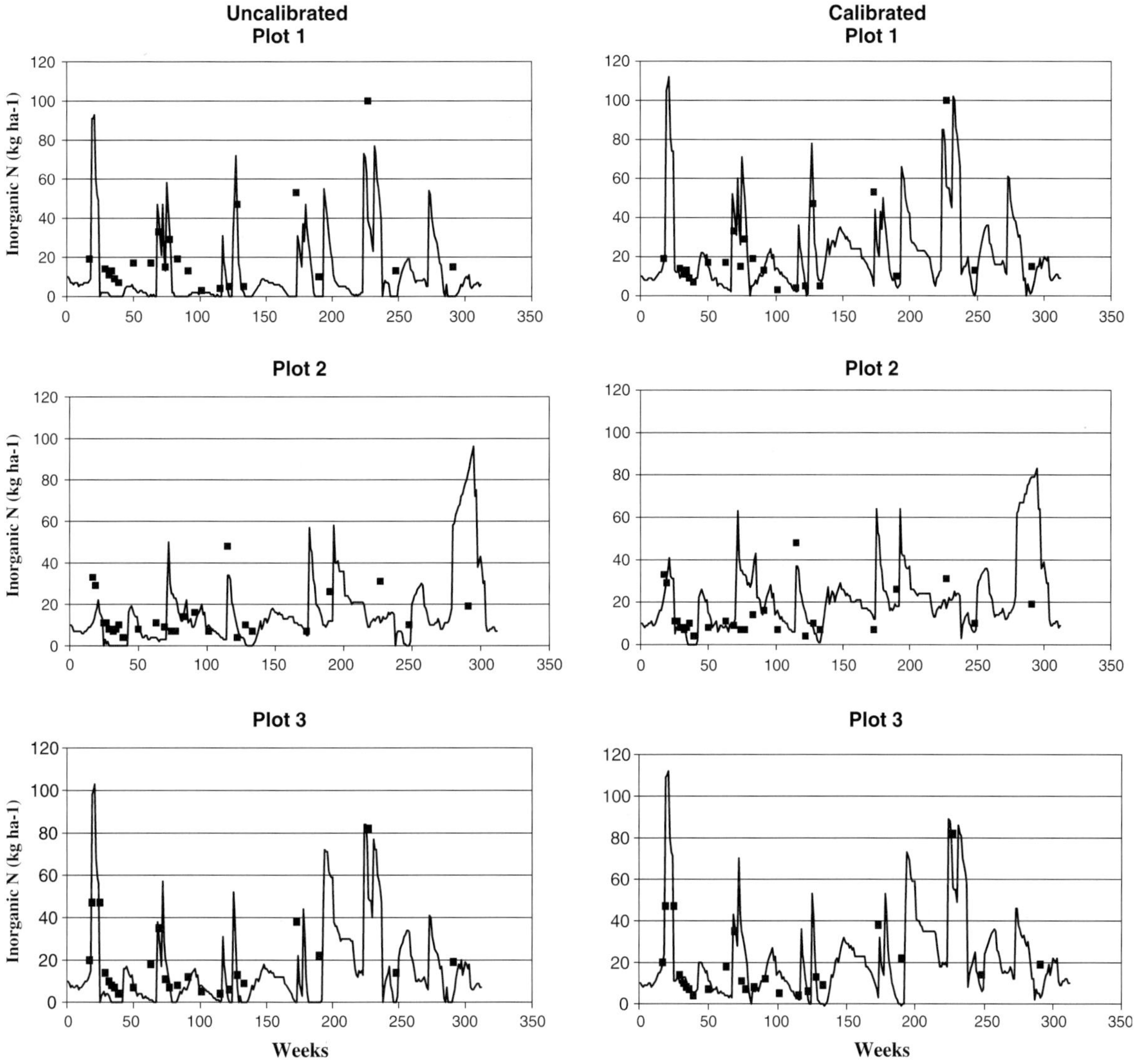

Figure 1. Simulated (*drawn line*) and observed (*symbols*) dynamics of the amount of inorganic nitrogen (kg ha^{-1}) in the topsoil layer (0–30 cm in depth). The *time axis* (weeks) starts on 1 January 1993.

the full soil profile was assessed, but did not increase consistently for each layer separately (Table 5). This result is caused by the fact that calibration success was evaluated with a composite objective function in which inorganic nitrogen in a single layer was only one of the constituents. The absolute number of accurate predictions [i.e. within the envelope of 20 kg (N) ha^{-1}] was much smaller for the full soil profile than for each of the layers. This result is caused by an accumulation of errors in going from two layers to one layer. If there are underestimations or overestimations for both layers, the summed error is higher than the errors of the separate soil layers. If the two soil layers exhibit opposite errors, the summed error will be smaller. The latter effect apparently occurred less frequently than the former.

Calibration decreased the difference between measured and modelled total crop nitrogen uptake during the 6-year experimental period from 187 to 2 kg ha^{-1} in plot 1, from 103 to 7 kg ha^{-1} in plot 2 and from 229 to 63 kg ha^{-1} in plot 3.

The three quantitative measures of model performance, index of agreement, RMSE and

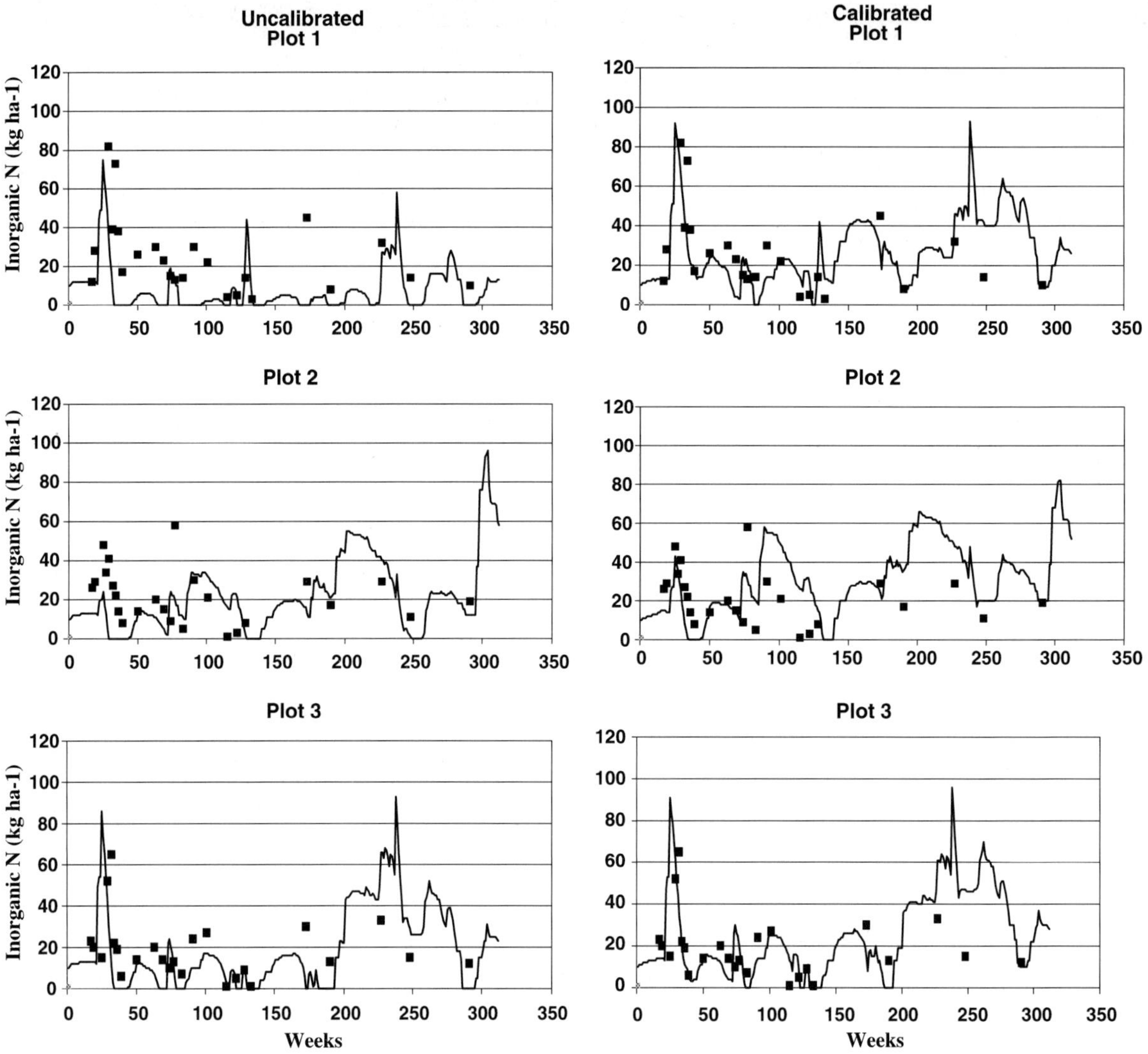

Figure 2. Simulated (*drawn line*) and observed (*symbols*) dynamics of the amount of inorganic nitrogen (kg ha^{-1}) in the subsoil layer (30–90 cm in depth). The *time axis* (weeks) starts on 1 January 1993.

percentage results within a 20 kg (N) ha^{-1} envelope were also used to assess the results of the independent evaluation of the model on data collected during the last 3 years using an increasing number of observations for calibration. Results for the index of agreement revealed an increasing trend when additional data was used for calibration, with the exception of plot 3 for which the values were high from the start (data not shown). RMSE values for the top soil slightly increased in plots 1 and 3 and substantially decreased in plot 2 when more plot-specific information was used for calibration (Table 6). RMSE consistently decreased for the subsoil, implying that more information decreased the average error of the model. RMSE results for the soil profile on a whole improved as a result of calibration using more plot-specific information.

The percentage of predictions within the range of a 20 kg (N) ha^{-1} prediction error was not affected for the top soil but increased for the subsoil (Table 7). Only for plot 3 did the quality of predictions improve when extending the calibration data from week 83 to week 138. The poor results for plot 2, topsoil layer 0–30 cm, were influenced by the weed-overgrown wheat crop in the final year.

Table 3. Index of agreement (–) (Willmott 1982) for the three datasets[a] (plot 1–plot 3) before and after calibration.

Soil layer (cm)	Plot 1		Plot 2		Plot 3	
	Before calibration	After calibration	Before calibration	After calibration	Before calibration	After calibration
0–30	0.67	0.80	0.52	0.68	0.72	0.83
30–90	0.14	0.77	0.43	0.64	0.48	0.74
0–90	0.33	0.80	0.49	0.63	0.77	0.62

[a]Number of data points for plot 1 is 23 (24 for subsoil); for plot 2, 26; for plot 3, 25.

Table 4. Values of RMSE [kg (N) ha^{-1}] (Wallach and Goffinet 1989) for the three datasets[a] (plot 1–plot 3) before and after calibration.

Soil layer (cm)	Plot 1		Plot 2		Plot 3	
	Before calibration	After calibration	Before calibration	After calibration	Before calibration	After calibration
0–30	20	16	17	15	19	19
30–90	26	15	17	16	22	20
0–90	41	23	25	21	27	23

[a]Number of data points for plot 1 is 23 (24 for sub soil); for plot 2, 26; for plot 3, 25.

Table 5. Percentage of observations with an absolute prediction error of less than 20 kg (N) ha^{-1}.

Soil layer (cm)	Plot 1		Plot 2		Plot 3	
	Before calibration	After calibration	Before calibration	After calibration	Before calibration	After calibration
0–30	91	87	88	77	80	80
30–90	58	92	88	73	76	84
0–90	39	65	50	69	56	64

Table 6. RMSE [kg (N) ha^{-1}] obtained following calibration using observations covering weeks 1–42, 1–83 and 1–138, respectively.

Soil layer (cm)	Plot 1			Plot 2			Plot 3		
	Week 1–42	Week 1–83	Week 1–138	Week 1–42	Week 1–83	Week 1–138	Week 1–42	Week 1–83	Week 1–138
0–30	25 (6)[a]	37 (12)	34 (18)	47 (9)	26 (15)	28 (21)	22 (8)	26 (14)	27 (20)
30–90	42 (7)	18 (13)	15 (19)	41 (9)	14 (15)	14 (21)	28 (8)	17 (14)	16 (20)
0–90	49 (13)	47 (25)	44 (37)	71 (18)	25 (30)	26 (42)	36 (16)	33 (28)	33 (40)

[a]Numbers in parenthesis represent number of data points used for calibration.

Table 7. Percentage of independent observations[a] with an absolute prediction error of less than 20 kg (N) ha^{-1} that were determined following calibration of the observations covering weeks 1–42, 1–83 and 1–138, respectively.

Soil layer (cm)	Plot 1			Plot 2			Plot 3		
	Week 1–42	Week 1–83	Week 1–138	Week 1–42	Week 1–83	Week 1–138	Week 1–42	Week 1–83	Week 1–138
0–30	60	60	60	40	40	40	80	80	80
30–90	20	80	80	40	80	80	20	40	80
0–90	40	60	40	0	20	80	40	40	40

[a]For number of data points used for calibration see Table 6.

The nitrogen dynamics can be analysed from the calibrated plots. In Table 8, the nitrogen balance per plot as calculated using results of NDICEA is presented as an average over the 6-year period. For plots 1 and 2, total inputs and outputs are approximately equal. For plot 3, modelled nitrogen uptake over the same 6-year period was 63 kg less than measured, resulting in a gap of 10 kg ha^{-1} year^{-1}.

Simulated leaching (Figure 3a) and denitrification (Figure 3b) are shown for plot 1 to illustrate these capabilities of NDICEA. In the same graphs, crop nitrogen uptake is shown, as a means to relate the losses to periods of crop presence and bare soil. A peak in leaching occurred in week 24, shortly after fertilizer application and sowing of the sugar beets. Major leaching further took place mainly from December until March, except for the period with the green manure crop (weeks 192–218) when leaching was suppressed.

Denitrification peaks occur when inorganic N in the topsoil is relatively high (see Figure 1) and temperature is high enough. In general, this is after fertilizer application in the spring. The highest modelled peaks occurred in week 194, following fertilizer application in the form of green manure, and in weeks 226 and 233; in both cases, the denitrification peaks occurred 2 weeks after fertilizer application in sugar beets.

Discussion

Model evaluation

A mix of methods for model evaluation was selected, which represent qualitative and quantitative measures of model behaviour as well as theoretical (index of agreement) and more practically oriented (RMSE, envelope of acceptability) measures. This type of evaluation of a rich model has the objective of overcoming the limitations of single measures and demonstrates various strong and weak aspects of the model (van der Werf et al. 1999). We evaluated the NDICEA model in two different manners: (1) by using it as a sophisticated regression model to be fitted to all available data; (2) by comparing predictions based on calibration of a portion of the dataset to independent observations. By using the model as a regression model, we demonstrated that the dynamics of inorganic nitrogen in the topsoil in the experiments could be described with modelling efficiency index values of 0.52–0.83 (Table 3), a maximum mean squared error of 20 kg (N) ha^{-1} (Table 4) and over 77% of the predictions with a prediction error of less than 20 kg (N) ha^{-1} (Table 5). The results for the topsoil and the subsoil together showed wider range

Table 8. Nitrogen (kg ha^{-1} year^{-1}) balance derived from NDICEA calculations with calibration on all data.

	Plot 1	Plot 2	Plot 3
Fertilizer	120	116	169
Atmospheric deposition	20	20	20
Net decay soil organic matter	37	37	5
Change in inorganic nitrogen	−3	−7	−3
Total in	174	166	191
Product	113	62	132
Leaching	37	49	31
Denitrification	26	56	28
Total out	176	167	201

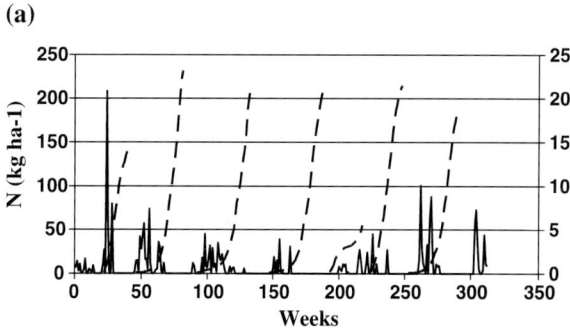
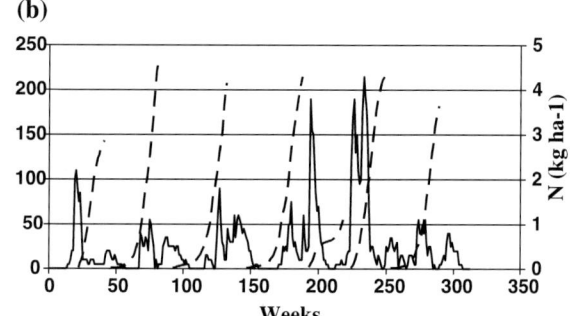

Figure 3. Simulated crop nitrogen uptake (*broken line, left Y-axis*), nitrogen leaching (*drawn lines* in **a**, *right Y-axis*) and denitrification (*drawn lines* in **b**, *right Y-axis*) in plot 1 based on calibration on all data [kg (N) ha^{-1}].

than those for the topsoil only. These results represent the maximum ability of the model to *describe* the observations.

Our comparison with independent data had the objective of showing whether the behaviour of the system could be *explained* from the modelled processes. Ten observations over the last 3 years of the study period were available for this purpose. The results for the topsoil indicated average prediction errors (RMSE) that ranged between 22 and 47 kg (N) ha^{-1} (Table 6) depending on the amount of local information used for calibration and the soil layer. Between 40 and 80% of the predictions demonstrated a prediction error of less than 20 kg (N) ha^{-1} (Table 7). The results further indicate that 25–30 inorganic nitrogen samples for topsoil and subsoil together over a 2-year period provided sufficient information to explain the nitrogen dynamics during the last 3 years of the experiment, represented by ten observations, five for each soil layer. The additional 12 samples taken in the third year of the experiment did not significantly improve the predictive capabilities of the model in any of the plots. The results were similar for the three experimental plots, indicating that the model deals well with very different types of nitrogen input, from inorganic fertilizer to organic and green manures. These results corroborate those of Koopmans and Bokhorst (2002), who found a good correlation between measured and calculated inorganic nitrogen in the topsoil using uncalibrated datasets of farms with different soil types and different fertilizer use. The researchers found that inorganic nitrogen level in the subsoil was described less well and concluded that site-specific calibration would be useful.

Table 8 summarizes the results of the dynamic simulations as nitrogen balances. One striking result is the low net decay of soil organic matter in plot 3 compared to plot 1 and plot 2. This difference may be explained by both a high production in plot 3, leading to a high input of crop residual N, and by manure applications. Since no data are available on changes in soil organic matter, this can not be verified. The pool of soil organic matter, due to its mere size, can have a considerable influence on results of a simulation of nitrogen dynamics. If data would have been available and could have been included in the calibration procedure, more certainty would have been obtained on the losses by leaching and denitrification. For any research on nitrogen dynamics, we recommend the incorporation of measurements on soil organic matter.

Environmentally and economically important losses by leaching and denitrification are not visible, and are difficult or expensive to measure. Nevertheless, they are important to farmers and can be influenced by farm practice. A model can provide insight into the processes by which these losses occur, and can be used to evaluate measures to decrease them. In Figure 3, leaching predominantly occurred from December to March, the period of the largest rainfall surplus. Possible manners of reducing leaching may be by removing the crop residues of sugar beets, earlier sowing of cereals and by growing an intercrop. Denitrification occurred mainly in the spring and once in August, following fertilizer application with the green manure, at times when relatively high amounts of inorganic soil nitrogen were present during periods of high soil moisture and with substantial decomposition rates of OM (see Eq. 20). Split applications of fertilizers might reduce the losses by denitrification. Using NDICEA, the effects of these measures can be tested. The question of when and how much fertilizer has to be applied to obtain target yields can also be assessed using this model.

Model implementation

The research version of the model evolved from a Pascal version running under the DOS operating system, to a Windows based Delphi-Pascal version. Although useful for research and used in this article, this version (NDICEA 4.59.2) is not user friendly and cannot be used by the intended audience, farmers and extension workers. A farmer's version has been developed – NDICEA Stikstofplanner 5.3 (NDICEA Nitrogen planner) – which is available at http://www.ndicea.nl in Dutch or English. This version 5.3 has two modules: one for the modelling of cropping systems spanning several years, and one suited for single year-single field evaluations. The interface combines most crop, manure and soil parameters in a limited number of choices that are meaningful for the intended users. Most soil, crop and manure parameters are fixed and cannot be changed by the user, and only default regional environments are used with regional data of 1985, a normal Dutch year. The modelling procedure for the rotation module is adapted on a few points. Calculations are done for two full cropping cycles, and the results of the second full crop rotation cycle are shown. In this way, uncertainties about initial soil organic matter are reduced, resulting in a better impression of the nitrogen dynamics and the effects of the cropping system on soil organic matter. The single year-single field module has the same interface and shows 'advanced' buttons in the main screens, thereby offering access to a range of soil-, crop- and manure-related parameters. It is linked to the world wide web for downloads of actual weather data.

NDICEA was developed for well-drained mineral soils. In principle, its process-based concept provides transportability across climates, as long as the required location-specific inputs are provided (Table 9). Prudent application under practical farming conditions should nevertheless involve careful assessment of the application domain of the component relations for water, nitrogen, and organic matter dynamics. Particular attention is needed for the quantification of losses of nitrogen from organic sources during application. This constitutes an area that warrants extension of the model with existing calculation procedures (e.g. Sogaard et al. 2002).

Modelling philosophy

Decisions about the structure of NDICEA were strongly inspired by the applied purpose of the model – i.e. to arrive at a representation of the system that can be used as a tool for management support. Local relevance was constantly weighed against detail in the representation of the mechanistic aspects of processes. This resulted in the target-oriented approach, and in replacing detailed descriptions of complex processes by summary functions, calibration factors, and calibrations on local data as an inherent aspect of the approach. The target-oriented approach, in which observed or expected crop production is fixed as a target, from

Table 9. Location-specific information required to run the NDICEA model.

	Description	Unit
Environment	Weekly average air temperature	°C
	Reference evapotranspiration (ET_0)	mm week^{-1}
	Precipitation	mm week^{-1}
	Irrigation	mm week^{-1}
	Atmospheric deposition	kg (N) ha^{-1} year^{-1}
Soil[a]	Thickness of soil layer i (thl_i)	cm
	Minimum depth ground water level	cm
	Maximum depth ground water level	cm
Topsoil and subsoil each[a]	Saturated soil moisture content (smo)	m^3 (water) m^{-3} (soil)
	Texture-specific coefficient related to soil pF curve (gam)	–
	Texture-specific coefficients related to capillary rise (cr_c and cr_x)	–
Crops[a]	Date of sowing or planting (week number)	–
	Date of harvest (week number)	–
	Total crop nitrogen uptake calculated, for example, from above ground, and root dry matter production and respective nitrogen contents ($N_{crop,\ tar}$)	kg (N) ha^{-1}
	Maximum rooting depth of the crop (rd)	cm
	Apparent age of crop, residue and root mass when incorporated in the soil (A_0)	yr
Fertilizer[a]	Date of application (week number)	–
	Amount applied	kg ha^{-1}
	Dry matter content	–
	Organic matter content of dry matter	–
	Organic nitrogen content of dry matter	–
	Inorganic nitrogen content of dry matter	–
	Apparent age of manure organic matter (A_0)	year

[a]For all soil, crop and fertilizer parameters, default values are given in the model.

which dynamic water and nitrogen requirements are derived, is also found in the SUNDIAL model (Smith et al. 1996). In many published models, however, crop dynamics are simulated from an initial condition, which renders results much more sensitive to accumulating errors.

Of all the parameters included in NDICEA, some have to be measured on-farm while others can be derived from literature. The inputs will vary in accuracy from high – for example, yield – to low – for example, rooting depth – due to inherent variability. A number of spatially and temporally complex processes have been simplified greatly in NDICEA and replaced by calibration factors. These calibration factors, which pertain to soil processes such as preferential flow and microbial activity, can only be given a site-specific value through calibration, although default values are given in the model. Experience has shown that calibration on experimental data, which vary greatly (e.g. Koopmans and Bokhorst 2002; van der Burgt 2004), results in relatively conservative ranges of optimal values.

The application of NDICEA to the Müncheberg dataset has shown the importance of site-specific calibration. The general trend was that model performance improved with an increase in the number of measurements used for calibration. Three years of calibration data are not sufficient to address the trade-off between the costs of additional measurements and the benefits of more accurate predictions of nitrogen dynamics. The tackling of this question represents a logical next step in the process of creating a reliable management and decision-making tool for organic and conventional farmers.

References

Allison L.E. 1965. Organic carbon. In: Black C.A. (ed.), Methods of soil analysis, part 2, no. 9. Series: Agronomy, American Society of Agronomy, Madison, Wis., pp. 1367–1378.

Anonymous 1989. Handboek voor de akkerbouw en de groenteteelt in de vollegrond 1989. Publicatie nr 47. Proefstation voor de Akkerbouw en de Groenteteelt in de Vollegrond. Lelystad, The Netherlands.

Baht K.K.S., Flowers T.H. and O'Callaghan J.R. 1980. A model for the simulation of the fate of nitrogen in farm wastes on land application. J. Agric. Sci. 94: 183–193.

Bouma J. and Dekker L.W. 1978. A case study on infiltration into dry clay soil. I. Morphological observations. Geoderma 20: 27–40.

Bradbury N.J., Whitmore A.P., Hart P.B.S. and Jenkinson D.S. 1993. Modelling the fate of nitrogen in crop and soil in the years following application of 15N-labelled fertilizer to winter wheat. J. Agric. Sci. 121: 363–379.

Bronswijk J.J.B., Dekkers L.W. and Ritsema C.J. 1990. Preferent transport van water en opgeloste stoffen in de Nederlandse bodem: meer regel dan uitzondering. H_2O. 22: 594–597.

de Bruin H.A.R. 1987. From Penman to Makkink. In: Hooghart C. (ed.),Evaporation and weather. Proceedings and Information. Comm. Hydrol Res-TNO Proc. Inform. 39: 5–32.

Brussaard L. 1998. Soil fauna, guilds, functional groups and ecosystem processes. Appl. Soil Ecol. 9: 123–135.

van der Burgt G.J. 2004. Use of the NDICEA model in analysing nitrogen efficiency. In: Hatch D.J. et al. (eds), Controlling nitrogen flows and losses. Proc 12th Nitrogen Workshop. Wageningen Academic Publishers, Wageningen, pp. 242–243.

Cassman K.G. and Munns D.N. 1980. Nitrogen mineralization as affected by soil moisture, temperature and depth. Soil Sci. Soc. Am. J. 44: 1233–1237.

Driessen P.M. 1988. The QLE primer. A first introduction to Quantified Land Evaluation procedures. Department of Soil Science and Geology, Wageningen Agricultural University, Wageningen.

Ennik G.C. 1960. De concurrentie tussen witte klaver en engels raaigras bij verschillen in lichtintensiteit en vochtvoorziening, Mededeling 109, I.B.S. Wageningen, The Netherlands.

Goudriaan J. and van Laar H.H. 1994. Modelling potential crop growth processes. Current issues in production ecology, vol. 2. Kluwer, Dordrecht.

Hassink J., Bouwman I.A., Zwart K.B., Bloem J. and Brussaard L. 1993. Relationships between soil texture, soil structure, physical protection of organic matter, soil biota, and C and N mineralization in grassland soils. Geoderma 57: 105–128.

Hendriks C.J.M. and Oomen G.J.M. 2000. Mest, stro en voer, het gemengd bedrijf op afstand als optie voor een zelfstandige biologische landbouw in de regio West- en Midden-Nederland. Rapport 158, Leerstoelgroep Biologische Bedrijfssystemen/ Afdeling Kennisbemiddeling, Wageningen University, Wageningen, The Netherlands.

Hofstad E.G. andSchröder J.J. 2002. Stikstof en fosfaat stromen in de Nederlandse biologische landbouw. Report 48. Plant Research International, Wageningen, The Netherlands.

Huinink J.T.M. 1998. Neerslag, verdamping en neerslagoverschotten: regionale verschillen binnen Nederland. Informatie-en Kenniscentrum Landbouw (IKC), Ede, The Netherlands.

Janssen B.H. 1984. A simple method for calculating decomposition and accumulation of 'young' soil organic matter. Plant Soil 76: 297–304.

Janssen B.H. 1996. Nitrogen mineralisation in relation to C:N ratio and decomposability of organic materials. Plant Soil 181: 39–45.

Jones J.W., Hoogenboom G., Porter C.H., Boote K.J., Batchelor W.D., Hunt L.A., Wilkens P.W., Singh U., Gijsman A.J. and Ritchie J.T. 2003. The DSSAT cropping system model. Eur. J. Agron. 18: 235–265.

Keating B.A., Carberrya P.S., Hammer G.L., Probert M.E., Robertson M.J., Holzworth D., Huth N.I., Hargreaves J.N.G.,

Meinke H., Hochman Z., McLeab G., Verburg K., Snow V., Dimes J.P., Silburn M., Wang E., Brown S., Bristow K.L., Asseng S., Chapman S., McCown R.L., Freebairn D.M. and Smith C.J. 2003. An overview of APSIM, a model designed for farming systems simulation. Eur. J. Agron. 18: 267–288.

Klepper O. and Rouse D.I. 1991. A procedure to reduce parameter uncertainty for complex models by comparison with real system output, illustrated on a potato growth model. Agric. Syst. 36: 375–395.

Koopmans C.J. and Bokhorst J. 2002. Nitrogen mineralization in organic farming systems: a test of the NDICEA model. Agronomie 22: 855–862.

Mäder P., Fliebach A., Dubois D., Gunst L., Fried P. and Niggli U. 2002. Soil fertility and biodiversity in organic farming. Science 296: 1694–1697.

Makkink G.F. 1957. Testing the Penman formula by means of lysimeters. J. Int. Water Eng. 11: 277–288.

Mayer D.G. 2002. Evolutionary algorithms and agricultural systems. Kluwer, Boston.

Meynard J.M., Cerf M., Guichard L., Jeuffroy M.-H. and Makowski D. 2002. Which decision support tools for the environmental management of nitrogen?. Agronomie 22: 817–822.

van Mil M. 1981. Actual and potential nitrogen fixation in pea and field bean as affected by combined nitrogen. Thesis, Wageningen Agricultural University, Wageningen, The Netherlands.

Mirschel W., Wenkel K.-O., Wegehenkel M., Kersebaum K.C., Schindler U. and Hecker J.-M. 2007. Müncheberg field trial data set for agro-ecosystem model validation. In: Kersebaum K.C., Hecker J-M, Mirschel W, Wegehenkel M (eds), Modelling water and nutrient dynamics in soil-crop systems. Springer, Dordrecht, pp. 219–243.

Mitchell P. and Sheehy J.E. 1996. Comparison of predictions and observations to assess model performance: a method of empirical validation. In: Kropff M.J., Teng P.S., Aggarwal P.K., Bouma J., Bouman B.A. et al. (eds), Applications of systems approaches at field level. Kluwer, Dordrecht, pp. 437–451.

Oomen G.J.M. 1995. Nitrogen cycling and nitrogen dynamics in ecological agriculture. Biol. Agric. Hortic. 11: 183–192.

Press W.H., Flannery B.P., Teukolsky S.A. and Vetterling W.T. 1986. Numerical recipes: The art of scientific computing. Cambridge University Press, Cambridge.

Price W.L. 1979. A controlled random search procedure for global optimization. Comput. J. 20: 367–370.

Rijtema P.E. 1980. Nitrogen emission from grassland farms – a model approach. Technical Bulletin 119. ICW, Wageningen, The Netherlands.

Rijtema P.E. and Kroes J.G. 1991. Some results of nitrogen simulations with the model ANIMO. Fertiliser Res. 27: 189–198.

Smith J.U., Bradbury N.J. and Addiscott T.M. 1996. SUNDIAL: A PC-based system for simulating nitrogen dynamics in arable land. Agron. J. 88: 38–43.

Sogaard H.T., Sommer S.G., Hutchings N.J., Huijsmans J.F.M., Bussink D.W. and Nicholson F. 2002. Ammonia volatilization from field-applied animal slurry – the ALFAM model. Atmos. Environ. 36: 3309–3319.

Verveda H.W. 1983. Opbouw en afbraak van jonge organische stof inde grond en de stikstofhuishouding onder een vierjarige vruchtwisseling met grasgroenbemester. Internal Bulletin 58. Department of Soil Science and Plant Nutrition, Agricultural University, Wageningen, The Netherlands.

Wallach D. and Goffinet B. 1989. Mean squared error of prediction as a criterion for evaluating and comaring system models. Ecol. Modell. 44: 209–306.

van der Werf W., Leeuwis C. and Rossing W.A.H. 1999. Quality of modelling in fruit research and orchard management: issues for discussion. In: Wagenmakers P.S., van der Werf W. and Blaise Ph. (eds), Proc. 5th Int. Symp. Comput Modell Fruit Res Orchard Manage. Acta Hortic 499: 151–160.

Willmott C.J. 1982. Some comments on the evaluation of model performance. Bull. Am. Met. Soc. 64: 1309–1313.

Woli P.B. 2000. Nitrogen and organic matter balances as calculated by the target-oriented models Quadmod PLUS and NDICEA to evaluate crop rotations. MSc Thesis, Dept. Theoretical Production Ecology, Wageningen Agricultural University, The Netherlands.

Wösten J.H.M., Bannink M.H. and Beuving J. 1987. Waterretentie doorlatendheidskarakteristieken van boven – en ondergronden in Nederland: de Staringreeks. Rapport 1932. Stiboka, rapport 18 ICW, Wageningen, The Netherlands.

Yang H.S. 1996. Modelling organic matter mineralization and exploring options for organic matter management in arable farming in northern China. PhD thesis, Wageningen Agricultural University, The Netherlands.

CHAPTER NINE

Simulation of water and nitrogen flows on field scale: application of the SWAP–ANIMO model for the Müncheberg data set

Joop Kroes and Jan Roelsma

Abstract The model combination SWAP–ANIMO was evaluated using a data set from Müncheberg to simulate water and nitrogen dynamics at field scale. During 1993–1998 arable crop rotations were grown on poor sandy soils with deep groundwater levels. Measurements were available with high temporal resolution and soil horizons were monitored at several depths. Comparisons were made between simulated and measured soil moisture content, pressure head, temperature, mineral nitrogen, and nitrate concentrations. An uncertainty analyses and calibration was carried out using soil hydraulic properties, soil temperatures, and hysteresis. Calibration results showed that simulated soil temperatures are lower than measured. Soil mineral nitrogen and nitrate concentrations were simulated with deviations within acceptable ranges. In general, it was concluded that the model performance on these three fields was acceptable for the tested domain of the hydrological and nitrogen cycle. Future weather for the period 2000–2050 was simulated with the calibrated model and compared with an extension of actual weather. Boundary conditions were a reduced precipitation and an increased air temperature. Results showed a reduced leaching of water and an increase of nitrate concentrations in the leachate.

Keywords ANIMO, Field, Hydrology, Hysteresis, Model, Mineral nitrogen, Nitrate, Soil, SWAP, Weather

Introduction

Leaching of nutrients to groundwater systems causes a threat to present and future generations of people. Nutrients leach downward with the precipitation excess and may percolate into groundwater, which may be used for drinking water purposes. It is therefore required to take all preventive measures that are possible to prevent this leaching potential. Nowadays, models are often used to analyse leaching conditions and estimate alternative strategies or measures. These measures should best start as close to the source as possible and that is in most cases the upper part of the soil system. It is in the upper part of the soil system that most of the transformation processes occur. Since nitrogen transformations are part of the nitrogen cycle, anyone who wishes to analyse nitrogen transformations will study the nitrogen cycle. Since it is hard to monitor all required process components of the nitrogen cycle, it is common to apply comprehensive simulation models to fill gaps in the measurements and to analyse strategies to reduce leaching. An overview of

Joop Kores and Jan Roelsma
Alterra, Green World Research, P.O. Box 47,
6700 AA Wageningen, The Netherlands
e-mail: joop.kroes@wur.nl, tel.: +31 317 474372,
fax: +31 317 41900

the use of comprehensive unsaturated flow and transport models is given by Vanclooster et al (2004).

Especially, when dynamics of processes are important, e.g., shift of application timing or adjustment of irrigation strategies, one may consider use of deterministic models. One such set of deterministic models was developed in the Netherlands and exists out of the models SWAP and ANIMO, respectively for water and nutrient management. Both models address the upper part of the soil and have a long history. Soil–water–atmosphere–plant–system (SWAP) describes the hydrological cycle and was developed to quantify evapotranspiration of different crops. Agricultural nutrient model (ANIMO) describes the carbon and nitrogen cycle and was developed to quantify the relation between fertilizer management, soil management, and the leaching of nutrients to groundwater and surface water systems for a wide range of soil types and different hydrological conditions.

The main purpose of this article is the evaluation of the performance of the model combination SWAP–ANIMO to simulate water and nitrogen dynamics at field scale using a German data set to calibrate the model. The calibrated model combination was applied to a climate scenario with predicted weather for the period 2000–2050.

Materials and methods

Data set

The simulations were carried using a data set from Müncheberg (Mirschel et al. 2007).

The data set describes three field plots on poor sandy soils with different agricultural management practices. It covers a time period from 1993 to 1998.

Time series of daily meteorological data were available for the period 1 January 1992–31 December 1998.

The available soil temperatures (°C), from depths of 5, 20 and 50 cm, were used to validate the simulated soil temperatures.

Data were available for: winter wheat, winter rye, winter barley, and sugar beets.

Management data (sowing, harvest, fertilization, tillage) were given for all treatments.

To describe the soil hydraulic behaviour ($\theta(h)$ and $K(h)$) a set of van Genuchten parameters was given for each soil horizon.

Daily values of pressure heads were measured by tensiometers at different depths. Values of soil water contents were incidentally measured gravimetric and daily by TDR probes.

Mineral nitrogen was monitored in soil samples at different depths during 6–8 sampling dates per year.

Nitrate concentrations were monitored at different depths by tension-driven suction cups; raw data were supplied by Mirschel et al. (2007).

Model tools – SWAP

A mathematical model which describes soil water movement is the model SWAP. SWAP is the successor of the agrohydrological model SWATR (Feddes et al. 1978) and some of its numerous derivatives. The experiences gained with the existing SWATR versions were combined into SWAP, which integrates water flow, solute transport, and crop growth according to current modelling concepts and simulation techniques. Alterra and Wageningen Agricultural University have developed the computer model SWAP in close cooperation. The theory of the processes simulated by the model is extensively described by Van Dam (2000).

This study was carried out with SWAP version 3.0.3 (Kroes and Van Dam 2003). Information about the SWAP model can be found on www.swap.alterra.nl.

SWAP is a computer model that simulates transport of water, solutes and heat in variably saturated top soils. Transport processes at field scale level and during whole growing seasons are considered. System boundaries at the top are defined by the soil surface with or without a crop and the atmospheric conditions. The lateral boundary can be used to simulate interaction with surface water systems. The bottom boundary is located in the unsaturated zone or in the upper part of the groundwater and describes the interaction with a regional groundwater system (Fig. 1).

The model applies a numerical solution of the basic equation for soil water flow, known as the Richards' equation:

$$\frac{\partial \theta}{\partial t} = \frac{\partial \left[K(h) \left(\frac{\partial h}{\partial z} + 1 \right) \right]}{\partial z} - S(h) \qquad (1)$$

where θ is the volumetric water content (m^3 m^{-3}), t is time (day), K is hydraulic conductivity (m day^{-1}), h is

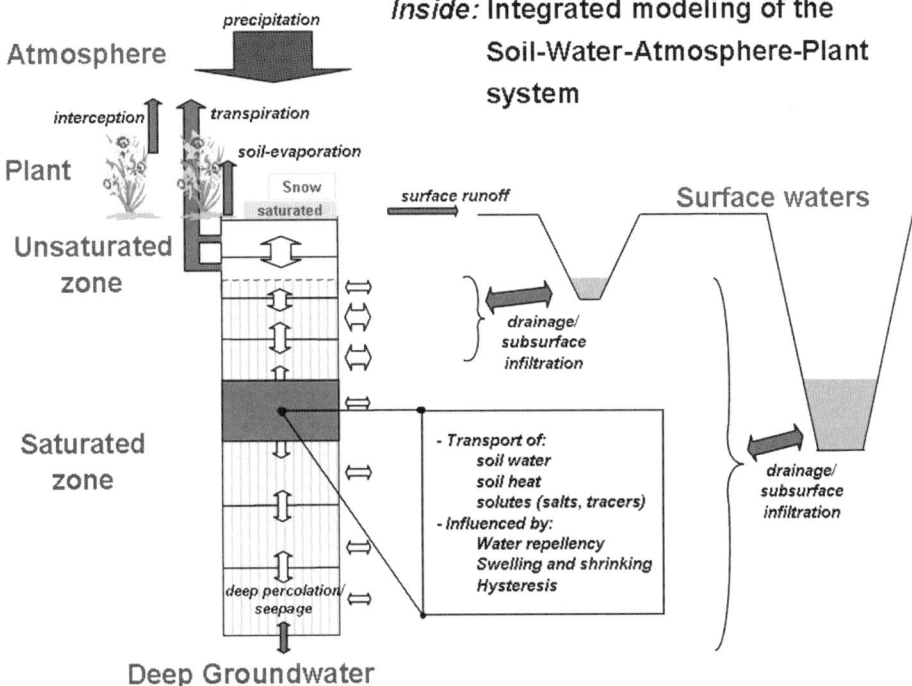

Fig. 1 Schematic overview of modelled system

soil water pressure head (m) and z is the vertical coordinate (m), taken positively upward and S is soil water extraction rate by plant roots (m^3 m^{-3} day^{-1}).

Richards' equation has a clear physical basis at a scale, where the soil can be considered to be a continuum of soil, air, and water. SWAP solves the non-linear Richards' equation numerically, subject to specified initial and boundary conditions and with known relations between θ, h, and K. These soil hydraulic relations are described by the analytical expressions of Van Genuchten and Mualem. These relationships can be measured directly in the soil, determined in the laboratory, or might be obtained from basic soil data. SWAP applies Richards' equation integrally for the unsaturated–saturated zone, including possible transient and perched groundwater levels.

Physical and empirical methods determine actual soil evaporation. Frost periods are accounted for with a simple module which reduces water flow and accumulates snowfall.

Potential evapotranspiration was calculated using Penman-Monteith equation and procedures as recommended by Allen et al. (1998). This implied an application of Penman-Monteith equation to reference grass in combination with crop coefficients. Interception is modelled as a separate process. Water stress is a function of soil water pressure.

The interaction between soil water and surface water may be described using several options, similar as the interaction between soil water and deep groundwater. However, for these fields in Müncheberg, without groundwater, there is no interaction with a surface water system. The interaction with deep groundwater is only in the downward direction and, therefore, defined as a free-draining lower boundary condition.

SWAP contains a simple and a detailed crop module. In the simple crop model, the crop development with time is prescribed as leaf area index (or soil cover fraction), crop height, and rooting depth as function of development stage. The detailed crop module is based on the crop growth model WOFOST. Due to a lack of time, required to apply detailed crop modelling, and due to the focus of current applications, the simple crop module was applied in this study.

Solute transport mechanisms are included in the model and allow simulation of ordinary pesticide and salt transport, including the effect of salinity on crop growth. Since salinity problems did not occur in Müncheberg, solute transport was not simulated with the SWAP model. In this study, where a detailed modelling of nutrient transport and transformation is required, daily water fluxes and water contents were generated by SWAP and used as input for the nutrient model ANIMO (Renaud et al. 2004).

Soil heat flow is solved with a numerical solution of the differential equation for soil heat flow:

$$C_{heat}\frac{\partial T}{\partial t} = \frac{\partial}{\partial z}\left(\lambda_{heat}\frac{\partial T}{\partial z}\right) \qquad (2)$$

where C_{heat} is the soil heat capacity (J m^{-3} °C^{-1}) and λ_{heat} is the thermal conductivity (J m^{-1} °C^{-1} day^{-1}) and T is the soil temperature (°C), t is time (day) and z is depth (m).

Thermal conductivity and soil heat capacity are calculated from the soil texture and the volume fractions of water and air as described by De Vries (1975). At the soil surface the daily average temperature is used as boundary condition.

Model tools – ANIMO

The ANIMO model simulates the main processes of the carbon, nitrogen, and phosphorus cycles (Groenendijk and, Kroes, 1999; Groenendijk et al. 2004, Rijtema et al. 1999). The organic matter cycle plays an important role for long-term effects of land use changes and fertilization strategies. Currently, the ANIMO model serves as one of the parent models for the development of the Dutch consensus leaching model STONE (Wolf et al. 2003). For this study the carbon and nitrogen cycle have been used.

In this study ANIMO version 4.0 was applied (Renaud et al. 2004).

The ANIMO model requires hydrological and soil temperature data supplied by an external field plot model (e.g. SWAP) or a regional groundwater flow model (e.g. SIMGRO). The vertical schematization resulting from the spatial discretization as applied in the water quantity model forms part of the input of the model (Fig. 2).

The simulated transformation processes are all part of the carbon, nitrogen, and phosphorus cycle. In the organic carbon cycle the following processes are described: (a) application of organic materials (e.g. manure); (b) decomposition of root materials; (c) decomposition of fresh organic materials and transformation to humus; (d) turnover of humus. For each kind of organic material, fractions are specified, which have different decomposition rates, and N and P contents. The organic part of both the nitrogen and phosphorus cycle in the soil runs largely parallel to the organic carbon cycle.

In the inorganic part of the nitrogen cycle the following processes are described: (a) addition of mineral

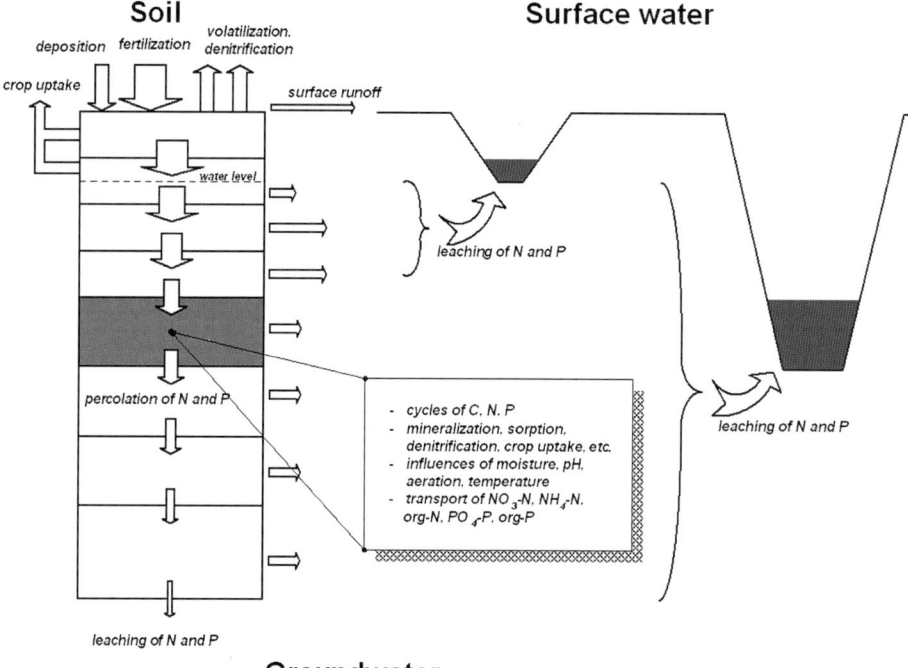

Fig. 2 Schematic overview of modelled system in the ANIMO model

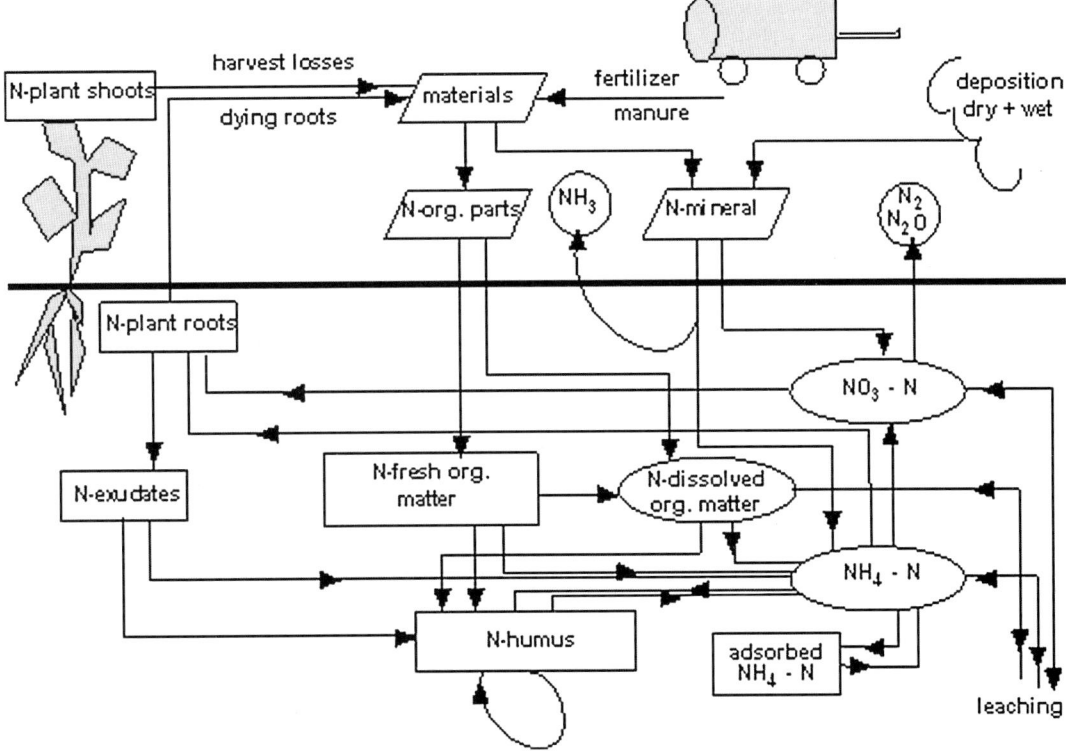

Fig. 3 Relation diagram of the nitrogen cycle in the ANIMO model

nitrogen (e.g. fertilizers); (b) ammonium volatilization; (c) ammonium sorption; (d) nitrification; (e) denitrification; (f) nitrogen uptake by crop (Fig. 3). In the inorganic part of the phosphorus cycle the following processes are described: (a) addition of mineral phosphorus; (b) phosphorus sorption; (c) precipitation; (d) phosphorus uptake by crop. In the ANIMO model the rate variables for organic matter transformation are corrected for the influences of temperature, moisture, pH, and oxygen demand. The nitrification rate is corrected for influences of temperature, moisture, and pH.

Substances that can be transported with water fluxes are: NH_4, NO_3, PO_4, and dissolved organic matter fractions. For this transport, combined with production or consumption, a transport and conservation equation is used. The calculation procedure follows the flow direction in the schematic column. For the first soil compartment, the boundary condition for the incoming flux from above is the precipitation with a concentration of the precipitation flux. For the last compartment, the boundary condition of the incoming flux is the seepage flux with a concentration of the soil solution below the described profile. Physical dispersion is simulated by the thickness of the model compartments and the length of the time step. For additions to the soil system, the model has an extra reservoir on top of the compartment division. The additions can be added to this reservoir and infiltrate into the soil system with the precipitation flux. Overland flow (surface run-off) will be discharged from this reservoir. The reduction factor for crop uptake is determined on base of the summarized crop uptake during previous time steps. For grassland the uptake includes diffusion.

From data set to model input

Meteorological data

Nearly all meteorological data could be applied without conversion. Mean relative air humidity, given as percentage, was converted to actual vapour pressure (Pa), using:

$$e_{act} = 0.01\, e_{sat} H \qquad (3)$$

where e_{act} is the actual vapour pressure (Pa), e_{sat} is the saturated vapour pressure (Pa) and H is the mean relative air humidity (%).

Saturated vapour pressure was derived from daily mean air temperatures (Tetens 1930):

$$e_{sat} = 610.78 \exp\left(17.2694 \frac{T_{mean}}{T_{mean} + 238.3}\right) \quad (4)$$

where T_{mean} is the given mean daily mean temperature (°C) at 2 m height.

The values given for global solar radiation on 8 November 1993 seemed too low and were multiplied by 10 (from 64 to 640).

The missing values for 27 June 1998 were generated by linear interpolation between the values for the previous and the next day; precipitation for that day was set to zero.

Soil

The soil characteristics were converted and extended to a schematized model profile of 3 m thickness, which was divided into 30 model compartments increasing in size from a thickness of 1 cm near the soil surface to 40 cm at a depth of 3 m. The properties of soil horizons at certain depths were assigned to model compartments at corresponding depths.

Land use

From analyses of the available crop data for each plot, the following parameters were derived and applied as driving forces for the simplified crop model: dates of sowing, emergence, ploughing, and harvest; and number of growing days.

LAI was directly related to the measured dry matter content in stem and leaves.

Soil water stress parameters were assumed to be similar to Dutch conditions and taken from a Dutch data set (Kroes and Van Dam 2003).

Irrigation was only applied in 1993 at the beginning of the growing season in three applications of 6 mm each.

Soil and fertilizer management were derived from the data set and resulted in management events with specific timing, types, and amounts of different kind of fertilizer. An overview of the management events for plot 1 is given as an example in Table 1.

Boundary conditions

The hydrological simulations used meteorological data as top boundary and a free-draining bottom boundary

Table 1 Fertilizer and soil management events of plot 1

Event	No. of events
Ammonium nitrate lime 50 kg/ha N	1
Ammonium nitrate lime 60 kg/ha N	2
Ammonium urea solution 30 kg/ha N	3
Ammonium urea solution 35 kg/ha N	1
Ammonium urea solution 40 kg/ha N	4
Ammonium urea solution 50 kg/ha N	2
Ammonium urea solution 70 kg/ha N	1
Ammonium urea solution 80 kg/ha N	1
Cultivator 10 cm	1
Cultivator 15 cm	3
Cultivator 8 cm	2
Cultivator drill 10 cm	3
Green manure (oil radish) ploughed in	2
Leaves sugar beet ploughed in	2
Ploughing 18 cm	1
Ploughing 23 cm	1
Rotary tiller 10 cm	2
Straw winter barley ploughed in	1
Straw winter rye ploughed in	1
Straw winter wheat ploughed in	2

at a depth of 3 m. All simulations were initialized with an additional simulation period of 1 year (1992), which minimized disturbances in the initial moisture profile and initial water fluxes.

The nutrient simulations used dry and wet deposition as top boundary, next to management events. Values for deposition were derived from long-term data from stations (http://www.emep.int/) in Germany and estimated to be 20 kg N a^{-1}. Deposition was evenly distributed over wet and dry deposition and over NO_3-N and NH_4-N. All nutrient simulations were initialized with organic matter distributions in the soil profile that were derived from the measured data sets. A pre-processing simulation period of 12 years was carried out to enable a proper approach of initial conditions in the carbon and nitrogen cycle and to achieve initial mineral nitrogen contents.

Uncertainty analyses

From analyses of field data it was concluded that there are uncertainties in available soil hydraulic data. Soil moisture retention curves were supplied as a parameter set to describe relation between soil pressure head and soil moisture ($\theta(h)$ and $K(h)$) content using the Mualem-Van-Genuchten equations. The detailed monitoring made it possible to compare soil pressure heads and soil moisture contents measured at the same time and place. From these data it was concluded

that: some of the measurements were not plausible; e.g., very low pressure heads do not always indicate (nearly) saturated conditions; the supplied parameter set was not always suitable to describe $\theta(h)$. The drawn curves do not represent the average relation between soil moisture content (θ) and soil hydraulic head (h).

The uncertainty in hydraulic parameters tends to increase with depth.

To estimate the impact of these uncertainties, a simplified analysis was carried out for the deeper soil layers (> 0.9 m depth) of plot 1: increase saturated soil moisture content (θ_s) with steps of 0.055 between 0.265 and 0.43; and increase saturated hydraulic conductivity (K_s) with steps of 25 cm day^{-1} between 1 and 175 cm day^{-1}.

Drying and wetting of the soil influence the relation between soil moisture (θ) and hydraulic head (h). The main drying curve was supplied with the data set. To analyse the impact of these drying and wetting conditions on the retention curve (hysteresis) another analysis was carried out. Linear scaling (Kroes and Van Dam 2003) was applied to the given main drying curve, by increasing a factor (f_{hyst}), which describes the relation between the shape parameter for main drying and wetting:

$$f_{hyst} = \frac{\alpha_{wet}}{\alpha_{dry}} \qquad (5)$$

where α_{wet} and α_{dry} are shape parameters, respectively for the wet and drying curve (–).

The factor (f_{hyst}) received four values (1, 1.5, 2 and 4) resulting in four different curves for wetting conditions (Fig. 4). The drying curves were left in their original shape.

An additional analysis was carried out in the context of the climate scenario. This scenario predicted increasing air temperature. An analysis was carried out for the impact of an increasing change of 1 °C of the air temperature on leaching of water and nitrate.

Results of uncertainty analyses focused on average leaching of water and nitrate at a depth of 0.9 m below the soil of plot 1.

Statistical criteria

Model results for three different plots were analysed by: verifying the plausibility of the mass balances; and graphical and statistical comparisons between

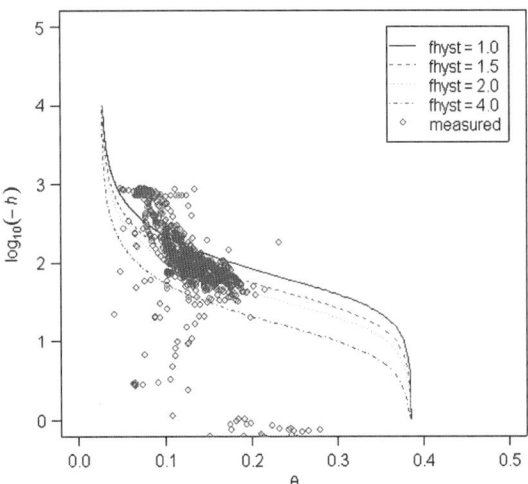

Fig. 4 Prescribed soil moisture retention curve $h(\theta)$ for different drying and wetting conditions (f_{hyst} = 1, 1.5, 2, 4) with measured soil hydraulic heads as function of measured soil moisture contents at a depth of 0–30 cm below plot 1 are included

simulated and measured values at different depths of: soil moisture contents, soil pressure heads, mineral nitrogen contents, and nitrate concentrations.

Results of the hydrological model SWAP were analysed using ordinary statistics, mean error, and root mean square error (Table 2). The nitrogen results from ANIMO were analysed using additional coefficients of determination and residual mass (Table 2; after Loague and Green 1991).

Impact of weather changes

To analyse the impact of weather changes on the leaching of water and nitrate, two scenarios were analysed for the period 2000–2050. One scenario ("predicted") used a set of weather parameters achieved from Gerstengarbe from the Potsdam Institute for Climate Impact Research. The other scenario ("extended") was an extension (multiplication) of the weather from the period 1993–1998. The land from the period 1993–1996 was maintained throughout the whole period and was identical in both scenarios. A comparison of both scenarios showed that "predicted" has an air temperature which is 1.5°C higher and a precipitation, which is about 18% lower than "extended" (Table 3). Humidity and radiation are less affected, whereas wind speed has a mean and maximum value, which is respectively lower and higher in "predicted" compared to "extended".

Table 2 Statistical criteria used in the model evaluation

Criteria	Symbol	Description, measure for	Equation	Range	Optimum
Mean error	ME	Mean error between predicted and observed values (>0: underestimate)	$\sum_{i}^{n}(O_i - P_i)$	$> -\infty$	0
Root mean square error	RMSE	Average difference between predicted and observed values	$\sqrt{\dfrac{\sum_{i}^{n}(P_i - O_i)^2}{n}}$	≥ 0	0
Coefficient of determination	CD	Relation in spreading of observed versus spreading of modelled values; >1: predicted deviates less, <1: predicted deviate more	$\dfrac{\sum_{i}^{n}(O_i - O_a)^2}{\sum_{i}^{n}(P_i - O_a)^2}$	≥ 0	1
Coefficient of residual mass	CRM	Relation between observed and predicted values; >0: systematic underestimation <0: overestimation	$\dfrac{\sum_{i}^{n} O_i - \sum_{i}^{n} P_i}{\sum_{i}^{n} O_i}$	≤ 1	0

P = Predicted (modelled), O = Observed (measured), i = a date with modelled and measured results, a = average

Table 3 Characteristics of two weather scenarios for the Muencheberg region

Weather	Parameter	Unit	Min.	Median	Mean	Max.
Predicted	Radiation	kJ m^{-2} day^{-1}	829	8,711	10,120	30,210
	T_{min}	°C	−23	6	6	23
	T_{max}	°C	−16	14	14	41
	Humidity	kPa	0.1	0.9	1.0	3.1
	Wind speed	m s^{-1}	0.0	1.3	1.6	9.5
	Precipitation	mm day^{-1}	0.0	0.0	1.3	87.7
Extended	Radiation	kJ m^{-2} day^{-1}	64	8,008	9,857	30,040
	T_{min}	°C	−22	5	4	21
	T_{max}	°C	−15	12	12	38
	Humidity	kPa	0.1	0.9	0.9	2.1
	WindSpeed	m s^{-1}	0.0	2.3	2.5	8.7
	Precipitation	mm day^{-1}	0.0	0.0	1.6	42.1

Results

Simulations were carried out for the three plots. Results will be discussed in detail for plot 1, and a summary will be given of all three plots.

Results were achieved after calibration of the model using the following parameters: saturated moisture content was reduced for the soil layers deeper than 0.9 m; saturated soil hydraulic conductivity was reduced for the deepest soil layers; organic matter content of soil layers deeper than 0.9 m.

An uncertainty analysis was carried out to analyse the impact of additional changes.

The impact of weather changes results from an analysis with a long-term future weather scenario.

Water balance

The hydrological simulations for three plots resulted in sound water balances. An example of such a balance is given for the soil layer from soil surface to 1 m depth of plot 1 (Table 4).

Precipitation reached the soil–plant system as rainfall and snow; on average between 2% and 8% of the precipitation was snowfall (Table 4). The total average evapotranspiration was 70% of the precipitation, which was divided over interception, soil, and crop evaporation as respectively 9%, 18% and 43%. Some run-off occurred (<1%), and the remaining part (29% of the precipitation) leached as bottom flux at 1 m depth.

Table 4 Water balance of the soil layer 0–1 m below plot 1; all terms in mm year^{-1}

Year	Rain-fall	Snow-fall	Irrigation	Interception	Soil evaporation	Transpiration	Surface run-off	Bottom flux	Delta storage
1993	570	55	18	48	122	187	8	233	45
1994	699	14	0	45	131	249	1	321	−34
1995	473	34	0	71	78	247	2	119	−10
1996	464	22	0	64	68	290	0	74	−10
1997	526	16	0	34	116	307	0	68	17
1998	643	12	0	54	125	249	1	217	9
Average	563	25	3	53	107	255	2	172	3

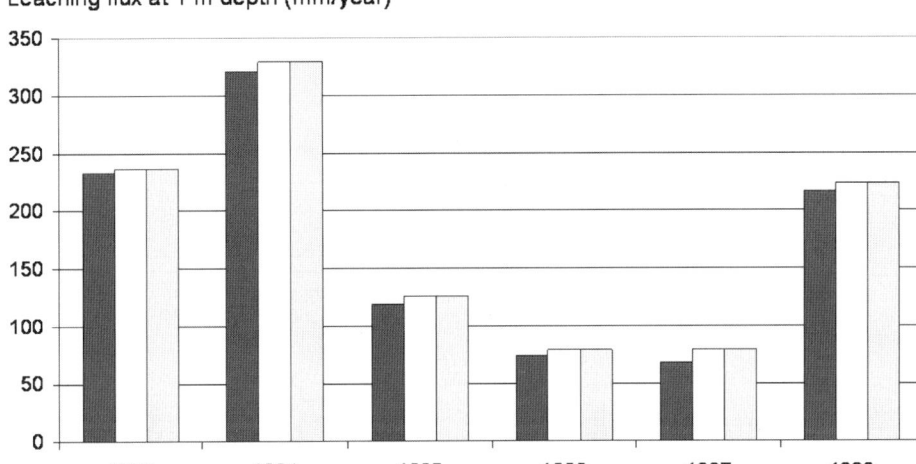

Fig. 5 Leaching flux (mm year^{-1}) at 1 m depth below three plots during 1993–1998

The leaching flux at 1 m depth varied little between the 3 different plots (Fig. 5), because the cropping pattern was similar. The variation between years was much stronger and showed a variation with a minimum leaching of 68 mm (13% of the precipitation) in the year 1996 and a maximum leaching of 329 mm (45% of the precipitation) during the wet year 1994. Highest leaching occurred during the first 2 wet years.

Soil moisture

Simulated soil moisture contents were compared with measured gravimetric and TDR values at three different depths: 0–30, 30–60, 60–90 and compared with TDR values at 90–120 and 120–150 cm. Results are summarized in Table 5 for all three plots. Average simulated values corresponded well with mean measured values. The RMSE indicates that simulated values were closer to gravimetric than to TDR measurements, although the differences were relatively small. The average RMSE was 0.04 m^3 m^{-3}, which seems reasonable. An example of a time course (Fig. 6) shows that simulated moisture contents in the upper part were about 5% too high in the wet autumn of 1993 and too dry in the winter of 1994/95.

Soil water pressure head

Simulated and measured soil water pressure heads were compared with measured using data from three different depths (Table 6). In the upper part of the soil (0–30 cm), the values corresponded relatively well, but the deviation increased with depth. RMSE values confirmed this values between 114 and 199 hPa, between 0 and 30 cm and over 5000 hPa at greater depth.

Table 5 Soil moisture content (m³ m⁻³) below three plots at different depths; mean values as measured (gravimetric and TDR) and simulated, as well as ME and RMSE based on a comparison between simulated values and gravimetric and TDR measurements

Plot	Depth (cm)	Mean value Gravimetric	TDR	ME Simulated	Gravimetric	TDR	RMSE Gravimetric	TDR
1	0–30	0.13	0.11	0.14	−0.01	−0.03	0.02	0.04
1	30–60	0.11	0.11	0.09	0.02	−0.02	0.03	0.04
1	60–90	0.13	0.12	0.13	0.01	0.01	0.02	0.04
1	90–120		0.12	0.12		0.01		0.04
1	120–150		0.13	0.12		−0.01		0.02
2	0–30	0.13	0.11	0.13	−0.01	0.02	0.03	0.04
2	30–60	0.11	0.11	0.08	0.03	−0.03	0.04	0.04
2	60–90	0.14	0.16	0.08	0.06	−0.07	0.06	0.08
2	90–120		0.15	0.13		−0.01		0.04
2	120–150		0.13	0.11		−0.02		0.02
3	0–30	0.14	0.11	0.14	−0.01	0.04	0.02	0.03
3	30–60	0.11	0.11	0.08	0.03	−0.02	0.04	0.04
3	60–90	0.13	0.13	0.09	0.04	−0.04	0.05	0.05
3	90–120		0.14	0.18		0.04		0.04
3	120–150		0.17	0.20		0.03		0.03

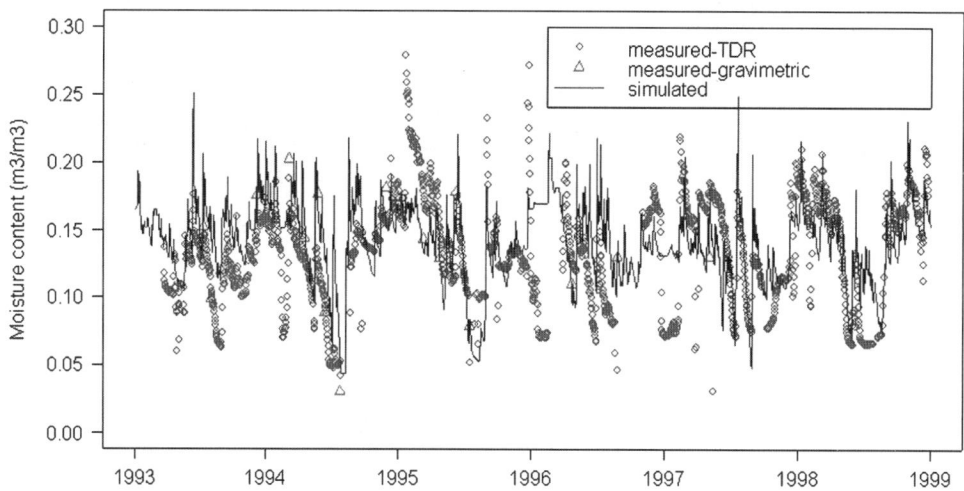

Fig. 6 Soil moisture content (m³ m⁻³) at 0–30 cm depth below plot 1 during 1993–1998

Table 6 Soil water pressure heads below three plots at different depths; mean values as measured and simulated, as well as the ME and RMSE (hPa)

Plot	Depth (cm)	Measured	Simulated	ME	RMSE
1	30	−163	−154	−9	199
1	60	−154	−869	716	3190
1	90	−70	−1629	1559	5175
2	30	−108	−171	63	182
2	60	−64	−1140	1076	5034
2	90	−63	−2004	1941	6027
3	30	−132	−146	14	114
3	60	−94	−1069	975	3803
3	90	−63	−1512	1449	5146

A time course of the pressure heads in the layer 0–30 cm indicates (Fig. 7) that it was hard to get a good fit between simulated and measured values.

Soil temperatures

Soil temperatures were simulated with thermal conductivity and soil heat capacity calculated from soil texture and volume fractions of water and air simulated by SWAP. At the soil surface the daily average temperature were used as boundary condition. Simulated average values were lower than measured values (Table 7),

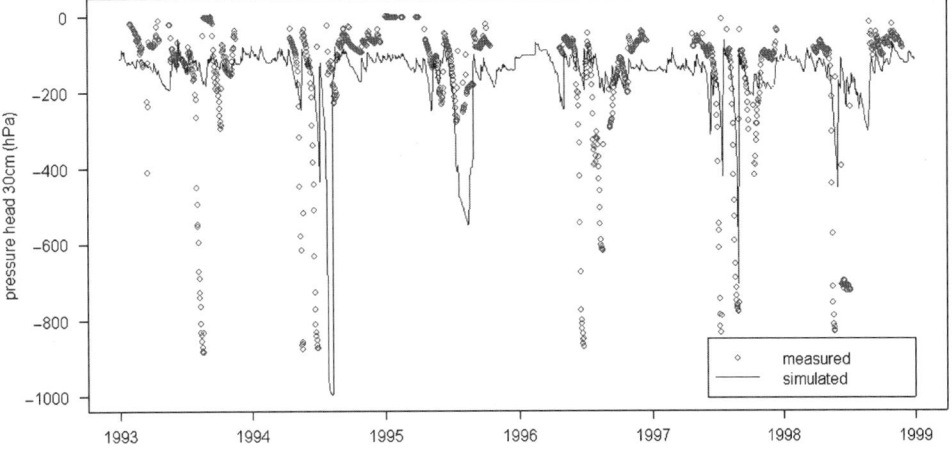

Fig. 7 Soil water pressure head at 30 cm depth below plot 1 during 1993–1998

Table 7 Soil temperature (°C) below plot 1 at three different depths; mean values as measured and simulated, as well as the RMSE

Depth (cm)	Measured	Simulated	ME	RMSE
5	9.38	8.24	1.14	2.76
20	10.02	8.23	1.79	2.61
50	9.64	8.17	1.47	1.65

which was mainly caused by a simulation of too low soil temperatures during winter periods (Fig. 8).

Nitrogen balance

The simulations with the ANIMO model resulted for the three different fields in carbon and nitrogen balances for the simulation period from 1993 to 1998. Mass balances are given for plots 1 and 3 (Table 8 and Fig. 9).

Supply of nutrients to the soil system is about 200–208 kg N ha^{-1} year^{-1} and originated mainly (>90%) from fertilizer applications and crop residues (Fig. 9: average of 181 kg N ha^{-1} for plot 1). Atmospheric wet and dry deposition added 19 kg N ha^{-1}. Crop uptake took out most of the nitrogen (90%) and most of the remaining part leached downward (21 kg N ha^{-1} for plot 1).

The nitrogen leaching flux of plot 1 varied between different years from a minimum value 4.3 kg ha^{-1} in the dry year of 1996 to a maximum of 44.7 kg ha^{-1} (Table 8). In plot 3 the last year 1998 showed a high leaching of 129 kg ha^{-1}, which resulted from a high storage during dry years 1996 and 1997.

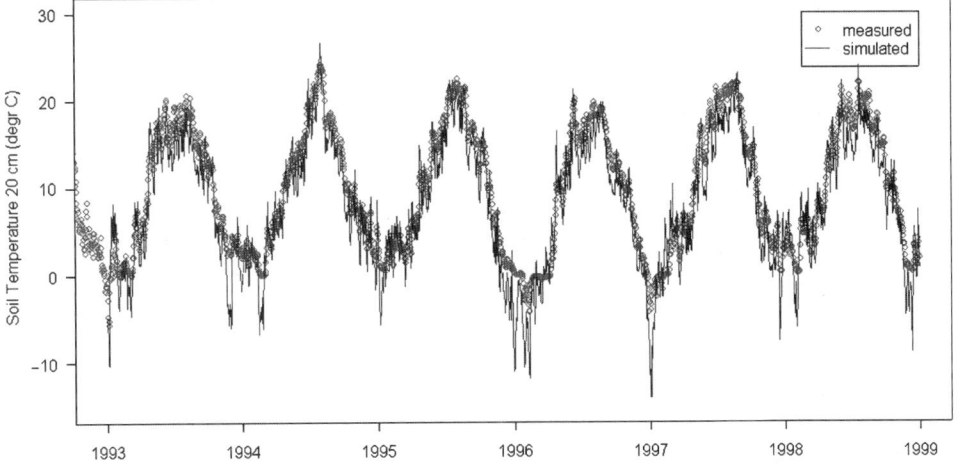

Fig. 8 Soil temperature at 20 cm depth below plot 1 during 1993–1998

Table 8 Nitrogen balance of soil layer 0–1 m for plots 1 and 3

Plot	year	Atmospheric deposition	Addition + crop residues	Total IN	Mineralization	Nitrification	Crop uptake	Nitrate leaching	Total OUT	Storage
1	1993	20	249.7	269.7	126.3	147.5	131.8	21.6	153.4	116.3
	1994	21.6	167.9	189.5	118.6	190	178.3	44.7	223	−33.5
	1995	18	146.3	164.3	83.6	148.9	147.3	14.3	161.6	2.7
	1996	17.8	200.2	218	66.2	164.5	192.7	4.3	197	21
	1997	18.6	247.9	266.5	90.5	154.2	195.1	7.9	203	63.5
	1998	20.4	71.6	92	111.7	120.4	119.5	33.7	153.2	−61.2
	Average	19	181	200	99	154	161	21	182	18
3	1993	20	274.3	294.3	145.5	200.8	140.2	31.1	171.3	123
	1994	21.6	90	111.6	121.8	180.5	156.5	74.4	230.9	−119.3
	1995	18	95	113	70	135.1	143.1	13.7	156.8	−43.8
	1996	17.8	313	330.8	95	265.2	195.2	6.1	201.3	129.5
	1997	18.6	310.1	328.7	107.5	208.4	196	42.5	238.5	90.2
	1998	20.4	35	55.4	122.5	132.6	138	129.2	267.2	−211.8
	Average	19	186	206	110	187	162	50	211	−5

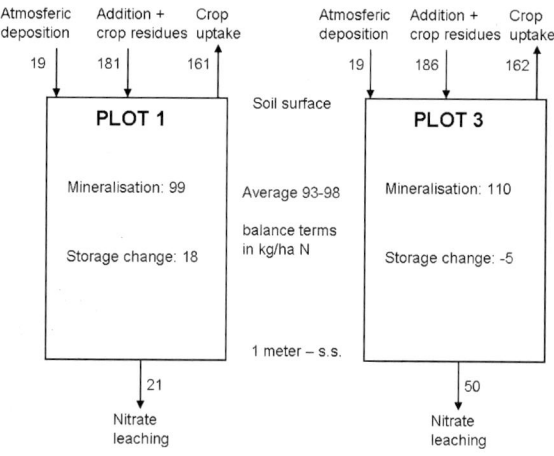

Fig. 9 Nitrogen balance (kg ha^{-1} a^{-1}) of plots 1 and 3, average values for the period 1993–1998

Soil mineral nitrogen was modelled and results for all plots (Table 9) indicate that model performance is better for plots 2 and 3 than for plot 1. This is largely based on a relatively high deviation in the layer 0–30 cm, with an RMSE of 42 kg ha^{-1} and a CD > 6. The deviations are caused by high measured mineral nitrogen content of 201.1 kg ha^{-1} on 21 April 1993, which is not modelled. Leaving out this value reduces the RMSE from 41.6 to 16.2 kg ha^{-1}. In general, simulated results were lower than measured results; this is confirmed by ME and CRM values that are positive and indicate an underestimation.

The time course of soil mineral nitrogen for plot 1 (Fig. 10) shows that high peaks in modelled and measured results corresponded relatively well. Unfortunately, the amount of measurements was limited in the years 1996–1998, where highest values were modelled.

Nitrate concentrations were measured with suction cups at different depths in three plots. Results (Table 10) indicate high average measured values of around 100 mg l^{-1} NO$_3$ at depths of 1.5–2.0 m. Negative ME and CRM values indicated that the modelled results systematically overestimated measured results. Available measured nitrate concentrations at lower depth were higher with average values of 145.6 mg l^{-1} NO$_3$ at a depth of 0.9 m below the soil of plot 1. Simulated results corresponded relatively well when comparing average values (Table 10). A comparison with the time course (Fig. 11) shows that the exact timing was no always accurate, which cause relatively large deviations. The period in 1997 with high modelled concentrations was unfortunately not monitored.

Uncertainty analyses

The uncertainty in saturated soil moisture contents and saturated hydraulic conductivity at a depth greater than 0.9 m had little influence (<3% changes) on the leaching of water and nitrate (Table 11). A reduction of the saturated hydraulic conductivity from a refer-

Table 9 Soil mineral nitrogen below plots 1–3 at different depths; mean values as measured and simulated, as well as ME, RMSE (kg ha^{-1} a^{-1}), CD and CRM (–)

Plot	Depth (cm)	Measured	Simulated	ME	RMSE	CD	CRM
1	0–30	28.0	14.9	13.1	41.6	6.3	0.5
1	30–60	11.4	3.8	7.6	11.4	7.7	0.7
1	60–90	13.4	5.2	8.2	12.9	5.3	0.6
1	0–90	52.7	23.9	28.8	54.6	6.4	0.6
2	0–30	13.9	7.2	6.7	12.5	1.6	0.5
2	30–60	10.5	3.6	6.9	12.6	6.8	0.7
2	60–90	10.2	3.1	7.1	9.6	3.7	0.7
2	0–90	34.6	14.0	20.6	28.4	1.9	0.6
3	0–30	22.8	15.0	7.8	27.7	3.5	0.3
3	30–60	9.3	6.8	2.5	10.8	0.6	0.3
3	60–90	10.8	8.1	2.7	11.7	0.9	0.3
3	0–90	42.8	29.9	12.9	34.5	1.8	0.3

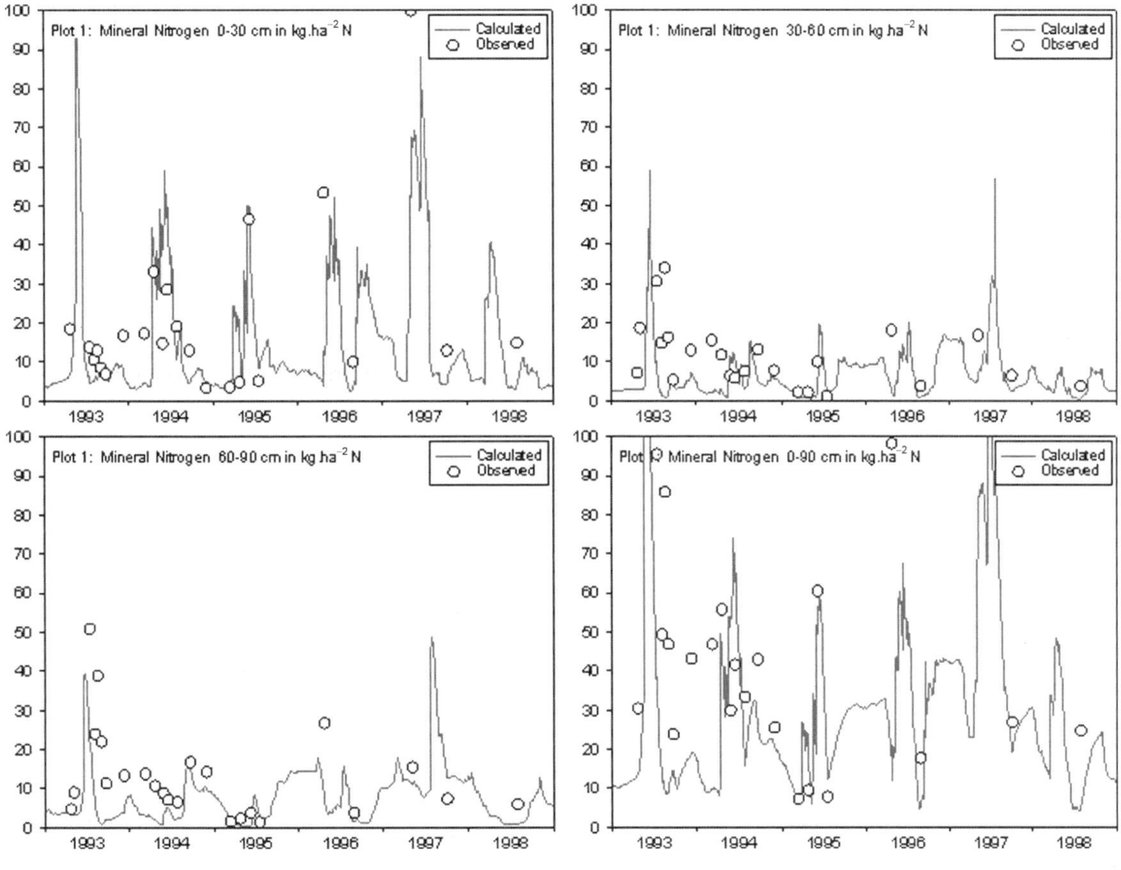

Fig. 10 Soil mineral nitrogen (kg ha^{-1}) for plot 1 for the period 1993–1998; results are given for four different soil layers (a) 0–30 cm, (b) 30–60 cm, (c) 60–90 cm, and (d) 0–90 cm

ence value of 23 cm day^{-1} to a value of 1 cm day^{-1}, caused a reduction of water leaching by 11%, but hardly affected nitrate leaching.

When the difference between drying and wetting scanning curve increased, the water leaching increased from 1% to a maximum of 9%. The impact of hysteresis on nitrate leaching increased from 11% to a maximum of 41% (Table 11).

A stepwise change of air temperatures shows that water leaching increased when air temperatures

Table 10 Nitrate concentrations below plots 1–3 at different depths; mean values as measured and simulated (mg l^{-1} NO$_3$), as well ME and RMSE (mg l^{-1} NO$_3$), CD and CRM (–)

Plot	Depth (cm)	Measured	Simulated	ME	RMSE	CD	CRM
1	90	145.6	106.0	39.6	72.5	1.4	0.3
1	150	104.4	82.1	22.3	49.7	2.9	0.2
2	200	85.5	115.9	−30.4	46.7	0.8	−0.4
3	200	120.3	129.2	−8.9	87.6	1.1	−0.1

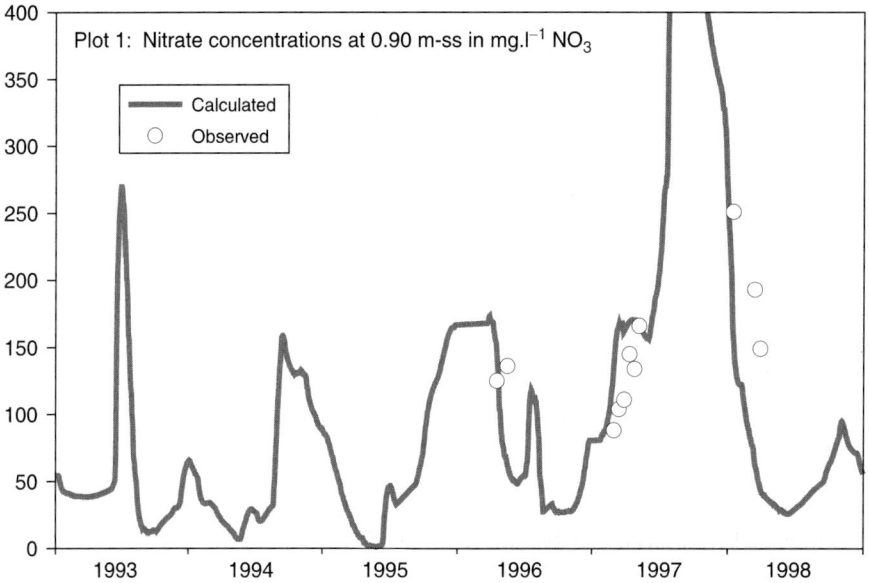

Fig. 11 Nitrate concentrations (mg l^{-1}) at a depth of 0.9 m of plot 1 for the period 1993–1998

decreased and water leaching decreased when temperatures rose (Table 11). An increase of 1°C resulted in a decrease of the water leaching of about 10% or 17 mm year^{-1}. The impact on nitrate leaching was reverse; an increase of 1°C resulted in an increase of 7% or 3 kg ha^{-1} year^{-1} NO$_3$-N.

Impact of weather changes

Results of the two scenarios with different weather showed that water leaching was reduced due to reduced precipitation (Table 12 and Fig. 12a, b). At a depth of 0.9 m below ground the average leaching was reduced from 178 mm for the "extended" to 116 mm year^{-1} for the predicted weather scenario. The same comparison showed that average nitrate leaching increased from 48 to 56 kg N ha^{-1} year^{-1} (Fig. 12c). Nitrate concentrations showed the largest increase: on average they nearly doubled from an average of 122–233 mg l^{-1} NO$_3$ (Fig. 12d).

Discussion and conclusions

Evaluating the performance of a comprehensive deterministic set of models, like SWAP–ANIMO, requires analyses using measurements of dominant terms of the water and nitrogen balances and most relevant process parameters. Such comprehensive data sets are costly and there are not many available. The data set from the three experimental plots in Müncheberg has the advantage that the flow direction is mainly vertical and well suitable to test one-dimensional models such as SWAP–ANIMO. The absence of a tracer experiment is unfortunate, but the presence of detailed (daily and at different depths) measurements

Table 11 Results of uncertainty analysis with data from plot 1: effect of variations in θ_{sat}, K_{sat}, hysteresis and air temperature on average leaching of water (mm) and nitrate (kg N ha^{-1}) at a depth of 0.9 m during the period 1993–1998. Percentage are given with respect to the values of the given data set (=100%)

Parameter	Value	Unit m^3 m^{-3}	Water leaching (mm)	Water leaching (%)	Nitrate leaching (kg N ha^{-1})	Nitrate leaching (%)
θ_{sat}	0.27		174.0	100	33.6	99
	0.32		173.4	100	33.9	100
	0.38		172.6	100	34.5	102
	0.43		171.9	99	35.1	103
K_{sat}	1.0	cm day^{-1}	157.9	89	34.9	98
	23.4		171.9	97	35.1	98
	50.0		174.3	98	35.2	98
	75.0		175.5	99	35.3	99
	100.0		176.2	99	35.5	99
	125.0		177.0	100	35.6	100
	150.0		177.3	100	35.7	100
	163.0		177.5	100	35.8	100
	175.0		177.8	100	35.8	100
f_{hyst}	1	–	171.9	100	35.1	100
	1.5		173.7	101	39.0	111
	2		179.4	104	42.5	121
	4		187.2	109	49.4	141
Increase of air temperature	−1	°C	192.2	112	32.7	93
	0		171.9	100	35.1	100
	1		150.8	88	39.2	112
	2		132.5	77	42.8	122
	3		114.6	67	45.5	130

Table 12 Results of weather changes on the average leaching of water (mm) and nitrate (kg N ha^{-1}) at a depth of 0.9 m during the period 1993–2050. Percentage are given with respect to the values of the given data set (=100%)

Scenario	Balans term	Unit	Minimum	Median	Mean	Maximum
Predicted	Precipitation	mm year^{-1}	330	497	505	838
	Water leaching 0.9 m	mm year^{-1}	22	116	132	326
	Nitrate leaching 0.9 m	kg N ha^{-1} year^{-1}	9	54	56	153
	Nitrate concentrations 0.9 m	mg/l NO$_3$	32	182	233	892
Extended	Precipitation	mm year^{-1}	486	625	600	712
	Water leaching 0.9 m	mm year^{-1}	45	178	186	389
	Nitrate leaching 0.9 m	kg N ha^{-1} year^{-1}	10	49	48	85
	Nitrate concentrations 0.9 m	mg l^{-1} NO$_3$	38	122	122	215

of soil moisture contents and soil hydraulic pressure heads is a great advantage. The data set includes main components of the nitrogen cycle, including fertilizer and soil management events, which are crucial for a proper modelling of nitrogen behaviour under field conditions. Comparisons between measured and modelled moisture contents showed acceptable agreement. A contour graph of moisture contents versus time and depth of plot 1 (Fig. 13) shows drying out of the soil profile up to a depth of about 0.3–0.6 m and to greater depth, especially during the drier years 1995–1997.

The contour graph of nitrate concentrations (Fig. 14) shows that the highest concentrations were found during or after periods of fertilizer management. When comparing the contour graphs of nitrate concentrations and moisture contents, it can be seen that the highest nitrate concentrations were found (modelled and measured) during the second half of the year 1997, which corresponded to the driest period. High nitrate

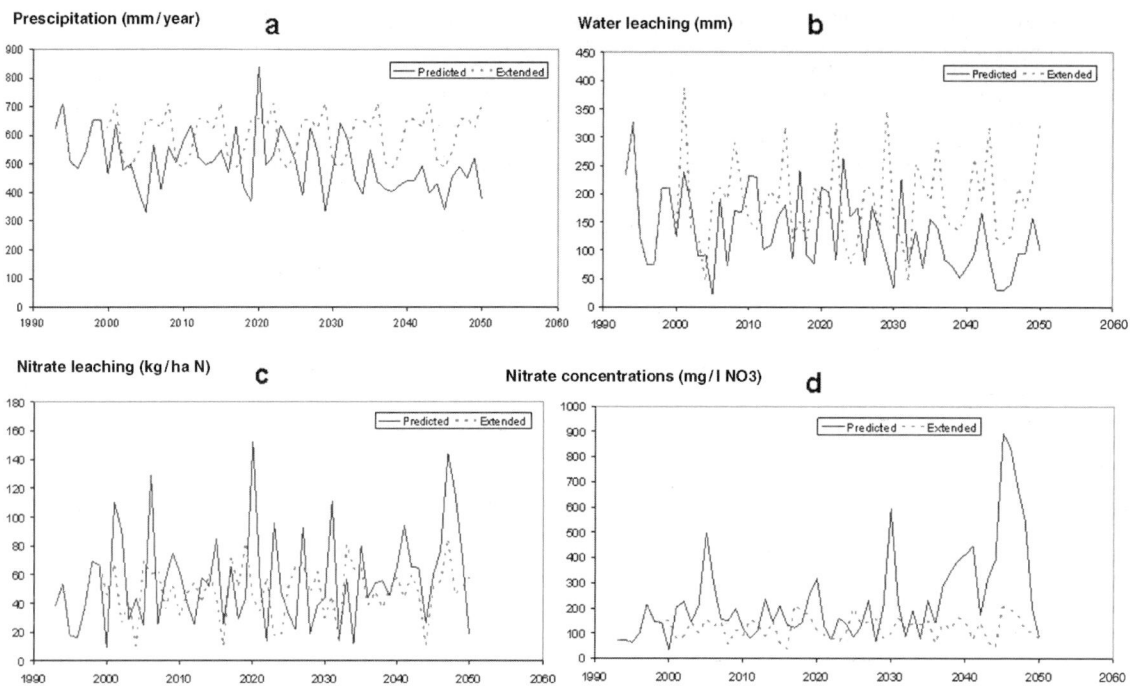

Fig. 12 Results of predicted and extended weather scenario for the period 1993–2050; (a) precipitation (mm year^{-1}); (b) water leaching (mm year^{-1}) at 0.9 m depth; (c) nitrate leaching (kg ha^{-1} year^{-1} NO$_3$-N) at 0.9 m depth; (d) nitrate concentrations (mg l^{-1} NO$_3$) at 0.9 m depth

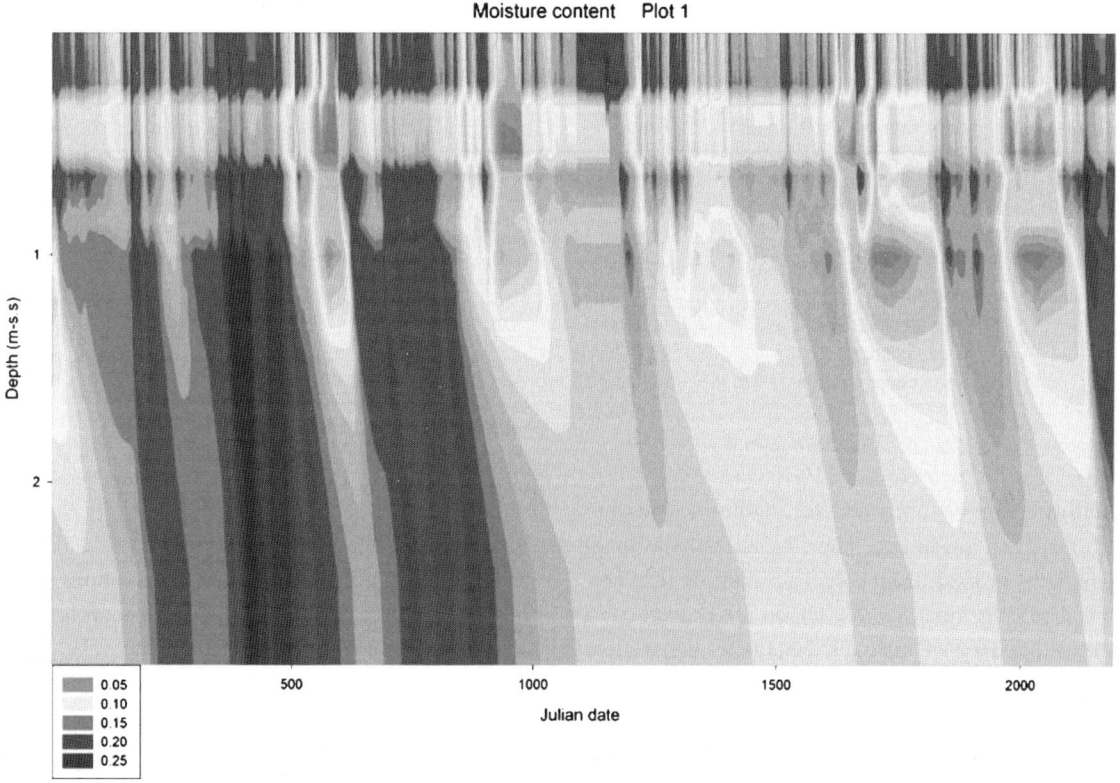

Fig. 13 Soil moisture content (m^3 m^{-3}) as function of time and depth below plot 1

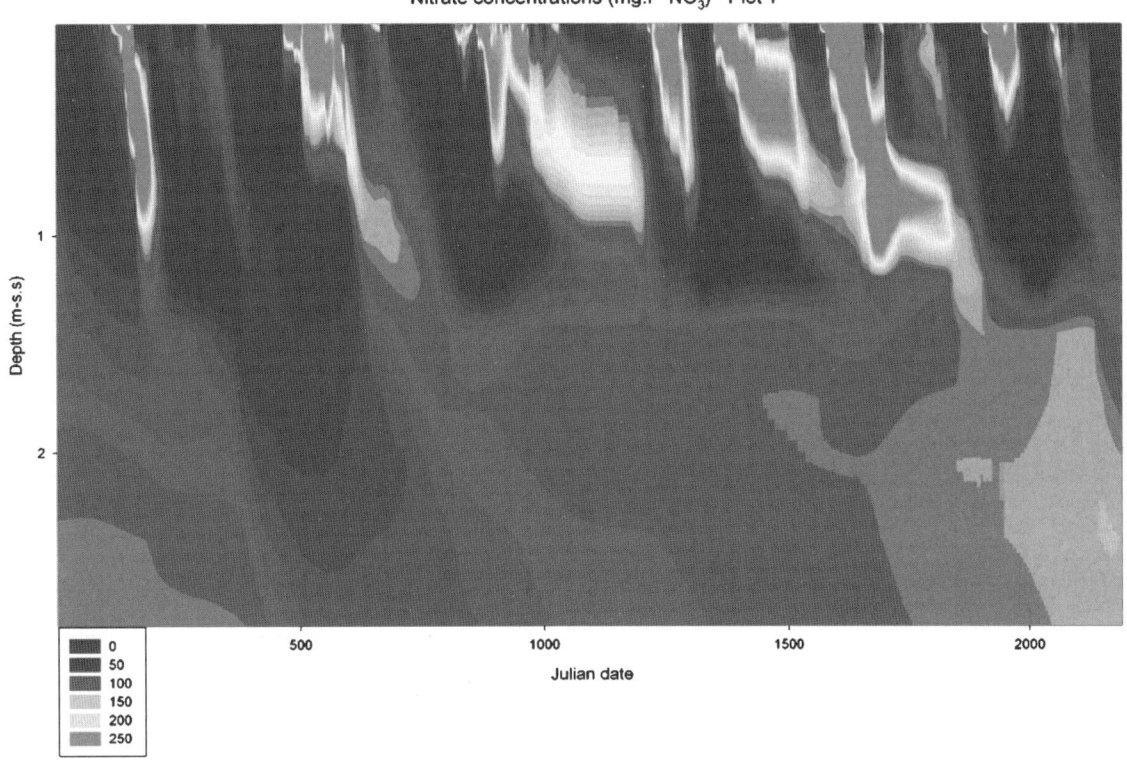

Fig. 14 Nitrate concentrations (mg l^{-1}) as function of time and depth below plot 1

concentrations in this area were caused by a combination of fertilizer application and dry periods (low moisture content) following a leaching event of precipitation excess.

The comparison of simulated and measured soil moisture contents and soil pressure heads showed deviations, which could be explained by the uncertainty in the soil moisture retention parameters and neglecting of hysteresis. The comparison between measured soil moisture and pressure head and the given retention curves showed deviations which indicate relatively large uncertainties in measured values and given retentions parameters.

The introduction of hysteresis influenced both leaching of water and nitrogen.

The uncertainties in soil hydraulic properties allowed adjustment of these properties, which resulted in improved results.

Crop development was a boundary condition and, therefore, a comparison of crop parameters was not appropriate.

Soil temperatures were simulated lower than measured, which had an influence on the transformation processes of the nitrogen cycle. Nitrification, decomposition of organic matter and the resulting mineralization were reduced. Low organic matter contents minimized the impact of this deviation.

Soil mineral nitrogen and nitrate concentrations were generally simulated lower than measured values. Deviations were within acceptable ranges.

In general, it may be concluded that the model performance on these three plots was acceptable for the tested domain of the hydrological and nitrogen cycle.

Future weather changes were simulated with the calibrated model. Boundary conditions were a reduced precipitation and an increased air temperature. Model results showed a reduced leaching of water, but an increased leaching of nitrate and a strong increase of nitrate concentrations.

Acknowledgement This study was only possible, thanks to financial support from the EU-COST718 action on AgroMeteorology and the Dutch Ministry of Agriculture and Fishery.

We thank professor Gerstengarbe from the Potsdam Institute for Climate Impact Research for the future weather data we were able to use in this study. Special thanks to the organizers from the ZALF Institute in Müncheberg who let us use their data set.

References

Allen RG, Pereira LS, Raes D, Smith M (1998) Crop evapotranspiration. Guidelines for computing crop water requirements. Irrigation and Drainage Paper 56. FAO, Rome, Italy.

De Vries DA (1975) Heat transfer in soils. In: De Vries DA, Afgan NH (eds) Heat and mass transfer in the biosphere I. Transfer processes in plant environment. Scripts Book Company, Washington, DC, pp 5–28

Feddes RA, Kowalik PJ, Zaradny H (1978) Simulation of field water use and crop yield. Simulation Monographs. Pudoc, Wageningen, The Netherlands

Hijmans RJ, Guiking-Lens IM, Diepen CA van (1994) User's guide for the WOFOST 6.0 crop growth simulation model. Technical Document 12, Winand Staring Centre, Wageningen, The Netherlands

Groenendijk P, Kroes JG (1999) Modelling the nitrogen and phosphorus leaching to groundwater and surface water; ANIMO 3.5. Report 144, DLO Winand Staring Centre, Wageningen, The Netherlands

Groenendijk P, Renaud LV, Roelsma J (2004) Prediction of Nitrogen and Phosphorus leaching to groundwater and surfacewaters; Process descriptions of the ANIMO 4.0 model. Report 983. Alterra, Wageningen, The Netherlands

Groenenberg JE, Salm C van der, Westein E, Groenendijk P (2000) Sensitivity analyses and limited uncertainty analyses of the ANIMO model (published in Dutch: Gevoeligheidsanalyse en beperkte onzekerheidsanalyse van het model ANIMO). Report 200, 138 pp, Alterra, Wageningen, The Netherlands

Kroes JG, Dam JC van (eds) (2003) Reference Manual SWAP version 3.0.4. Wageningen, Alterra, Green World Research, Alterra-report 773, 211 pp, Wageningen, The Netherlands

Loague K, Green RE (1991) Statistical and graphical methods for evaluating solute transport models: Overview and application. J Contam Hydrol 7:51–73

Mirschel W, Wenkel K-O, Wegehenkel M, Kersebaum KC, Schindler U, Hecker J-M (2007) Müncheberg field trial data set for agro-ecosystem model validation. In: Kersebaum KC, Hecker J-M, Mirschel W, Wegehenkel M (eds) Modelling water and nutrient dynamics in soil-crop systems. Springer, Dordrecht, pp 219–243

Renaud LV, Roelsma J, Groenendijk P (2004) User's guide of the ANIMO 4.0 nutrient leaching model. Wageningen, Alterra Report 224. 154 pp, Wageningen, The Netherlands

Rijtema PE, Groenendijk P, Kroes JG (1999) Environmental impact of land use in rural regions. The development, validation and application of model tools for management and policy analysis. Series on environmental science and management, Vol. 1. Imperial College Press, London

Tetens O (1930) Uber einige meteorologische Begriffe. Z Geophys 6:297–309

Vanclooster M, Boesten J, Tiktak A, Jarvis N, Kroes JG, Muñoz-Carpena R, Clothier BE, Green SR (2004). On the use of unsaturated flow and transport models in nutrient and pesticide management. In: Feddes RA, Rooij GH de, Dam JC van (eds) Unsaturated-zone modeling: progress, challenges and applications, Series : Wageningen Ur Frontis Series, Vol. 6. xxii, 364p. ISBN: 1-4020-2917-9, Kluwer/Springer, New York. Also available on http://www.wageningen-ur.nl/frontis/

Dam JC van (2000) Field-scale water flow and solute transport. SWAP model concepts, parameter estimation, and case studies. Ph.D. thesis, Wageningen University, Wageningen, The Netherlands, 167 p, English and Dutch summaries

Wesseling JG, Kroes JG, Metselaar K (1998) Global sensitivity analysis of the Soil-Water-Atmosphere-Plant (SWAP) model. Report 160, DLO-Winand Staring Centre, Wageningen, The Netherlands

Wolf J, Beusen AHW, Groenendijk P, Kroon T, Rötter R, Zeijts H van (2003) The integrated modeling system STONE for calculating nutrient emissions from agriculture in the Netherlands. Environmental Modelling & Software 18:597–617

CHAPTER TEN

Evaluation of water and nutrient dynamics in soil–crop systems using the eco-hydrological catchment model SWIM

Joachim Post[1], Anja Habeck[2], Fred Hattermann[2], Valentina Krysanova[2], Frank Wechsung[2] and Felicitas Suckow[2]

Abstract The process-based eco-hydrological spatially distributed catchment model SWIM (Soil and Water Integrated Model) was used to model water and nutrient dynamics in soil–crop systems. SWIM integrates hydrological processes, vegetation/crop growth, erosion, soil carbon, phosphorous and nitrogen dynamics at the river basin scale. A module for the turnover of soil organic matter was recently added by integrating the soil organic matter module of the forest growth model 4C into SWIM. As part of the model evaluation exercise described in this paper, SWIM was evaluated against data on soil temperature, soil hydrology, crop yield, soil nitrogen and long-term soil organic matter dynamics at the plot scale. The model was run predominantly without calibration except for parameterization data provided for the field plots (e.g. soil physical parameters).

Soil temperature and soil water were simulated well, with modelling efficiency index (IA) 0.87–0.96 for soil temperature and 0.54–0.92 for soil water. Simulated crop yield compared satisfactory well to measured yield, with IA values ranging from 0.37 to 0.87. Some problems occurred for long-term simulations (51 years) due to the fact that SWIM does not consider technical management changes such as seed quality improvements. Soil nitrogen dynamics were represented satisfactory under the different crop rotation and fertilization regimes at the measurement sites. Modelling efficiency index varied between 0.18 and 0.79. The simulation of long-term soil carbon dynamics resulted in a good representation of the measurements under the different fertilization treatments. The long-term trend of soil organic carbon (SOC) could be successfully represented by SWIM with a modelling efficiency index between 0.46 and 0.69.

Although SWIM was not designed as a plot or field scale agro-ecosystem model, it was able to reproduce the measured data for different plots, representing different edaphic, climatic and management conditions, in their temporal dynamics and magnitudes.

Keywords Agro-ecosystems, Crop growth, Crop yield, Eco-hydrological modelling, Plot scale validation, Soil carbon, Soil hydrology, Soil nitrogen, Soil organic matter, Soil temperature

Introduction

Regional scale dynamic environmental modelling investigating the impacts of regional environmental change requires modelling systems that are able to

---

Joachim Post, Anja Habeck, Fred Hattermann, Valentina Krysanova, Frank Wechsung and Felicitas Suckow
[1]Corresponding author: Joachim Post Potsdam, Institute for Climate Impact Research, Dept., of Global Change and Natural Systems, P.O. Box 60 12 03, 14412 Potsdam, Germany.,
tel.: +49 (0) 331-288-2417; fax: +49 (0) 331-288-2695;
e-mail: joachim.post@pik-potsdam.de
[2]Potsdam Institute for Climate Impact Research Dept., of Global Change & Natural Systems, P.O. Box 60 12 03, 14412 Potsdam, Germany

simulate the relevant processes adequately under the constraints of moderate data availability and model parameterization. A possibility to assess the impact of regional environmental change and management practices on ecosystem dynamics is the use of integrated process-based eco-hydrological catchment models, which are spatially distributed and operate at time steps appropriate for the natural processes described. Catchments integrate many forces, including land use and climate, fluxes of water, nutrients and pollutants (Krysanova et al. 1999). An integrated consideration of hydrological processes, vegetation dynamics, biogeochemical cycles and interactions between these quantities driven by soil, climate, land use and land management information is, therefore, necessary to assess regional environmental change and the consequences thereof. One prerequisite for the application of regional scale environmental models is a successful evaluation against observed data.

Agro-ecosystems are highly altered ecosystems through human interactions and are of high importance in both providing nutrients for humans and being a source of environmental pollution. In this context, coupled modelling of soil temperature and water dynamics, crop growth and yield, soil nitrogen and carbon transformations and transport in lateral and vertical dimensions can be useful to assess agro-ecosystem dynamics and their impacts on the environment.

This paper describes the results of the model evaluation exercise during the international workshop "Modelling water and nutrient dynamics in soil-crop systems" at the Leibniz-Centre for Agricultural Landscape Research (ZALF) e. V. Müncheberg, Germany in June 2004. In the frame of this workshop, observed data for soil temperature, soil hydrology, soil nitrogen contents, crop yield and long-term soil organic matter dynamics have been provided for three sites in Germany (Bad Lauchstädt in Saxony-Anhalt, Müncheberg in Brandenburg and Berlin), representing different crop rotations, fertilization regimes, soil management and soil types. This comprehensive data set of documented site conditions offers the opportunity to test the model capability in simulating relevant processes and interactions between them on the field scale for agro-ecosystems.

The soil and water integrated model (SWIM) (Krysanova et al. 1998) was used in this model evaluation exercise. It was developed mainly for meso- to macro-scale (100–100,000 km^2) catchment modelling of water quality, water quantity and vegetation growth (e.g. crop yield) assessments under the control of land use, land management and climate (Krysanova et al. 1998, 1999, 2000; Hattermann et al. 2005; Wechsung et al. 2000). In this context a comprehensive field scale evaluation of relevant processes is fundamental. But it has to be mentioned that SWIM was not designed to be used as a field scale agro-ecosystem model for crop yield prediction or a farm scale fertilization recommendation tool. SWIM was developed for regional environmental change assessment and, therefore, the relevant processes and interactions have to be represented in the correct magnitudes and with correct temporal behaviour under the various agricultural management practices relevant at that scale.

Testing ecosystem behaviour under documented site conditions is fundamental to assessing ecosystem behaviour under hypothesized conditions (Grant 2001). Based on a successful validation, useful information on questions related to soil C sequestration and soil quality, soil nutrient and water uptake by plants, soil nutrient loss and water quality issues, soil disturbance and land management impacts on ecosystem can be gained through the application of ecosystem models (Grant 2001).

Materials and methods

Experimental sites

The data sets provided by the workshop organizers are described in detail for the Müncheberg plots in Mirschel et al. (2007), for the Bad Lauchstädt plots in Franko et al. (2007) and for the Berlin data sets in Diestel et al. (2007). Here we provide an overview of the data sets used for simulation with SWIM (Table 1).

Our aim was to provide an integrated model evaluation by including soil temperature, soil water, soil nitrogen and carbon and crop growth dynamics rather than focusing on a single process. Therefore, we present here representative examples for each process for all three measurement sites.

Model description

SWIM (Krysanova et al. 1998) is a continuous-time, spatially distributed model. SWIM works on a daily

Table 1 Overview of simulated sites used for model evaluation

Field site	Plot	Time frame (year)	Soil temperature	Soil hydrology	Crop yield	N dynamics	C dynamics
Muencheberg	Plot 1	6		✓	✓	✓	–
	Plot 2		✓	✓	✓	✓	–
	Plot 3			✓	✓	✓	–
Bad Lauchstaedt	Crop Rotation	4	✓	✓	✓	✓	✓
	Black Fallow		–	✓	–	✓	✓
	Plot 1	100	–	–	✓	–	✓
	Plot 6		–	–	✓	–	✓
	Plot 13	51	–	–	✓	–	✓
	Plot 18	simulated	–	–	✓	–	✓
Berlin -Lysimeter	1–4	3	●	✓	–	–	–
	5–10		●	✓	–	–	–
	11–12		●	✓	–	–	–

✓ Simulated, – no measurement, ● not simulated

time step and integrates hydrology, vegetation, erosion, nutrients (nitrogen (N) and phosphorus (P)) and sediment fluxes at the river basin scale (Fig. 1). The spatial aggregation units for river basin modelling are subbasins, which are delineated from digital elevation data. The subbasins are further disaggregated into so-called hydrotopes, hydrologically homogenous areas. The hydrotopes are delineated by overlaying of subbasin, land use and soil maps (Krysanova et al. 2000). The model is connected to meteorological, land use, soil and agricultural management data (Fig. 1). For detailed process descriptions, validation studies and data see Krysanova et al. (1998, 1999, 2000). An extensive hydrological multiscale and multicriterial

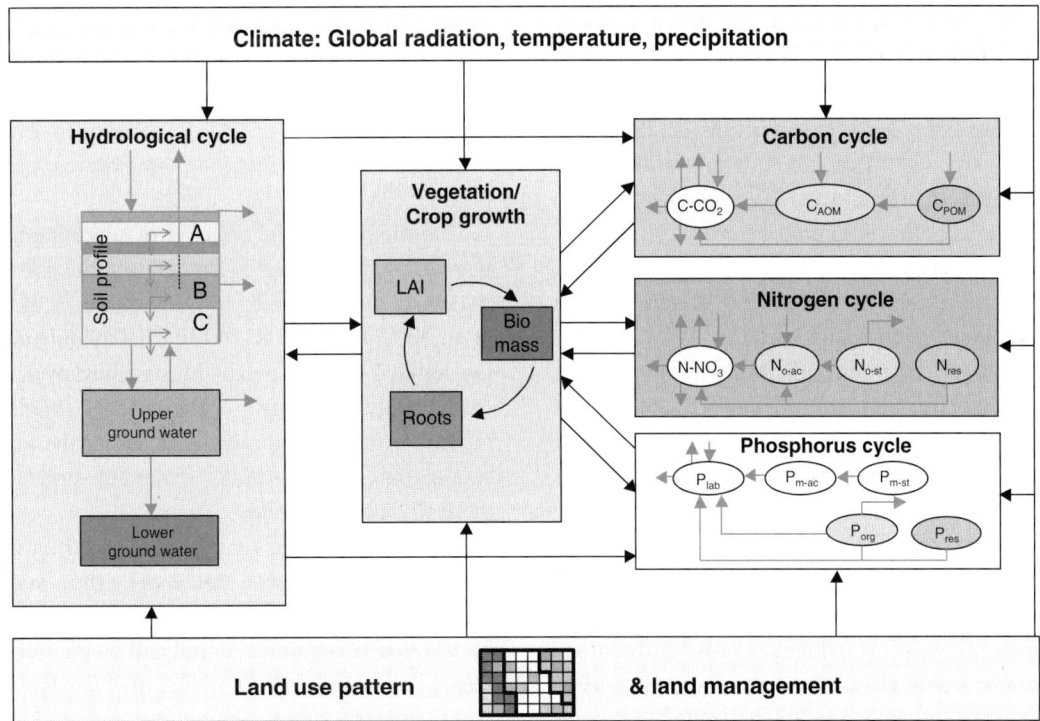

Fig. 1 Flow chart of the SWIM model, integrating hydrological processes, crop/vegetation growth, nutrient (nitrogen, N and phosphorus, P) dynamics and soil carbon turnover

validation of the model in the Elbe basin (Germany) including sensitivity and uncertainty analyses is described in Hattermann et al. (2005). Point scale validation using lysimeter data for soil hydrological and nitrogen processes are described in Beblik et al. (2001). The relevant processes for the presented work are described briefly.

The hydrology module is based on the water balance equation, taking into account precipitation, evapotranspiration, percolation, surface run-off and subsurface run-off for the soil column, which is subdivided into several layers according to the soil database (Fig. 1). The water balance for the shallow aquifer includes ground water recharge, capillary rise to the soil profile, lateral flow and percolation to deep aquifer (Krysanova et al. 1998).

The Priestley–Taylor (1972) or Penman–Monteith (Monteith and Unsworth 1990) methods are used (depending on input data availability) to estimate the potential evapotranspiration. Soil evaporation and plant transpiration are calculated as functions of leaf area index (LAI) using the approach of Ritchie (1972).

Surface run-off is calculated using a modification of the Soil Conservation Service (SCS) curve number technique. Water, which has infiltrated into the soil, percolates through the soil layers using a storage routing technique (Arnold et al. 1990). The water percolated from the bottom soil layer is defined as groundwater recharge (Hattermann et al. 2005). Lateral subsurface flow or interflow is calculated simultaneously with percolation using a cinematic storage model. Interflow occurs in a given soil layer, if the soil layer below is saturated.

Soil temperature is calculated on a daily basis at the centre of each soil layer. The calculation is based on an empirical relationship between daily average, minimum and maximum air temperature and a damping factor for soil depth. The effect of current weather conditions and land cover (snow, aboveground biomass) are considered (Krysanova et al. 2000; Neitsch et al. 2002).

The module representing crops and natural vegetation is an important interface between hydrology and nutrients (Fig. 1). A simplified EPIC approach (Williams et al. 1984) is included in SWIM for simulating arable crops (like wheat, barley, rye, maize, potatoes), using specific parameter values for each crop type. The model uses a concept of phenological crop/plant development based on daily accumulated heat units, Monteith's approach (1977) for potential biomass, water, temperature and nutrient stress factors and harvest index for yield partitioning.

The nitrogen module includes the following pools: nitrate nitrogen ($N\text{-}NO_3$), active ($N_{o\text{-}ac}$) and stable ($N_{o\text{-}st}$) organic nitrogen, organic nitrogen in the plant residue (N_{res}) and the processes: mineralization, denitrification, plant uptake, fertilization, input by precipitation, wash-off with surface and subsurface flows and leaching to groundwater (Fig. 1). The nitrogen mineralization model is a modification of the PAPRAN mineralization model (Seligman and van Keulen 1991). Mineralization of fresh organic nitrogen and active organic nitrogen pool depends on the C:N ratio, soil temperature and water content. Denitrification occurs only under the conditions of oxygen deficit and is described as a function of soil water content, soil temperature, organic matter content and mineral nitrogen content. Plant uptake of nitrogen is estimated using a supply and demand approach. The daily plant demand of nutrients is estimated as the product of biomass growth and optimal concentration in the plants. Actual nitrogen uptake is the minimum of supply and demand. The plant is allowed to take nutrients from any soil layer that has roots. The main purpose of the nitrogen module within SWIM is to assess catchment scale water quality issues like nitrate pollution of groundwater and surface water bodies under regional environmental change. All soil nitrogen evaluation results presented here were performed using the original SWIM nitrogen module.

The module for the turnover of soil organic matter was recently extended by integrating the soil organic matter module of the forest growth model 4C (Lasch et al. 2002; Grote et al. 1999). This module describes the coupled soil carbon and nitrogen turnover. For this model evaluation exercise, the recently extended soil organic matter module is used only for the long-term soil organic carbon (SOC) dynamics simulations at the Bad Lauchstädt site.

The soil carbon (C) turnover is based on the tight relationship between soil and the vegetation processes. On the one hand, an input exists into the soil by addition of organic material and on the other hand, there is a release of CO_2 into the atmosphere. To describe the C budget, organic matter is differentiated into active organic matter (AOM) as humus pool and

primary organic matter (POM) as litter pool. The latter is separated in up to five fractions for each vegetation and crop type as stems, twigs and branches, foliage, fine roots and coarse roots.

The carbon turnover into different pools is represented by first-order reaction kinetics (Chertov and Komarov 1997; Franko 1990; Parton et al. 1987; Jenkinson 1990). The carbon change in the POM pool (C_{POM}) is controlled by matter (plant type) and plant litter fraction-specific reaction coefficients, which control the rate of turnover. The transformation of C_{POM} to AOM (C_{AOM}) is controlled by a matter (plant type) and plant litter fraction-specific synthesis coefficient. The turnover of C_{AOM} is made from the synthesized portion and the carbon used in the process of mineralization driven by the mineralization coefficient.

Model parametrization and initialization

It is beyond the scope of this paper to provide a full description on input data and parameterization necessary to run the model. Except for the recently extended description for soil organic matter turnover, a detailed description of SWIM including all information on necessary input data and model parameterization can be found in Krysanova et al. (2000). Any changes from standard model parameterizations are provided in the respective sections.

Main parameters necessary for soil parameterization relevant to soil hydrological processes are soil textural information (percentages of clay, silt and sand), bulk density, soil porosity, available water capacity, field capacity and saturated soil conductivity. If saturated soil conductivity is not specified, it is calculated by standard pedotransfer functions (PTFs) (e.g. Rawls and Brakensiek 1982; Van Genuchten 1980). These data should be available for the entire layered soil profile. Related to soil C and N processes, initial organic carbon and nitrogen contents and initial nitrate content have to be provided with agricultural management practices (crop rotation, crop management, fertilization). General information was provided for the three field sites under study, for which necessary soil parameters have been assigned from standard soil science textbooks or the standard configuration in SWIM for the respective soil type has been used.

The model parameterization for soil organic matter turnover was done to simulate the relevant processes for agro-ecosystems under eastern German conditions. Therefore, related environmental studies in the region and literature were used for parameterization. Determination of main parameters and coefficients was mainly done either by field experiments (litter bag experiments) or under laboratory conditions (incubation experiments) cited in literature. Main source of these parameters are for agricultural plants investigations by Klimanek (1990 a, b), McGechan and Wu (2001) and Franko (1990).

Statistical evaluation

One way to evaluate model simulation is by visual/graphical comparison of the simulation values produced by the model with actual measured values from the field experiments. Besides this qualitative way of assessing the goodness of the simulation, a statistical assessment of the residuals (the differences between the observed and the simulated values) was performed. Addiscott and Whitmore (1987) stated that using one statistical method alone to quantify the discrepancy between model simulations and measured data can be misleading. It is hence necessary to use a set of statistical methods to determine common strengths and weaknesses in the simulation through distinct statistics to describe different aspects of the accuracy of the simulation (Smith et al. 1996). Therefore, we adopted a quantitative method described by Smith et al. (1996) and Smith et al. (1997).

As most observed data are without replicate measurements, we considered altogether eight statistical methods, seven as proposed in the statistical procedure described in Smith et al. (1996) and the IA (Willmott 1982). Table 2 provides an overview of the applied statistics and the respective references. Each of these statistics provides a partial insight into model performance. By balancing different aspects of the used statistics allows an appropriate evaluation of model performance.

The root mean square error (RMSE) provides a term for the total difference between the predicted and the observed values. The lower limit is 0, which indicates no difference between measured and simulated values (Table 2). The modelling efficiency (EF) value compares the variance of predicted from the observed values to the variance of the observed values from the mean of the observations. A value of 1 denotes a

Table 2 Overview of statistics used in the model evaluation procedure

Method	Name	Equation	Perfect fit [range]	Purpose of method	Reference
RMSE	Root mean square error	$\sqrt{\sum_{i=1}^{n}(P_i - O_i)^2 / n}$	0 $[0 : +\infty]$	Total difference and coincidence between measured and simulated values	Loague and Green (1991)
EF	Modeling efficiency	$\dfrac{\left(\sum_{i=1}^{n}(O_i - \overline{O})^2 - \sum_{i=1}^{n}(P_i - O_i)^2\right)}{\sum_{i=1}^{n}(O_i - \overline{O})^2}$	+1 $[-\infty : +1]$		Smith et al. (1996)
CD	Coefficient of determination	$\dfrac{\sum_{i=1}^{n}(O_i - \overline{O})^2}{\sum_{i=1}^{n}(P_i - \overline{O})^2}$	+1 $[0 : +\infty]$		Loague and Green (1991)
E	Relative error	$\dfrac{100}{n}\sum_{i=1}^{n}\dfrac{(O_i - P_i)}{O_i}$	0 $[-\infty : +\infty]$	Consistency of errors	Addiscott and Whitmore (1987)
M	Mean difference	$\sum_{i=1}^{n}\dfrac{(O_i - P_i)}{n}$	0 $[-\infty : +\infty]$		
t(m)	T value of M with critical 2.5% levels (two-tailed)	$\dfrac{M}{s_d / \sqrt{n}}$	–	Student's t statistic of M	Chatfield (1983), Smith et al. (1996)
IA	Modelling efficiency index	$1 - \dfrac{\sum_{i=1}^{n}(P_i - O_i)^2}{\sum_{i=1}^{n}(|P_i - \overline{O}| + |O_i - \overline{O}|)^2}$	+1 $[0 : +1]$	Measure of agreement between measured and simulated values	Willmott (1982)
r	Sample correlation coefficient	$\dfrac{\sum_{i=1}^{n}(O_i - \overline{O}) \cdot (P_i - \overline{P})}{\sqrt{\sum_{i=1}^{n}(O_i - \overline{O})^2} \cdot \sqrt{\sum_{i=1}^{n}(P_i - \overline{P})^2}}$	1 $[-1 : +1]$	Association between simulated and measured values	Draper and Smith (1966)

O_i = observed (measured) values, P_i = predicted (simulated) values, $\overline{O}$ = mean of observed values, $\overline{P}$ = mean of predicted values, n = number of samples.

perfect match of predicted and measured values. The coefficient of determination (CD) is a measure of the proportion of the total variance in the observed data that is explained by the predicted data (Table 2). A value of 1 again indicates a perfect fit. RMSE, EF and CD are measures to prove how closely the simulated values correspond to the measured values and, therefore, assess the coincidence of two data sets (Smith et al. 1996).

The sample correlation coefficient (r) and the IA are measures of the association of two data sets, i.e. the similarity of the shape of plotted simulation values in respect to measured values. A value of 1 for both statistical methods indicates a perfect fit and the same pattern of observed and simulated values.

The relative error (E) and the mean difference (M) are indicators for the bias in the total difference between simulations and measurements. They can be used to assess consistent or inconsistent errors in the simulations in respect to observations. Values of 0 for both indicate a perfect fit between simulation and measurements. The significance of the coherency between simulation and observation can further be tested using Student's t of M. Although it has to be mentioned that using Student's t-test in this context fails for testing, if measured and simulated values are related since they are not completely independent (i.e. calibration of simulations regarding observations) (Smith et al. 1997). Here, we test, if the simulations and measurements values differ significantly from each other using a two-tailed 2.5% significance measure.

Results and discussion

Soil temperature

Soil temperature was simulated at the Bad Lauchstädt site for two soil depths (20 and 50 cm) and at the Müncheberg site at 5 and 20 cm soil depths over a period of 8 years for the Müncheberg site and 6 years for the Bad Lauchstädt site. Simulation was performed on a daily basis for the two sites. In general, the simulated soil temperature compared satisfactory with the measured data (Fig. 2), especially for the present purpose of soil temperature as an reduction factor for soil carbon and nitrogen turnover.

It can be noted that near soil surface simulation of soil temperature performed better than deeper soil depths, which is expressed in lower IA values and higher RMSE values for deeper soil depths (Table 3). SWIM uses a relatively simple formulation mainly driven by daily average, minimum and maximum air temperature and a damping factor for soil depth, which is dependent upon the bulk density and the soil water content. The influence of plant canopy or snow cover on soil temperature is incorporated with a weighting factor, which is dependent on total aboveground biomass and residue and the water content of the snow cover on the current day. This not a physically based description of soil heat fluxes and results in higher daily fluctuations of the simulations and decreasing goodness of simulation with soil depth compared to the measurements (Fig. 2).

For the Müncheberg plot, problems occurred mainly in spring and autumn seasons, where the sandy soil is warming or cooling quicker than soils with higher clay contents. Simulations are not representing

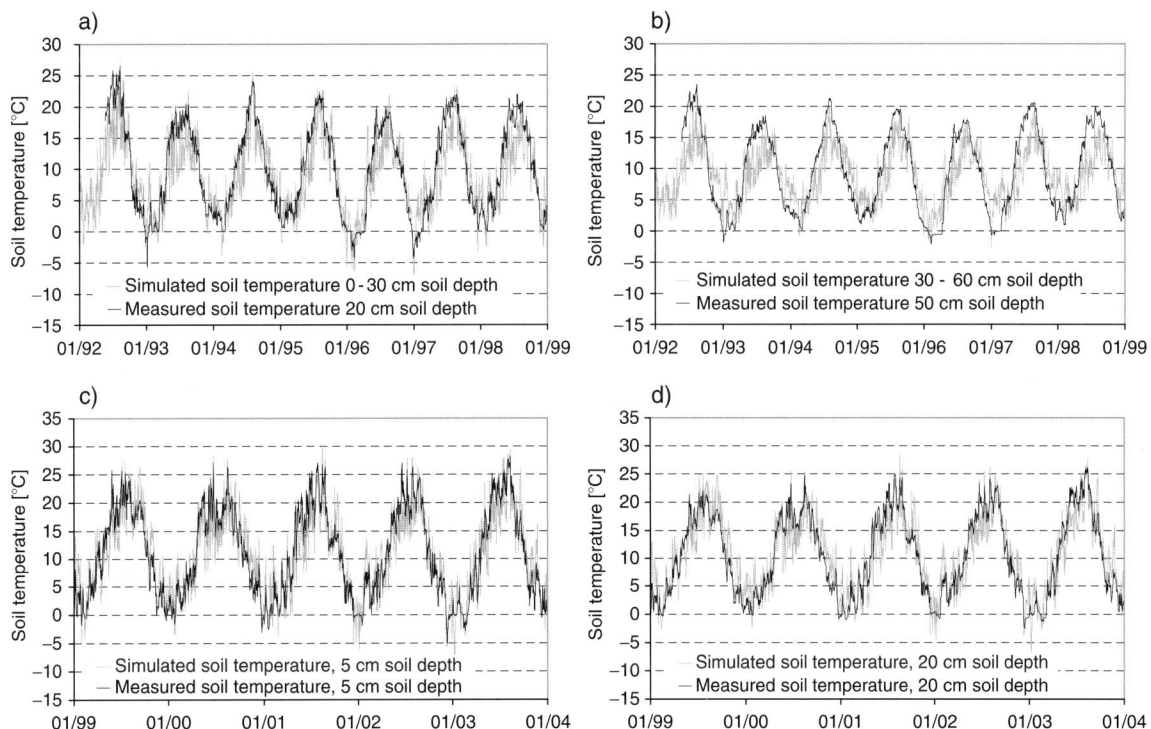

Fig. 2 Measured and simulated values of soil temperature for (a) Müncheberg 20 cm, (b) Müncheberg 50 cm (c) Bad Lauchstädt 5 cm, (d) Bad Lauchstädt 20 cm soil depth

Table 3 Representation of statistics describing the model performance in simulating soil temperature

Field site	RMSE	E	M	t(m)*	t_{crit}	EF	CD	IA	r
(a) 20 cm	3.3	−14.1	0.81	2.9**	1.98	0.85	1.75	0.94	0.94
(b) 50 cm	3.9	−57.4	0.61	1.61	1.98	0.69	2.82	0.87	0.87
(c) 5 cm	3.1	−19.6	−0.02	−0.09	2.0	0.93	1.39	0.96	0.97
(d) 20 cm	3.0	−41.9	−0.26	−0.96	2.0	0.91	1.46	0.95	0.97

(a) Müncheberg 20 cm, (b) Müncheberg 50 cm, (c) Bad Lauchstädt 5 cm, (d) Bad Lauchstädt 20 cm soil depth
*t-value, critical at 2.5% (two-tailed)
**significantly biased

this properly, resulting in a significantly biased simulation for the Müncheberg plot at 20 cm soil depth, where t values of M were greater than the critical two-tailed 2.5% t value (Table 3). For the 50 cm soil depth, this effect was the strongest, resulting in the lowest association between measured and simulated values (lowest r value of 0.87 of all simulations, Table 3).

The problems described here can partly be solved through calibrating the damping (for soil depth) and weighting (for land cover) factor used in the soil temperature description separately for the black earth and sandy soil. This would lead to a better representation of the measurements and the respective conditions (e.g. current plant cover, soil heat properties) at the two sites, but the described discrepancies are seen to be negligible for the purpose of soil temperature as a reduction factor for soil carbon and nitrogen turnover and soil temperature impacts on water movement.

Soil water content

Soil moisture simulations have been performed for all three Müncheberg plots (only plot 3 is shown here), for the two short-term Bad Lauchstädt plots, one representing a crop rotation and one black fallow plot (the latter is not shown) and for the *Cambisol* soil lysimeter at the Berlin site (lysimeters 9–12).

Different techniques have been applied for measuring soil moisture contents at various depths. The most reliable and accurate measurements are seen to be the gravimetric measured soil moisture contents. Additional techniques used were tensiometer measurements, field domain reflectometers (FDR) and time domain reflectometers (TDR) measurements. For the Bad Lauchstädt sites, three replicate measurements using FDR instruments were conducted. Standard deviations between replicate measurements varied between 0.7 and 2.6 Vol% for the crop rotation. At the Müncheberg sites, gravimetric measurements were available for comparison with simulations.

Figure 3 shows comparisons of soil moisture dynamics for Müncheberg, plot 3 (Fig. 3a), for Bad Lauchstädt at the crop rotation site (Fig. 3b) at two soil depths and for the Berlin site the lysimeters 9–12 (0–90 cm soil depth) representing a *Cambisol* soil for different groundwater distances (135 and 210 cm) as representative examples over all simulations.

The temporal dynamics and magnitudes of volumetric soil water contents at different soil depths are seen to be reliable represented by the simulations. Important to note is that interactions with soil water uptake by plants and with meteorological variables (precipitation and climatic water balance) are distinguishable and represented by the simulations (and the measurements). The general pattern of these mechanisms is met by the simulations (Fig. 3).

Discrepancies in the above-mentioned patterns (representation of plant water uptake, climatic water balance influences) are most pronounced for the Bad Lauchstädt comparisons. Here, the simulations show the influence of plant water uptake and water stress to be to great due to a negative climatic water balance (Fig. 3b, left and right). In times of plant water uptake during the growing season and of negative climatic water balance, an underestimation of soil moisture is seen. The strong underestimation for the 90 cm soil depth plot is likely to be due to an insufficient representation of root water uptake and the assignment of maximal rooting depth as an input parameter. Using a standard maximal root depth parameter for the specific crops as proposed by, e.g. Breuer et al. (2003) or Krysanova et al. (2000), might not be the appropriate value for the present soil type and properties. A calibration of this parameter would lead to a better representation, but that was not

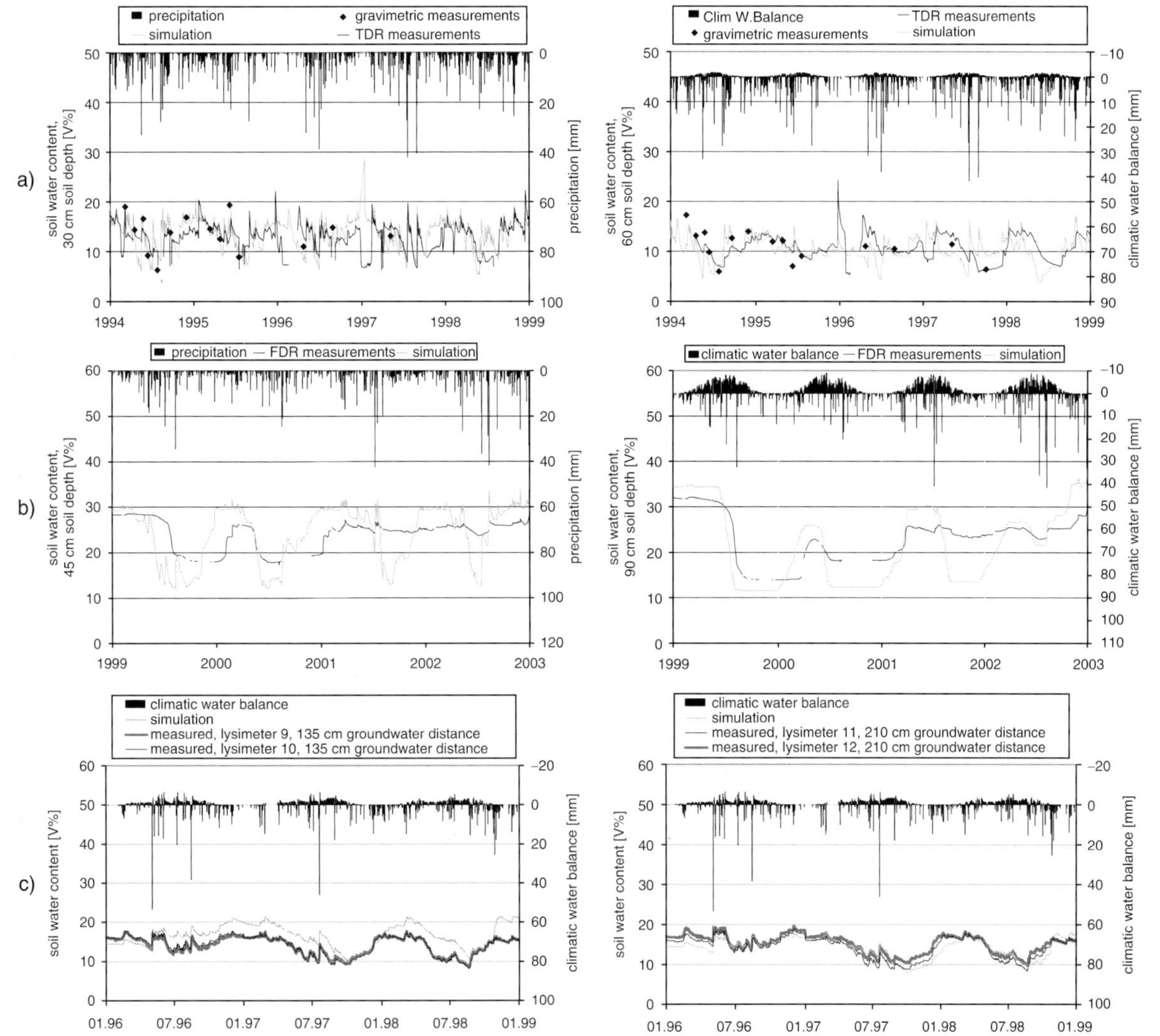

Evaluation of water and nutrient dynamics using model SWIM

Fig. 3 Measured and simulated values of soil water contents [Vol%] for (a) Müncheberg, plot 3 for 30 cm (left), 60 cm soil depth (right), (b) Bad Lauchstädt crop rotation plot for 45 cm (left) and 90 cm (right) soil depth (grey lines are simulations, black lines TDR or FDR measurements and black dots gravimetric measurements of soil water content) and (c) Berlin lysimeter 9 and 10 for 135 cm groundwater distance (affected by groundwater) and lysimeters 11 and 12 for 210 cm groundwater distance (not affected by groundwater) on a *Cambisol* soil. Black lines are lysimeter measurements and grey lines are the respective simulations for 0–90 cm soil depth. At the secondary axis values for precipitation [mm] (left) and climatic water balance [mm] (right) for the Müncheberg site (a) and Bad Lauchstädt site (b) are shown. For Berlin (c) only climatic water balance [mm] is given

the scope of this exercise. Additionally, water uptake by roots is overestimated for deeper soil depths (see Fig. 3b, right). For this case the root water uptake description might not be appropriate. The plant water use is estimated using the approach of Williams and Hann (1978), which is driven by the plant water transpiration rate. It is parameterized assuming that about 30% of the total water use comes from the top 10% of the root zone. Simulated soil moisture for the uppermost layer (not shown) is overestimated, indicating that for this case the model has to allow more water to be taken by the plants in these layers. Another factor at the Bad Lauchstädt site is the high loess content of the black earth soil, which causes problems assigning soil physical properties. This is due to the fact, that loess soils have an inherent high variability of soil physical properties like available field capacity or soil porosity. Using general soil parameterization and PTFs (e.g. Rawls and

Brakensiek 1982; Van Genuchten 1980) are one source of errors in simulating this site. Studies by Hattermann et al. (2005) using SWIM showed that uncertainties in simulating river discharge in loess regions are highest compared to mountain or lowland regions. This is mainly due to difficult loess soil parameterization (e.g. soil physical properties), high amounts of macro-pores in loess soils and high variability of soil physical properties derived from laboratory measurements.

Consequently simulation performance was weaker at the Bad Lauchstädt plot than the Müncheberg and Berlin plot having higher RMSE values and lower values for IA (except Müncheberg, 90 cm soil depth) and CD (Table 4). CD values at the Müncheberg plots are close to one or higher, which denote that the deviation of the simulations from the mean of the observed values is less than observed in the measurements, i.e. the model describes the measured data better than the mean of the measurements (Smith et al. 1996). The CD values for the Bad Lauchstädt site lie between 0 and 1, indicating that the deviation of the simulations from the mean observation is greater than observed in the measurements (Smith et al. 1996). This points to problems with model parameterization presumably related to the above-mentioned loess soil properties.

In consideration of the more reliable gravimetric soil moisture measurements at the Müncheberg site, the simulated soil water dynamics performed well. The temporal dynamics are more or less consistent with low bias (M) and good association (r values between 0.6 and 0.75, Table 4). The effects of plant water uptake and climatic water balance are better represented. The sandy soil leads to higher soil water dynamics than the Bad Lauchstädt soil, which is satisfactory reflected by the simulation.

The Berlin site shows a similar goodness of simulations in respect to measurements like the Müncheberg simulations. For lysimeters 9 and 10, which are affected by groundwater (distance of 135 cm, Fig. 3c, left), the simulation overestimate the measurements. The simulations further show a higher soil water content as for lysimeters 11 and 12 with no groundwater influence (distance of 210 cm, Fig. 3c, right). Higher soil water content for groundwater-influenced sites can be expected beforehand, but is not mirrored by measurements (Fig. 3c). The overestimation is also expressed in a higher RMSE value for groundwater-influenced lysimeters (Table 4c). The consistency of this error (the overestimation) is mirrored in low values of E and M, which are close to zero. This indicates a systematic bias, either in the measurements or in the simulation. For the 210 cm groundwater distance lysimeter (11 and 12), the simulations fit well to the measurements (Fig. 3c, left and right). The general course of simulated soil water dynamics, influenced by climatic water balance and plant water uptake, reflects the dynamics in the measurements. This site shows lowest RMSE values and highest IA of 0.92 and a high r value of 0.81 (Table 4c).

Crop yields

The performance of the crop growth model was evaluated using measured crop yields – as an indirect measure for crop growth – for the long-term and short-term field plots at Bad Lauchstädt and the field plots at Müncheberg.

The growth cycle of a plant in SWIM is controlled by plant attributes summarized in the plant growth database (Krysanova et al. 2000; Neitsch et al. 2002) and by the timing of operations listed in the management

Table 4 Representation of statistics describing the model performance in simulating soil water contents

Field site	RMSE	E	M	$t(m)$*	t_{crit}	EF	CD	IA	r
(a) 30 cm	2.7	0.3	0.53	0.86	2.14	0.54	1.7	0.82	0.75
(a) 60 cm	3.6	3.5	0.93	1.55	2.13	0.26	2.2	0.72	0.60
(b) 45 cm	5.6	−1.9	−0.47	−0.82	2.01	0.69	0.5	0.54	0.18
(b) 90 cm	5.3	1.8	0.28	0.47	2.01	0.58	0.4	0.81	0.27
(c) 135 cm	3.2	2.2	0.23	0.96	2.03	0.54	0.7	0.58	0.83
(c) 210 cm	1.4	3.9	0.42	1.8	2.03	0.42	0.6	0.92	0.81

(a) Müncheberg plot 3 with respective soil depths, (b) Bad Lauchstädt crop rotation with respective soil depths, (c) Berlin lysimeter 9, 10 (135 cm groundwater depth) and 11, 12 (210 cm groundwater depth), *Cambisol* soil

*t-value, critical at 2.5% (two-tailed)

file. The exact dates of sowing, harvesting and other management practices provided have been implemented into the management routines of SWIM. Related parameters remained unchanged and are set accordingly as for regional impact studies, described, e.g. in Krysanova (1998, 2000). According to Williams (1984), crop growth is mainly driven by solar radiation and adjusted daily considering for plant stress factors (water, temperature, nitrogen and phosphorus) using a simplification of the EPIC crop model. Therefore, crop growth and consequently yield simulations are seen to reflect mainly climatic and soil hydrological effects.

For the short-term assessments, the simulations are satisfactory comparing to the measurements. IA and r values are highest for all comparisons and M values range around 13 (12.6 for the Bad Lauchstädt site and − 13.6 for the Müncheberg site, Table 5a and c), indicating a mean difference of 13 dt ha^{-1} for both sites. Although showing considerable differences, the simulation can be evaluated as satisfactory. On one hand the simulations have not been calibrated by adjusting the harvest index or other relevant parameters, on the other SWIM is not designed as a plot scale yield prediction model. Points, which need further consideration, are a better representation of nutrient influences on crop growth, for long-term studies an introduction of a time-dependent harvest index, which reflects technical improvements (improving seed/sort quality), and a higher sensitivity of crop growth on soil properties. As it can be seen in Fig. 4a, b, the levels of crop yield in the simulations are less distinctive than in the measurements. The higher soil fertility of the Bad Lauchstädt plot, resulting in general higher crop productivity than the sandy soil at the Müncheberg site, is not sufficiently mirrored in the simulations.

For the long-term picture at the Bad Lauchstädt site, simulations are poor compared to a year-to-year comparison (Fig. 4c). The average yields, however, can be met fairly well. In the measurements, a clear trend of increasing yields can be seen (Fig. 4c, dotted black line), which should be probably assigned to technical and management improvements in agriculture during the last 50 years, especially in improving seed quality, higher efficiency of fertilization and in increasing the harvested plant part in comparison with shoots of a crop. SWIM allocates harvest from total biomass using a harvest index, which is crop-specific but does not change with time. Thus, the technological improvements cannot be mirrored in the simulations (grey dotted line in Fig. 4c). Only climatic constraints are influencing crop growth in the simulations. Therefore, a direct comparison with measured total biomass and an appropriate representation of technical and management advances in agriculture within the model would be necessary for a long-term assessment at the plot scale. An adjusted harvest index considering technical and management improvements over time would have increased the goodness of the simulations, but correct parameter values are difficult to obtain. Not considering these facts, the long-term comparison shows a low IA value for winter wheat (Table 5 b).

Soil nitrogen dynamics

Temporal dynamics of soil nitrogen contents have been evaluated at two sites. For the Müncheberg plots, results are shown for plot 3 in Fig. 5a and Table 6. Plots 1 and 2 lead to similar results and are not shown here. Plot 3 consists of a sugar beet/winter wheat/winter barley/winter rye rotation starting in 1993. Before each rotation, *Phacelia* was grown and ploughed under as green manure. Various amounts of inorganic and organic fertilization have been applied, for a detailed description see Mirschel et al. (2007). Results are shown for the years 1993, 1994 and 1995 because in this period most measurements were available providing a sufficient basis to perform an appropriate evaluation. Additionally, results are shown for 0–90 cm soil depth, not considering layers 0–30, 30–60 and 60–90 cm here. Evaluation of soil nitrate concentrations for the Bad Lauchstädt site is shown for the short-term crop rotation and for the black fallow plot in the compartments 0–20 cm at two sampling dates per year (1998–2002, Fig. 5b, c). The crop rotation consists of winter barley, winter wheat, sugar beet, spring barley, potato and winter wheat starting in 1997. Various amounts of organic manure and mineral N fertilizer have been applied (for details see Franko et al. (2007)). For both sites, the mineral N contents of each soil layer, with which the model was initiated, were estimated from the measurements. Sowing, harvesting, fertilization, etc. during the model run are on the same dates as those in the field experiments.

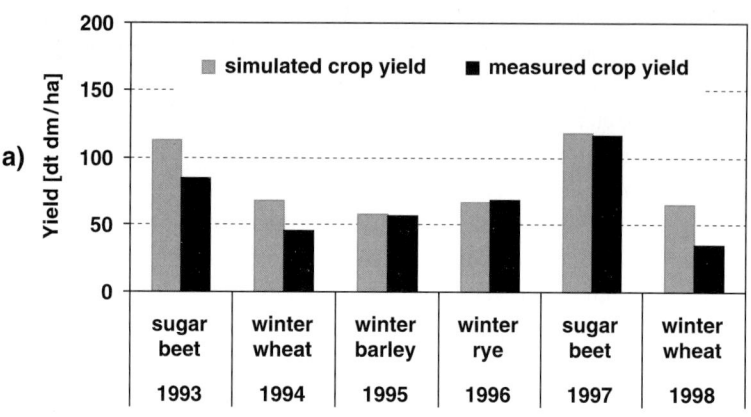
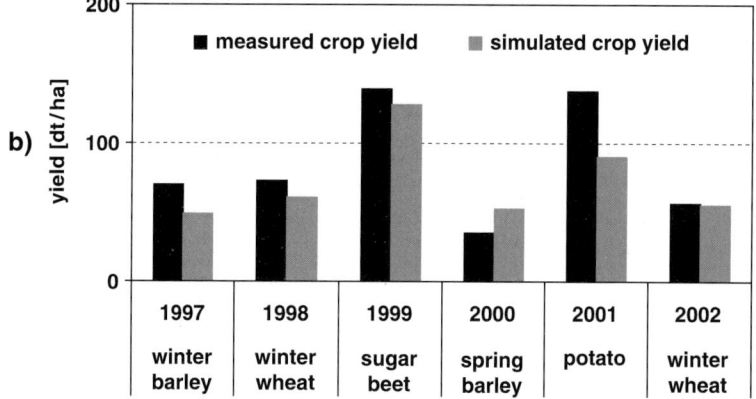
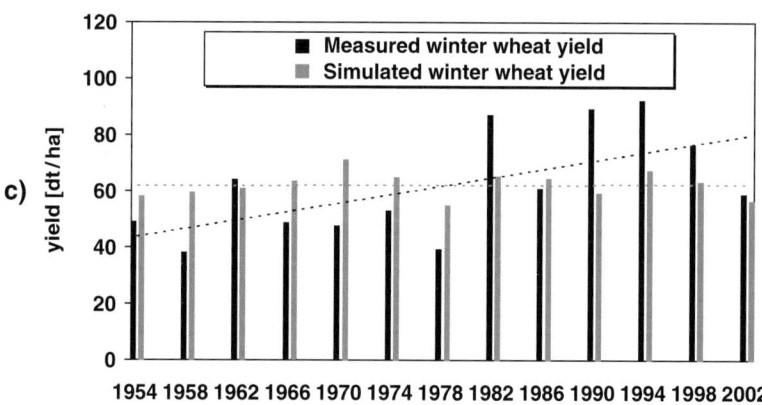

Fig. 4 Measured and simulated values of crop yield for (a) Müncheberg plot 1, (b) Bad Lauchstädt crop rotation plot and (c) Bad Lauchstädt long-term (plot 1) for winter wheat

Table 5 Representation of statistics describing the model performance in simulating crop yield

Field site	RMSE	E	M	$t(m)$*	t_{crit}	EF	CD	IA	r
(a)	19.1	−28.4	−13.6	−2.25	2.57	0.49	0.93	0.85	0.86
(b)	17.3	−8.3	−0.2	−0.04	2.18	0.08	16.9	0.36	0.28
(c)	23.4	6.8	12.6	1.42	2.57	0.65	1.63	0.87	0.88

(a) Müncheberg plot 1, (b) Bad Lauchstädt long-term plot only for winter wheat yields, (c) Bad Lauchstädt, short-term crop rotation plot
*t-value, critical at 2.5% (two-tailed)

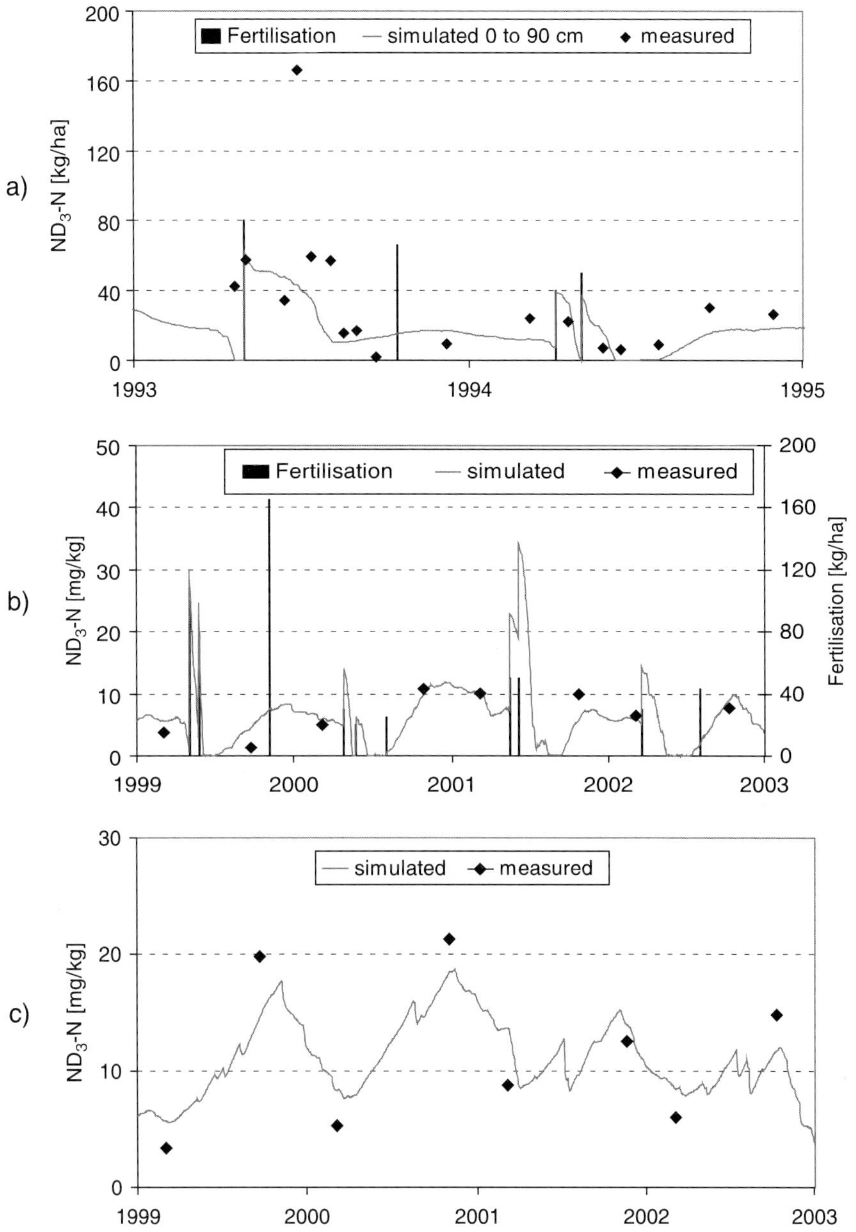

Fig. 5 Measured and simulated values of soil nitrogen (nitrate, NO$_3$-N [kg/ha]) dynamics for (a) Müncheberg plot 3, (b) Bad Lauchstädt crop rotation plot and (c) Bad Lauchstädt black fallow plot. Grey lines are simulated values, black dots are measurements and black bars are indicating the amount of fertilisation [kg/ha]

Table 6 Representation of statistics describing the model performance in simulating soil nitrogen (nitrate) dynamics

Field site	RMSE	E	M	t(m)*	t$_{crit}$	EF	CD	IA	r
(a) 0–90 cm	38	−419.13	0.03	0.37	2.07	−0.25	1.8	0.48	0.21
(b) 0–20 cm	6.4	−27.5	−0.11	−0.16	2.31	0.56	2.3	0.72	0.76
(c) 0–20 cm	16.5	−21.3	−0.37	−0.3	2.37	0.72	2.6	0.18	0.89

(a) Müncheberg plot 3, (b) Bad Lauchstädt crop rotation, (c) Bad Lauchstädt black fallow
*t-value, critical at 2.5% (two-tailed)

Simulated nitrate contents for plot 3 (Müncheberg) increase following each inorganic fertilization to comparable levels in the measured data, except for October 1993 where an ammonium urea solution was applied (Fig. 5a). This did not directly effect nitrate contents, because the solution entered the soil as ammonia and was then transformed to nitrate resulting in increasing nitrogen content thereafter (Fig. 5). The simulated values generally agree with the observation in the correct pattern and magnitudes. The value of CD is greater than 1 (1.8, Table 6a), indicating that the variation in the observed values is higher than in the simulated values. This means that either the model is not adequately describing extreme values or the experimental measurements are erroneous. The high value in June 1993 is most likely a measurement error. Under these instances, the high RMSE and the low E value have to be interpreted. Occurrence of extreme values enlarges E (and RMSE) in a way that does not truly reflect the accuracy of the simulation and should thus be excluded for interpretation. A low M value (0.03, Table 6a) and an IA value of 0.48, however, indicate a low bias of the simulation.

For the crop rotation field experiment (Bad Lauchstädt) a similar quality of simulation performance can be stated as for the Müncheberg plot. Inorganic fertilizer application led to a rapid increase in nitrate concentration in the simulation. But the correct order of magnitude could not be evaluated as measurements at these times are missing. Apart from that, the model correctly represents the general course of nitrate dynamics. The organic fertilization is not resulting in a rapid increase of nitrate concentration in the simulation, because the main form is ammonium being either transformed to nitrate, washed out or taken up by the plants or is subject to decomposition. IA and r values of 0.72 and 0.76, respectively (Table 6b) are indicating a high association of simulated and measured values. The total difference between the simulated and the measured values is at 6.4 (RMSE, Table 6b) and EF value is positive indicating a low coincidence between simulations and measurements. With an M value close to zero (−0.11), it can be stated that no consistent errors are within the model representation. Same statistical interpretation can be made for the black fallow plot. This plot shows a good temporal representation of the soil nitrate dynamics, but minima and maxima are not fully represented by the model. This can also be seen in the CD value (2.6, Table 6c), which is greater than 1, denoting that the deviation of the predictions from the mean observation is less than observed in the measurements.

Soil carbon dynamics

SOC dynamics have been simulated for the long-term static fertilizer experiment at Bad Lauchstädt (Franko et al. 2007). Data for four plots have been provided. Plot 1 received 30 t ha^{-1} farmyard manure every 2 years and varying rates of inorganic fertilizer (NPK), plot 6 received 30 t ha^{-1} farmyard manure every 2 years, plot 13 received varying rates of inorganic fertilizer (NPK) and plot 18 received no fertilization at all. A 4-year rotation of summer barley/potatoes/winter wheat/sugar beet was in use at all plots. Soil carbon levels (0–20 cm) have increased for plots 1 and 6 and have remained constant on plots 13 and 18 (Fig. 6a, b).

Simulation of SOC started 1951, because from 1951 onwards, all necessary meteorological input data have been provided. The model was initialized for each plot by simply using the value from 1951 derived from the linear trend of the measured data (1901–2002, dotted lines in Fig. 6). We considered only aboveground plant residuals (straw and shoots/stubbles) and belowground plant residuals (roots) as POM fractions, which have been allocated according to root to shoot ratios as proposed by, e.g. Klimanek (1990a, b).

A Monte Carlo-based sensitivity assessment of input parameters (not shown here) indicated, that the amount of dead plant material entering the soil (as evaluated indirectly through crop yield comparisons), the synthesis coefficient of POM and the turnover coefficient of AOM are most sensitive parameters of the model. Hence, a correct representation of these two parameters and the amount of dead plant material entering the soil column is very important.

Amount of dead plant material entering the soil was derived from simulated crop yields by using an empirical relationship proposed by Franko (1990) and Franko et al. (1997) to calculate the carbon inputs. Amount and timing of inorganic and organic fertilizer input, sowing and harvest times and atmospheric N deposition have been provided by the data-holders and have been fed into SWIM. The turnover

Fig. 6 Measured and simulated values of total soil organic carbon in the top 20 cm of soil at Bad Lauchstädt for (a) plot 6 (receiving 30 t/ha farmyard manure every 2 years) and plot 18 (no fertilization) and (b) plot 1 (receiving 30 t/ha farmyard manure every 2 years and various amounts of inorganic fertilization) and plot 13 (receiving various amounts of inorganic fertilization). Lines are showing the simulation, dots the measured values with standard deviation bars and dotted lines the long-term linear trend in the measurements

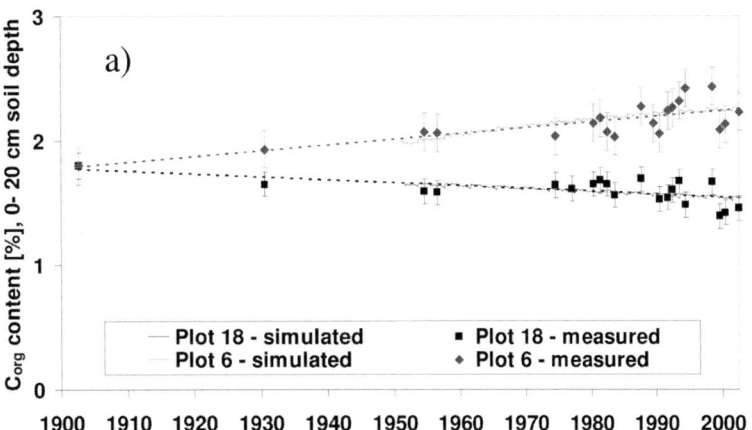
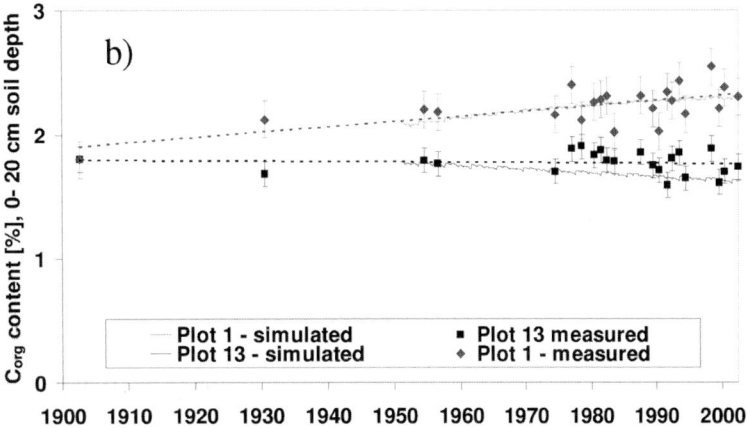

coefficient of AOM was calibrated on the long-term experiment in Müncheberg (not part of this evaluation exercise, Post et al. 2004) and remained unchanged for this study.

The simulated SOC contents of the four plots generally agree with the measured data. The simulations could reproduce well the impacts of organic fertilization on SOC dynamics and the pattern of the measurements was matched reasonable by the simulation. The simulated values lie between the standard error of most of the measured data. The long-term trends in measured SOC contents (dotted lines in Fig. 6a, b) are met by the simulations, except for plot 13, which show a slight decreasing trend in the simulation (Fig. 6b).

Table 7 Representation of statistics describing the model performance in simulating soil carbon dynamics for the Bad Lauchstädt long-term static fertilization experiment

Field site	RMSE	E	M	$t(m)$*	t_{crit}	EF	CD	IA	r
(a) plot 6	0.13	−0.29	−0.005	−0.53	2.0	0.99	0.36	0.69	0.36
(a) plot 18	0.08	−0.07	0.0005	0.08	2.0	0.99	0.34	0.53	−0.31
(b) plot 1	0.11	0.02	0.003	0.28	2.0	0.99	0.41	0.46	0.35
(b) plot 13	0.13	2.04	0.03	4.1**	2.0	0.98	0.44	0.49	−0.33

* t-value, critical at 2.5% (two-tailed)
** Significantly biased

Statistical analysis shows that for plot 13 the simulation shows a significant bias (greater student's t-value of M (4.1) than t_{crit} (2.0), see Table 7). RMSE values are in the same range as most of the models used in a soil organic matter model comparison at the Bad Lauchstädt site (Smith et al. 1997). EF values are positive for all plots, so the model accurately simulates soil organic carbon dynamics for that site.

Problems occurred at plot 13, where the simulation shows a decreasing trend, whereas the measurements remain more or less constant. Regular applications of NPK seem to stabilize the SOC content. The simulation could not represent this process adequately, hence inorganic fertilization effects on crop productivity are not considered correctly, at least for the Bad Lauchstädt site conditions. Therefore, the effects of increasing plant productivity due to inorganic fertilization on SOC dynamics can currently not be assessed satisfactorily. Although organic fertilization with farmyard manure and effects of crop rotation on SOC contents have been described correctly.

Conclusion

SWIM performed reasonably well in simulating soil temperature, soil water, soil nitrogen, soil carbon dynamics and crop yields at the plot scale using the standard model parameter values. For soil temperature, the empirical process formulation shows some disadvantages when evaluated at the plot scale. Probably, more physically based descriptions for soil heat flow (e.g. Grant 1995; de Vries 1963; Suckow 1984) would lead to better representation especially for deeper soil positions and under various land covers. But soil temperature is mainly used as impact factor on soil carbon and nitrogen turnover processes and impacts on water movement in SWIM and, for this purpose, the presented agreement is seen to be sufficient.

Comparisons of soil moisture showed that SWIM adequately represents plant uptake and water fluxes in the soil driven by meteorology. Problems occurred for the black earth soil (Bad Lauchstädt) where simulating soil hydrology was weakest mainly due to difficulties in parameterization of soil physical properties for soils with high loess and organic matter contents. Soil water dynamics play a central role for vegetation growth and soil carbon and nitrogen dynamics. Especially, soil nitrogen dynamics highly depend on soil water dynamics and errors in soil water simulations will propagate to soil nitrogen dynamics.

Modelling of crop yield for a long period showed the lowest performance in this exercise. The process description is clearly designed to assess crop growth at regional scales and is simplified mainly in the description of phenological processes in order to decrease requirements on input information. Taking this fact into consideration, the yield assessments performed satisfactory at the field scale and can be used to calculate plant biomass entering the soil. Especially, these quantities are important determinants for soil C and N turnover.

Simulation of soil nitrate contents showed reliable dynamics both in magnitudes and temporal behaviour under the different crop species and fertilization regimes. Although a detailed evaluation of soil nitrogen dynamics in soil–crop systems need to consider additional processes in more detail (e.g. ammonification, ammonium and nitrate plant uptake and effects on crop growth and transport with soil water). To assess these interactions more properly in SWIM, we are currently testing the recently extended soil carbon and nitrogen module for this purpose. Considering coupled soil carbon and nitrogen processes in this context is important as carbon turnover is providing energy for soil nitrogen processes and absence of sufficient inorganic nitrogen will inhibit the carbon turnover. However, since mineral nitrogen content in the soil is the result of multiple simultaneous processes and is highly dependent on crop growth, management practices and soil water dynamics, it is difficult to evaluate these processes (Diekkrüger et al. 1995).

Long-term soil carbon dynamics have been successfully simulated, especially the long-term trend of organic carbon. Impacts of organic fertilization and plant-derived organic matter inputs are well represented, whereas impacts of inorganic fertilization on crop growth and consequently SOM dynamics needs further consideration.

By evaluating the process dynamics as a whole, it can be stated that SWIM is able to reproduce the relevant interaction in soil–crop systems in respect to its purpose as a regional scale eco-hydrological model. A detailed validation of processes using field scale measurements is of importance both in checking the models capability and to detect problems and

deficits. Without field scale experimental plots and available data out of it, development and improvement of regional ecosystem models is impossible. Combined with evaluations of lateral fluxes of water and nutrients, it is then possible to perform global change assessment of agro-ecosystems at the regional scale using the eco-hydrological catchment model SWIM.

Acknowledgements We thank the European Science Foundation (ESF) for the support of Joachim Post with an exchange grant under the programme "The Role of Soils in the Terrestrial Carbon Balance (RSTCB)". We are grateful to Jo and Pete Smith (University of Aberdeen, UK) and the Modelling Research Group, School of Biological Sciences, Aberdeen University, UK. We also thank the workshop organizers for leading a very stimulating and interesting workshop.

References

Addiscott TM, Whitmore AP (1987) Computer simulations of changes in soil mineral nitrogen and crop nitrogen during autumn, winter and spring. J Agric Sci 109:141–157

Arnold JG, Williams JR, Nicks AD, Sammons NB (1990) SWRRB -a basin scale simulation model for soil and water resources management, Texas A&M University press, College Station

Beblik A, Cepuder P, Dreyhaupt J, Fank J, Feichtinger F, Franko U, Haberlandt U, Kersebaum KC, Krysanova V, Steinhadt U (2001) Stickstoffmodellierung für Lysimeter des Parthegebietes. Ergebnisse des Workshops "Stickstoffmodellierung" vom 8. bis 10. Juni 1999. In: Dreyhaupt J (ed.) Umweltforschungszentrum Leipzig-Halle GmbH, Leipzig, pp 25–57

Breuer L, Eckhardt K, Frede H-G (2003) Plant parameter values for models in temperate climates. Ecol Model 169:237–293

Chatfield C (1983) Statistics for technology. Chapman & Hall, London

Chertov OG, Komarov AS (1997) SOMM: a model of soil organic matter dynamics. Ecol Model 94:177–189

de Vries DA (1963) Thermal properties of soils. In: Wijk WR van (ed.) Physics of Plant Environment. North-Holland Publishing, Amsterdam, The Netherlands, pp 210–235

Diekkrüger B, Arning M (1995) Simulation of water fluxes using different methods for estimating soil parameters. Ecol Model 81:83–95

Diestel H, Zenkel T, Schwartengraeber R, Schmidt M (2007) The lysimeter station at Berlin-Dahlem. In: Kersebaum KC, Hecker J-M, Mirschel W, Wegehenkel M (eds) Modelling water and nutrient dynamics in soil-crop systems. Springer, Dordrecht, pp 259–266

Draper NR, Smith H (1966) Applied regression analysis. Wiley, New York

Franko U (1990) C-und N-Dynamik beim Umsatz organischer Substanz im Boden. Dissertation B thesis, Akademie der Landwirtschaftswissenschaften der DDR, Berlin

Franko U, Crocker GJ, Grace PR, Klir J, Korschens M, Poulton PR, Richter DD (1997) Simulating trends in soil organic carbon in long-term experiments using the CANDY model. Geoderma 81:109–120

Franko U, Puhlmann M, Kuka K, Böhme F, Merbach I (2007) Dynamics of water, carbon and nitrogen in an agricultural used Chernozem soil in Central Germany. In: Kersebaum KC, Hecker J-M, Mirschel W, Wegehenkel M (eds) Modelling water and nutrient dynamics in soil-crop systems. Springer, Dordrecht, pp 245–258

Genuchten MT van (1980) A closed-form equation for predicting the hydraulic conductivity of unsaturated soils. Soil Sci Soc Am J 44:892–898

Grant RF (1995) Dynamics of energy, water, carbon and nitrogen in agricultural ecosystems: simulation and experimental validation. Ecol Model 81:169–181

Grote R, Suckow F, Bellmann K (1999) Modelling of carbon-, nitrogen-, and water balances in pine stands under changing air pollution and deposition. In: Hüttl RF, Bellmann K (eds) Changes of atmospheric chemistry and effects on forest ecosystems. A roof experiment without roof. Nutrients in ecosystems. Kluwer, Dordrecht, pp 251–281

Hattermann FF, Wattenbach M, Krysanova V, Wechsung F (2005) Runoff simulations on the macroscale with the eco-hydrological model SWIM in the Elbe catchment-validation and uncertainty analysis. In: Fohrer N, Arnold JG (eds) SWAT 2000 Development and Application. Hydrological Processes 19:693–714

Jenkinson DS (1990) The turnover of organic carbon and nitrogen in soil. Phil Trans R Soc Lond B 329:361–368

Kartschall T, Döring P, Suckow F (1990) Simulation of nitrogen, water and temperature dynamics in soil. Syst Anal Model Simul 7:33–40

Klimanek EM (1987) Ernte-und Wurzelrückstände landwirtschaftlich genutzter Fruchtarten, Akademie der Landwirtschaftswissenschaften der DDR, Forschungszentrum für Bodenfruchtbarkeit Müncheberg, Müncheberg

Klimanek EM (1990a) Umsetzungsverhalten der Wurzeln landwirtschaftlich genutzter Pflanzenarten. Arch Acker-Pfl Boden 34:569–577

Klimanek EM (1990b) Umsetzungsverhalten von Ernterückständen. Arch Acker-Pfl Boden 34:559–567

Klimanek EM (1997) Bedeutung der Ernte-und Wurzelrückstände landwirtschaftlich genutzter Pflanzenarten für die organische Substanz des Bodens. Arch Acker-Pfl Boden 41:485–511

Krysanova V, Müller-Wohlfeil D-I, Becker A (1998) development and test of a spatially distributed hydrological/water quality model for mesoscale watersheds. Ecol Model 106:263–289

Krysanova V, Becker A (1999) Integrated modelling of hydrological processes and nutrient dynamics at the river basin scale. Hydrobiologia 410:131–138

Krysanova V, Wechsung F (2000) SWIM (Soil water integrated model) – User manual. 69, Potsdam Institute for Climate Impact Research (PIK), Potsdam

Lasch P, Badeck F-W, Lindner M, Suckow F (2002) Sensitivity of simulated forest growth to changes in climate and atmospheric CO_2. Forstwiss. Centralblatt 121:155–171

Loague K, Green RE (1991) Statistical and graphical methods for evaluating solute transport models: overview and application. J Contam Hydrol 7:51–73

McGechan MB, Wu L (2001) A review of carbon and nitrogen processes in European soil nitrogen dynamics models. In: Shaffer MJ, Ma L, Hansen S (eds) Modeling carbon and

nitrogen dynamics for soil management. Lewis Publishers, Boca Raton, FL, pp 103–171

Mirschel W, Wenkel K-O, Wegehenkel M, Kersebaum KC, Schindler U, Hecker J-M (2007) Müncheberg field trial data set for agro-ecosystem model validation. In: Kersebaum KC, Hecker J-M, Mirschel W, Wegehenkel M (eds) Modelling water and nutrient dynamics in soil-crop systems. Springer, Dordrecht, pp 219–243

Monteith JL (1977) Climate and the efficiency of crop production in Britain. Phil Trans R Soc Lond B 281:277–329

Monteith JL, Unsworth MH (1990) Principles of environmental physics. Edward Arnold, London

Neitsch SL, Arnold JG, Kiniry JR, Williams JR, King KW (2002) Soil and water assessment tool. Theoretical documentation, Texas Water Resources Institute, College Station, TX

Parton WJ, Schimel DS, CVC, Ojima DS (1987) Analysis of factors controlling soil organic matter levels in Great Plain grasslands. Soil Sci Soc Am J 51:1173–1179

Post J, Krysanova V, Suckow F (2004) Simulation of water and carbon fluxes in agro- and forest ecosystems at the regional scale. In: Pahl-Wostl C, Schmidt S, Rizzoli AE, Jakeman T (eds) Complexity and integrated resources management. Transactions of the 2nd biennial meeting of the International Environmental Modelling and Software Society (iEMSs), Manno, Switzerland, pp 730–736

Priestley CHB, Taylor RJ (1972) On the assessment of surface heat flux and evaporation using large scale parameters. Mon Weather Rev 100:81–92

Rawls WJ, LBD (1982) Estimating soil-water retention from soil properties. J Irrig Drain Div-ASCE 108:166–171

Ritchie JT (1972) A model for predicting evaporation from a row crop with incomplete cover. Water Resur Res 8:1204–1213

Seligman NG, Keulen H van (1991) PAPRAN – a simulation model of annual pasture production limited by rainfall and nitrogen. In: Frissel MJ, Veen JA van (eds) Simulation of nitrogen behaviour of soil-plant systems. Proc. Workshop, Wageningen, January–February 1980, pp 192–221

Smith JU, Smith P, Addiscott T (1996) Quantitative methods to evaluate and compare soil organic matter (SOM) models. In: Powlson DS, Smith P, Smith JU (eds) Evaluation of soil organic matter models–using existing long-term datasets. Series I, Global Environmental Change. Springer, Heidelberg, pp 181–199

Smith P, Smith JU, Powlson DS, McGill WB, Arah JRM, Chertov OG, Coleman K, Franko U, Frolking S, Jenkinson DS (1997) A comparison of the performance of nine soil organic matter models using datasets from seven long-term experiments. Geoderma 81:153–225

Suckow F (1984) Ein Modell zur Berechnung der Bodentemperatur. Zeitschrift für Meteorologie 35:66–70

Van't Hoff JH (1884) Etudes de dynamique chimique. Muller, Amsterdam

Wechsung F, Krysanova V, Flechsig M, Schaphoff S (2000) May land use change reduce the water deficiency problem caused by reduced brown coal mining in the state of Brandenburg. Landscape and Urban Planning 51:177–189

Williams JR, Hann RW (1972) HYMO, a problem-oriented computer language for building hydrological models. Water Resour Res 8:79–86

Williams JR, Renard KG, Dyke PT (1984) EPIC a new method for assessing erosion's effect on soil productivity. J Soil Water Conserv 38:381–383

Willmott CJ (1982) Some comments on the evaluation of model performance. Bull Am Meteorol Soc 64:1309–1313

CHAPTER ELEVEN

Modelling nitrogen dynamics in soil–crop systems with HERMES

Kurt Christian Kersebaum

Received: 25 August 2005 / Accepted: 5 July 2006
© Springer Science+Business Media B.V. 2006

Abstract Model runs with HERMES were performed over entire crop rotation cycles for two experimental sites on loamy and sandy soils in Germany with differently managed plots. The model was able to simulate soil water and nitrogen contents on the sandy plots of Müncheberg with an index of agreement (IA) >0.8 and >0.69. Crop growth and N-uptake was simulated well with IA values >0.89 and >0.75, respectively. For the loess site in Bad Lauchstädt model results for above-ground biomass, storage organ and N-uptake agreed well with observations over a 4-year rotation with IA values of 0.93, 0.94 and 0.71, respectively. Soil mineral nitrogen was significantly overestimated on the cropped plot (IA = 0.45) as well as on the black fallow plot (IA = 0.65) using the default initialization of the decomposable nitrogen pools from the organic matter content. Equilibration of the pools, using data from a neighbouring long term experiment, improved soil mineral nitrogen simulation to an IA of 0.72 for the cropped and 0.91 for the fallow plot. The long term model performance was investigated using data from 1903 to 2002 of four differently managed plots in Bad Lauchstädt. Soil organic carbon, derived from simulated nitrogen pools, showed acceptable results for the unfertilized plot, but a distinct underestimation on plots with farmyard manure application. Simulated historical winter wheat and potato yields were distinctly overestimated during the initial 50 years. Therefore, an adoption of crop parameters for older varieties is necessary. The index of agreement of 0.9 indicates that the annual weather impact on yield fluctuations was correctly reflected.

Keywords Crop growth · Nitrogen dynamics · Water dynamics

Introduction

The last century was characterized by a strong increase of agricultural production mainly due to improved nutrient availability, crop breeding and crop protection. However, these improvements were often linked to various problems like pollution of water resources by nutrients, trace gas emissions or decrease in soil fertility. Sustainability of agricultural management and land use is still a challenge. Increasingly, the evaluation of such production systems is supported by the use of agro-ecosystem models, in which the processes of water, carbon and

K. C. Kersebaum (✉)
Institute of Landscape Systems Analysis, Leibniz-Centre for Agricultural Landscape Research, (ZALF) Eberswalder Strasse 84, D-15374 Müncheberg, Germany
e-mail: ckersebaum@zalf.de

nitrogen dynamics in the soil–plant–atmosphere system play a central role. Although various models of different complexity have been developed, their application for practical purposes has been limited due to the high site specific input data requirements which are commonly not available. Additionally, there is an increasing demand for regional applications to develop regional management plans, which are required, e.g. for the EU Water Framework Directive. The input data requirements of state-of-the-art models and the scarcity and uncertainty of regional input data create a conflict, which is further exacerbated by the limitations of model validation at the regional scale. On the other hand, validation of models is required by decision makers who want to know the reliability of the model. Holling (1978) proposed to investigate the non-validity of models through the application on many different data sets without calibration. This implies the risk that the model results do not fit very well the observed data, but show also the limitations of the models as well as the uncertainty of model inputs. Within the framework of a workshop held in Müncheberg in 2004 (Kersebaum et al. 2007), the model HERMES (Kersebaum 1995) was applied using the data sets of Müncheberg and Bad Lauchstädt provided by Mirschel et al. (2007) and Franko et al. (2007) to simulate the effects of different site and management conditions on the nitrogen dynamics in the soil–crop system. The objective of this study was to evaluate the model behaviour on different temporal scales using standard parameter sets derived from basic input data. Model results achieved under these "operational" conditions were compared with those of different models, which participated at the workshop (Kersebaum et al. 2007). Model applications on a regional scale, e.g. in river basins, must consider longer historical time periods to explain the currently observed nutrient concentrations in water bodies. Hence, a second objective was to look at the long term behaviour of the model HERMES and to test how proxy inputs derived from long-term simulations can improve the model performance in short term simulations.

Material and methods

Model description

The HERMES model was developed for practical purposes (nitrogen fertilizer recommendations, nitrogen leaching assessment, land use evaluation) in the field of agriculture, water protection and land use planning. This implies that relatively simple model approaches were chosen to operate under restricted data availability. The model and the concept for model-based fertilizer recommendations has been described in detail by Kersebaum (1995) and Kersebaum and Beblik (2001). Therefore, only a brief characterization of the fundamentals will be given here.

The model simulates vertical one-dimensional processes only. The main processes considered are nitrogen mineralization, denitrification, transport of water and nitrogen, crop growth, and nitrogen uptake.

A capacity based approach is used to describe soil water dynamics (e.g. Burns 1974). The capacity parameters required by the model are attached to the model by external data files that are consistent with the German soil texture classification and their related capacity parameters (AG Boden 1994). Additionally, the AG Boden (1994) provides effective rooting depth for each textural class, which are used by the model as default values.

The daily precipitation data are corrected according to Richter (1995) for the systematic error of Hellmann measurements compared to soil surface precipitation. A crop specific potential evapotranspiration (PET) is calculated from daily weather data. Presently, two options of evapotranspiration formula are included in the model:

1. calculation of PET with the empirical HAUDE formula (Haude 1955) using crop specific monthly coefficients of Heger (1978) and the daily vapour pressure deficit at 2 o'clock p.m.,
2. calculation of a reference evapotranspiration for grass with the TURC-WENDLING formula (Wendling et al. 1991) using relative monthly kc factors to calculate crop specific

PET for other crops. The formula requires the diurnal average temperature and global radiation sum. Sunshine duration can be used instead of global radiation using the Ångström formula (Ångström 1924).

Daily net mineralization of nitrogen is simulated as a function of temperature and soil moisture content from two pools of potentially decomposable nitrogen (Richter et al. 1982), which are derived from soil organic matter and amount of crop residue related to the yield of the previous crop. As a "standard pool initialization", 13% of the total soil organic nitrogen is set as the initial value of the slowly decomposable nitrogen pool (Nuske 1983). Denitrification is modelled for the 0–30 cm depth of the soil profile using Michaelis–Menten kinetics, modified by reduction functions depending on water filled pore space and temperature. Details of the approach can be found in Kersebaum (1995).

The sub-model for crop growth was built on the basis of the SUCROS model (van Keulen et al. 1982). Its structure has been modified into a generic approach, which is able to simulate different crops using external crop parameter files. Up to 5 different crop organs and 10 development stages can be defined in the parameter file for dry matter partitioning. The yield is estimated at harvest from the weight of the storage organ, which is a dynamic result of the dry matter partitioning during crop development considering the specific growth conditions during the different crop development phases. Nitrogen recycling through crop residues is calculated automatically from the simulated crop nitrogen uptake minus the nitrogen exported at harvest with the yield and removed by-products (straw, leaves etc.).

Crop growth is limited by water and nitrogen stress. Water logging is considered by reducing transpiration and photosynthesis according to Supit et al. (1994). Water and nitrogen uptake is calculated from potential evaporation and crop nitrogen status, depending on the simulated root distribution and water and nitrogen availability in different soil layers (Kersebaum 1995). The concept of critical N concentration in plants (Greenwood et al. 1990) is used to consider nitrogen shortage. For cereals the critical N concentration in plants is described as a function of the crop development stage (Kersebaum and Beblik 2001). The generalized function for C3 crops of Greenwood et al. (1990) describing the critical N concentration as a function of the crop biomass had already been applied for potatoes by Kabat et al. (1995) and Greenwood and Draycott (1995) as well as for sugar beets (IIRB 2003). Therefore, we used this function unchanged in HERMES for both crops.

Data required by the model fall into three parts: weather data, soil information, and management data. Weather data required are daily precipitation, average air temperature, global radiation or alternatively sunshine duration, and the vapour pressure deficit at 2 o'clock p.m. if the HAUDE formula for potential evapotranspiration is used.

Soil information is required for each 10 cm of the profile. For the plough layer, the organic matter content and its C/N ratio are basic inputs. Textural class according to AG Boden (1994) for each depth increment is the most important soil input to the model. Information on groundwater depth and soil texture are used to calculate capillary rise, if applicable. Bulk density classes, organic matter content, ground water table or hydro-morphological indices and percentage of stones are used to modify the capacity parameters within the model according to AG Boden (1994). Surcharges to the capacity parameter are used with high organic matter content and shallow water table. Reductions are applied for low organic matter contents, a deep ground water table or stony soils. The hydro-morphological index, which ranges from 0 to 3 is directly added to the field capacity parameter to consider layers where water logging occurs.

Mandatory data for management are crop species, dates of sowing, harvest and soil tillage, mineral and organic nitrogen fertilization (kind of fertilizer, quantity and date of application/incorporation) and irrigation. Each fertilizer is defined by its total nitrogen content per application unit and its distribution to mineral (nitrate, ammonia) and the two organic pools (easily and slowly decomposable parts). A measured vertical distribution of mineral nitrogen content and soil moisture are used as initial values for the simulation.

Model applications to test sites

The model was applied using the data sets from the experimental sites at Müncheberg (Mirschel et al. 2007) and Bad Lauchstädt (Franko et al. 2007). The data sets focussed on different objectives: The data set from the Müncheberg experimental site and from the cropped and fallow plots of Bad Lauchstädt were used to evaluate the performance of the model's nitrogen and crop growth dynamics over relatively short (< 5 years) time periods. The long term experimental data set of Bad Lauchstädt (Franko et al. 2007) mainly aims at the long term carbon dynamics under different nutrient management. Although the HERMES model does not consider carbon dynamics and C/N interactions, it was also applied to the long term Bad Lauchstädt data set to investigate the long term behaviour of the model. The continuous simulation started in 1903 with the "standard pool initialization" based on the textural data and organic matter content. Additionally to the given data, the total specific nitrogen contents of the applied farmyard manure (Körschens 1994) were used and classified to consider the differences by three different manure types with average nitrogen contents. The different fertilization practices led to different developments of the decomposable nitrogen pools of the plots. The measurements of the organic carbon content since 1903 can be used to examine the long term nitrogen mineralization results from the model. Therefore, the total organic carbon content (C_{org}) was calculated for the upper 20 cm using the simulated nitrogen amount of the two decomposable organic nitrogen pools. The assumption of a constant C/N ratio over the whole time period is supported for this data set by Körschens (1994). A brief description of the experimental plots and the simulation variants is given in Table 1.

Basic soil data like soil texture and soil organic matter content were used to derive model parameters automatically (described above) to apply the model under "operative" conditions. Due to the availability of daily weather data, the simulations at Müncheberg were performed using the HAUDE formula for potential evapotranspiration, while in Bad Lauchstädt the TURC-WENDLING formula was applied. Daily weather data before 1956 were generated from 5-day averages by Franko et al. (2006). Because the required global radiation values were missing at Bad Lauchstädt between 1956 and 1991 the model calculated radiation from daily sunshine duration during this period. All model runs were performed throughout the whole crop rotation without any re-initialization.

The following calibrations were made for the simulation of the field trials: The given textural composition of plot 1 in Müncheberg resulted in a distinct overestimation of the water content. Due to the uncertainty caused by spatial variability and analysis error, an alternative profile was used that is similar to the profile description of the site from the medium scale soil map and to the profile descriptions of the other two plots. Compared to the "original" texture, the "calibrated" soil profile of plot 1 assumed a neighbouring texture class from AG Boden (1994), which implies in the upper depth (0–60 cm) a small shift in clay content of 1–3%. For the subsoil (60–90 cm), the "calibrated" texture was 9% lower in clay content compared to the original value. Model results, which were simulated based on this alternative profile are denoted as "calibrated". All other plots were simulated using the original soil data.

The short term data set of Bad Lauchstädt was obtained from a field neighbouring the long term data set. Soil data were identical to the long term experiment except the initial C_{org} content. For the simulations, which started in 1996, two different methods were applied to determine the initial size of the slowly decomposable organic nitrogen pool. First, the "standard pool initialization" of 13% of N_{total} was used. Because the given initial C_{org} content corresponds well to the C_{org} measurement in 1996 of the neighbouring long term experimental plot 13 (NPK), the simulated organic nitrogen in the slowly decomposable pool of this plot in 1996 (~4% of N_{total} after a 92 year simulation) was used for a second simulation, which is denoted as "equilibrated". Water contents were measured using FDR probes (Franko et al. 2007). Measurements in the upper 5 cm depth were recalibrated with gravimetric water contents, which led to significant higher water contents in the data set as without recalibration.

Table 1 Overview of investigation plots, their characteristic and simulation variants

Plot	Period	Clay/silt [%]	Management characteristics	Simulations
Müncheberg field trial (Eutric Cambisol, C_{org} 0–30 cm 0.62%)				
Plot 1 integrated	1992–1998	0–30 cm: 8/9 30–60 cm: 6/8 30–60 cm: 14/14	Cr: sb-ww-wb-wr-(or)-sb-ww mineral N fertilization	1. "original" 2. "calibrated"
Plot 2 organic	1992–1998	0–30 cm: 8/9 30–60 cm: 1/6 30–60 cm: 10/12	Cr: sb-ww-wb-wr-(or)-sb-ww Organic N fertilization (FYM and LM)	"original"
Plot 3 reduced	1992–1998	0–30 cm: 8/9 30–60 cm: 1/ 6 30–60 cm: 1/ 6	Cr: sb-ww-wb-wr-(or)-sb-ww Reduced mineral and organic N fertilization (FYM and LM)	"original"
Bad Lauchstädt long term experiment (Haplic Chernozem, C_{org} 0–20 cm 1.8 %)				
Plot 1 NPK+FYM	1903–2002	0–90 cm: 21/68	Cr: sb-spb-pt-ww straw /leaves removed 30 t FYM ha^{-1} every 2 years + mineral NPK fertilizer	"original"
Plot 6 FYM	1903–2002	0–90 cm: 21/68	Cr: sb-spb-pt-ww straw /leaves removed 30 t FYM ha^{-1} every 2 years	"original"
Plot 13 NPK	1903–2002	0–90 cm: 21/68	Cr: sb-spb-pt-ww straw /leaves removed mineral NPK fertilizer	"original"
Plot 18 no N	1903–2002	0–90 cm: 21/68	Cr: sb-spb-pt-ww straw /leaves removed no N fertilization	"original"
Bad Lauchstädt short term field trial (Haplic Chernozem, C_{org} 0–20 cm 1.85%)				
Cropped	1998–2002	0–90 cm: 21/68	Cr: sb-spb-pt-ww mineral N fertilization	1. "standard pool initialization" 2. "equilibrated"
Fallow	1998–2002	0–90 cm: 21/68	Bare soil No fertilization	1. "standard pool initialization" 2. "equilibrated"

Abbreviations: FYM = farmyard manure, LM = liquid manure, Cr = crop rotation, Sb = sugar beet (Beta vulgaris), ww = winter wheat (Triticum aestivum), wb = winter barley (Hordeum vulgare), wr = winter rye (Secale cereale), spb = spring barley (Hordeum vulgare), pt = potatoes (Solanum tuberosum), (or) = catch crop: oil radish (Raphanus sativus oleiformis)

For cereals, no changes in crop parameter were made compared to previous studies (e.g. Kersebaum and Beblik 2001; Kersebaum et al. 2002). However, the crop growth parameters of potatoes and sugar beets were calibrated because these crops were never simulated by the model before. Basic parameters, e.g. the specific leaf area per leaf mass and the light use efficiency, were taken from the literature (van Heemst 1988; Boons-Prins et al. 1993; Kooman and Spitters 1995). A manual calibration was made for the maximum photosynthesis at light saturation (A_{max}), the degree days of the development stages and the dry matter partitioning coefficients using the storage organ and above-ground biomass measurements of the short term Bad Lauchstädt crop rotation plot.

Evaluation of model performance

The following four statistical measures and indices were applied to evaluate the model: the mean bias error (*MBE*; Addiscott and Whitmore 1987)

$$\text{MBE} = \sum_{i=1}^{n} \frac{S_i - O_i}{n} \qquad (1)$$

the root mean square error (*RMSE*; Fox 1981):

$$\text{RMSE} = \sqrt{\frac{\sum_{i=1}^{n}(S_i - O_i)^2}{n}} \quad (2)$$

Index of agreement (IA) according to Willmott (1982)

$$\text{IA} = 1 - \frac{\sum_{i=1}^{n}(S_i - O_i)^2}{\sum_{i=1}^{n}\left(|S_i - \overline{O}| + |O_i - \overline{O}|\right)^2} \quad (3)$$

Modelling efficiency (ME) according to Nash and Sutcliffe (1970):

$$\text{ME} = 1 - \frac{\sum_{i=1}^{n}(S_i - O_i)^2}{\sum_{i=1}^{n}(O_i - \overline{O})^2} \quad (4)$$

where n is the number of samples, S_i and O_i are the simulated and the observed values, and $\overline{O}$ is the mean of the observed data.

RSME describes the average absolute deviation between simulated and observed values. The lower the RSME, the more accurate the simulation. The MBE considers positive and negative deviations, which make it suitable to indicate the bias of the model error. The index of agreement, IA, as an additional method for the evaluation of modelling performance results in a range between 0 and 1. The closer IA is to 1, the better the simulation quality, similar to the coefficient of determination. In contrast to IA, the modelling efficiency (ME) allows also negative values and compares the deviation between simulated and observed state variables with the variance of the observed values.

Results and discussion

Experimental field Müncheberg

Water dynamics are strongly interrelated to crop growth and processes of nutrient dynamics. Especially for sandy soils like those at the Müncheberg site, water holding capacity is very sensitive to the clay content, which is demonstrated by the relative narrow ranges of the soil texture classification and their corresponding large range of field capacity values (AG Boden 1994) of the sandy soils.

For plot 1, the use of the "calibrated" profile with a lower water holding capacity resulted in a better fit of model results to the measured water contents compared to the use of the "original" profile. Textural classes also provide information about the effective rooting depth. The shallower rooting depth of the "calibrated" profile had a strong influence on the simulation of water and nutrient uptake by crops and led to a better agreement between observations and simulation, especially for the subsoil (Table 2).

The use of the "calibrated" profile affected also the results of crop growth and nitrogen dynamics compared to the original parameter set through altered field capacity and rooting depth (Fig. 1 and Table 2). For simulated crop growth, the effects were strongest in 1994 and 1998. While in 1994 growth reduction of winter wheat was mainly due to drought stress during the summer months, the growth in 1998 was limited due to a lower simulated nitrogen content after the winter. This was caused by increased leaching through the use of the "calibrated" soil profile. In 1993, sugar beets took up less nitrogen due to the shallower rooting depth compared to the "original" soil data set, which results in an improved fit to the measured soil mineral nitrogen content (N_{min}) in the deepest layer (60–90 cm) and the whole profile (0–90 cm).

The statistical indices for the goodness-of-fit are summarized for all 3 plots on sand (Müncheberg) in Table 2. For plot 2 and 3 the model was applied without changing parameters, which were derived automatically from the given data sets. For both plots, simulated water contents in 0–90 cm agreed well with the observed values with an index of agreement (IA) of 0.8 for plot 1 and 0.89 for plot 3. However, the modelling efficiency (ME) showed low or negative values in some layers indicating a poor performance of the model at those depths, but positive values for the profile of 0–90 cm. Larger deviations (MBE) and negative values of the ME occurred in the upper

Table 2 Model performance indices (MBE = mean biased error, RMSE = root mean square error, IA = index of agreement and ME = model efficiency for different state variables of Müncheberg plot 1 (numbers in brackets: simulation with "original" soil data), plot 2 and plot 3 (* in brackets: results without first year sugar beet storage organ)

Plot		MBE	RMSE	IA	ME
1	Water content 0–30 cm [vol%]	0.9 (5.7)	2.9 (6.5)	0.88 (0.64)	0.48 (− 1.65)
	30–60 cm	−1.0 (6.8)	3.8 (8.4)	0.75 (0.46)	−0.12 (− 4.57)
	60–90 cm	−1.3 (8.4)	2.2 (10.8)	0.85 (0.40)	0.37 (− 13.72)
	0–90 cm [mm]	−4.0 (62.7)	21.5 (71.4)	0.86 (0.50)	0.39 (− 5.68)
	N_{min} 0–30 cm [kg N ha^{-1}]	−4.7 (− 5.2)	11.8 (12.1)	0.91 (0.91)	0.68 (0.67)
	30–60 cm	−2.2 (− 2.1)	10.4 (9.9)	0.51 (0.57)	−0.60 (− 0.44)
	60–90 cm	−4.5 (− 6.2)	8.9 (15.3)	0.78 (0.30)	0.43 (− 0.68)
	0–90 cm	−11.0 (− 13.2)	24.8 (31.1)	0.80 (0.72)	0.31 (− 0.01)
	Crop biomass above gr. [kg ha^{-1}]	214 (587)	1378 (1804)	0.96 (0.95)	0.86 (0.76)
	Storage organ	321 (827)	1123 (1625)	0.98 (0.96)	0.93 (0.86)
	N-uptake above gr. [kg N ha^{-1}]	13.1 (13.5)	36.5 (37.6)	0.90 (0.89)	0.53 (0.50)
2	Water content 0–30 cm [vol%]	5.5	6.8	0.61	−1.95
	30–60 cm	−1.7	2.9	0.71	0.00
	60–90 cm	−3.1	4.1	0.58	−0.73
	0–90 cm [mm]	2.0	21.6	0.80	0.38
	N_{min} 0–30 cm [kg N ha^{-1}]	−0.6	12.5	0.69	−0.38
	30–60 cm	−4.6	11.9	0.37	−0.38
	60–90 cm	−3.1	7.1	0.58	−0.04
	0–90 cm	−8.3	19.1	0.69	−0.23
	Crop biomass above gr. [kg ha^{-1}]	291	1736	0.89	0.52
	Storage organ	−1097 (384*)	2768 (1290*)	0.87 (0.97*)	0.61 (0.90*)
	N-uptake above gr. [kg N ha^{-1}]	3.1	31.2	0.75	−0.04
3	Water content 0–30 cm [vol%]	5.6	6.8	0.58	−2.62
	30–60 cm	−1.5	2.3	0.85	0.40
	60–90 cm	−2.0	2.6	0.70	−0.18
	0–90 cm [mm]	7.0	17.7	0.89	0.59
	N_{min} 0–30 cm [kg N ha^{-1}]	3.4	20.2	0.83	−0.26
	30–60 cm	0.0	8.2	0.69	−0.38
	60–90 cm	0.7	7.2	0.81	0.48
	0–90 cm	3.5	25.8	0.82	−0.04
	Crop biomass above gr. [kg ha^{-1}]	−336	1424	0.96	0.85
	Storage organ	−1830 (514*)	2965 (1317*)	0.90 (0.98*)	0.68 (0.93*)
	N-uptake above gr. [kg N ha^{-1}]	12.6	50.4	0.80	−0.23

30 cm, which were overestimated on average by 5.5 and 5.6 vol%, respectively. This indicates that the default reductions (AG Boden 1994) applied to the field capacity parameters of the plough layer did not sufficiently consider its very low organic matter contents. Soil mineral nitrogen simulation showed mostly negative ME values, but mainly IA values > 0.69. The large differences between the MBE and RMSE indicate the existence of outliers in the observed data, which is known to have a strong impact on both performance indicators (Legates and McCabe 1999). This holds also for the above-ground crop biomass and the above-ground nitrogen content. Above-ground biomass and separated results for the storage organs were in the order of magnitude, which were reported for other models (e.g. Kersebaum et al. 2007; Wegehenkel and Mirschel 2006) for the same dataset. Storage organ simulation showed distinct underestimations for plot 2 and 3, which received less nitrogen than plot 1. This was mainly caused by a drastic underestimation of the sugar beet storage organ in the first year of simulation due to nitrogen shortage simulated during beet growth. Leaving out this period, the results became much better (*values in brackets in Table 2) and achieved the same order of magnitude described for other crop growth models (e.g. Asseng et al. 2000; Bellocchi et al. 2002). Unfortunately, there were no data

Fig. 1 Comparison between simulations and measurements at Müncheberg plot 1 (1993–1998) for above-ground crop biomass, nitrogen uptake by plants and soil mineral nitrogen in 0 – 90 cm.
(dots = measurements; broken line = simulation "original" soil data; solid line = "calibrated" simulation, (or) = catch crop oil radish (Raphanus sativus oleiformis))

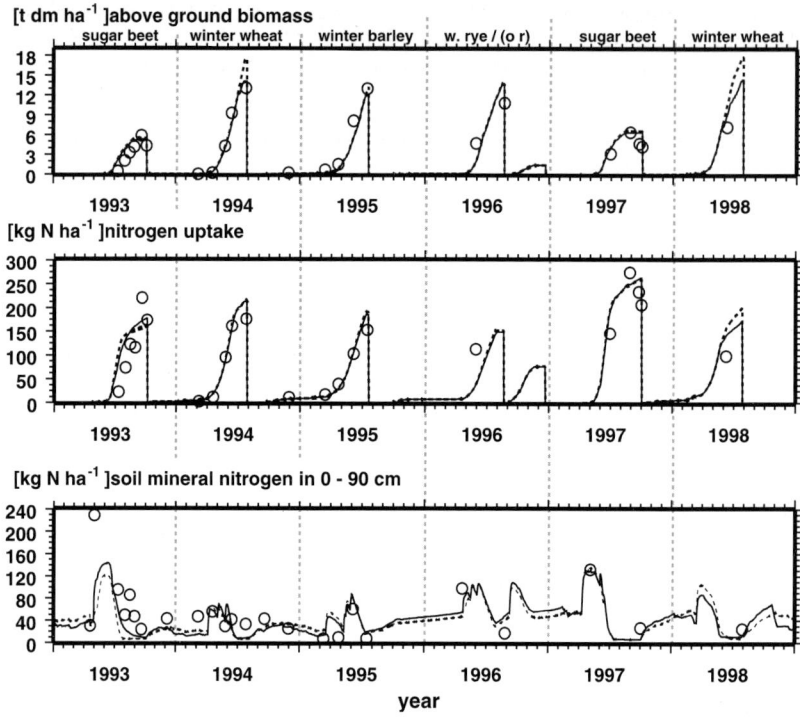

available concerning the management of the previous years to explain this deviation. High amounts of organic fertilizer or cultivation with legumes in the previous years could be a reason for a higher nitrogen release from organic matter compared to the simulation in the first year. The very high soil mineral nitrogen content (229 kg N ha^{-1}), which was measured in 1993 in plot 1 might be a further indicator to this plausible explanation.

The simulated nitrogen balances of the Müncheberg and Bad Lauchstädt plots are summarized in Table 3. Due to the limited nitrogen supply with organic fertilizers, simulated crop growth and nitrogen uptake of plot 2 was reduced compared to plot 1 (Table 3). Nevertheless, simulated nitrate leaching for those plots receiving organic fertilizers (plot 2 and 3) was higher due to the additional nitrogen release from organic N pools that happens partly during phases without

Table 3 Simulated elements of the nitrogen balance of the three Müncheberg plots (period 1 November 1993 to 31 October 1998) and the two short term plots at Bad Lauchstädt (1. November 1998 to 31 October 2002)

Process [kg N ha^{-1}]	Müncheberg Plot 1	Müncheberg Plot 2	Müncheberg Plot 3	B.Lauchstädt Short term cropped	B.Lauchstädt Short term fallow
Period	11/93–11/98	11/93–11/98	11/93–11/98	11/98–11/02	11/98–11/02
N-Mineralization	522	601	694	269	57
Mineral-N-application	625	340*	617*	360	0
N-Deposition	100	100	100	136	136
N-uptake	1051	701	995	791	0
N-leaching (90 cm)	174	224	298	3.6	131
Denitrification	0.05	0.12	0.14	1.0	2.1
Organic N-input**	485	582	682	386	0

*Including mineral portion of manure, **from manure and crop residues

crop cover. Although receiving less total nitrogen fertilization, the organic managed plot 2 showed an increase of the simulated nitrogen leaching compared to the integrated managed plot 1. This indicates that mineral fertilization allows a better temporal adaptation of the nitrogen supply on sites with low water holding capacity.

Long term experiment field Bad Lauchstädt

Simulation of long term yields

Figure 2 shows a comparison of the crop yields on plot 1 (NPK + farmyard manure) from 1905 to 2002 for winter wheat, spring barley, sugar beets and potatoes. The high fertilized plot 1 received enough nitrogen to avoid nitrogen stress for all crops. The observations show a clear trend of increasing yields for winter wheat and potatoes. For spring barley and sugar beets, the trend is less pronounced. The crop parameter sets were derived for current crop varieties. This can explain partly the overestimation of yields during the first 60 (potatoes) to 80 (winter wheat) years while at the end simulated yields agreed better with the observations.

For winter wheat, crop breeding has led to an increase of yields mainly through an increase of the harvest index (Dobben 1962; Austin 1989). Evans (1993), reviewing different investigations, showed that the harvest index for barley increased more slowly during the last century than for winter wheat. Increase of grain yields might be less pronounced for spring barley because it was mainly used for brewing. For sugar beets, crop breeding mainly aimed at a higher sugar content (Evans 1993). This might explain why there is no significant trend in the beet dry matter production. For potatoes, Henning (1988) reported an increase in production from 15 to more than 20 tons per ha between 1900 and 1980/85, which is thought to be mainly caused by an increasing availability of fertilizers and pesticides and partly by improved crop varieties. In the long term experiment always highest yielding varieties were used (Körschens 1994). Chemical crop protection was introduced in 1973, growth regulators for wheat and barley in 1975 and 1984, respectively (Körschens 1994).

Although Table 4 shows high deviations in terms of MBE and RMSE and negative ME values, the index of agreement was surprisingly high

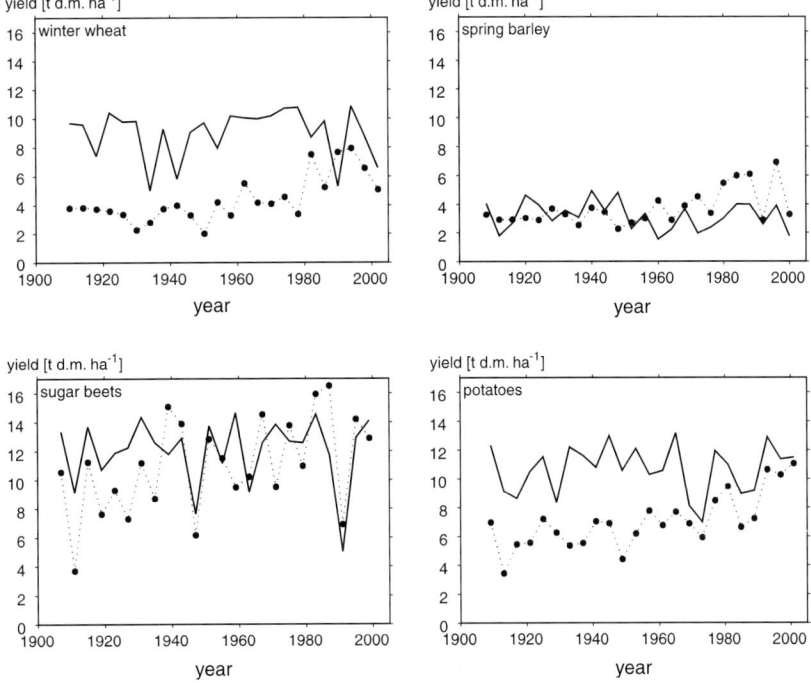

Fig. 2 Measured (dots) and simulated (solid lines) crop yields (d.m. = dry matter) of winter wheat, spring barley, sugar beets and potatoes for the long term field experiment Bad Lauchstädt on plot 1 (NPK+FYM) from 1905 to 2002

Table 4 Model performance indices for different state variables (organic carbon contents calculated from nitrogen simulation) of the long term and short term experimental plots at Bad Lauchstädt (number in brackets: model result considering the long term history, numbers with *indicate results after removal of FDR measurements during frost periods)

Plot	State variable	MBE	RMSE	IA	ME
Long term experiment					
1	C_{org} [%]	−0.41	0.42	0.33	−14.9
	Yield [kg FM ha^{-1}]	6453	11812	0.92	−3.43
6	C_{org} [%]	−0.34	0.37	0.38	−11,2
	Yield [kg FM ha^{-1}]	8719	14219	0.88	−4.06
13	C_{org} [%]	−0.12	0.15	0.43	−4.1
	Yield [kg FM ha^{-1}]	7313	12930	0.91	−3.74
18	C_{org} [%]	−0.03	0.09	0.32	−0.61
	Yield [kg FM ha^{-1}]	2332	7980	0.82	−3.35
All plots	C_{org} [%]	−0.29	0.21	0.35	−9.7
	Yield [kg FM ha^{-1}]	6190	11862	0.90	−3.69
Short term field trial					
Cropped	Water content 5 cm [vol%]	8.3 (6.2*)	10.5 (9.3*)	0.52 (0.58*)	−3.0 (− 2.5*)
	45 cm	4.4 (4.1*)	9.1 (8.8*)	0.50 (0.51*)	−7.1 (− 6.7*)
	90 cm	−2.1	4.8	0.83	0.02
	N_{min} 0–20 cm [kg N ha^{-1}]	21.3 (5.4)	27.3 (10.2)	0.45 (0.72)	10.2 (− 0.57)
	NO_3conc. 45 cm [mg NO_3 l^{-1}]	−50.6 (− 59.0)	52.2 (60.2)	0.28 (0.25)	−18.2 (− 24.6)
	90 cm	17.8 (12.2)	19.9 (13.9)	0.19 (0.18)	−38.4 (− 18.2)
	Crop biomass a.gr. [kg dm ha^{-1}]	736 (713)	1735 (1694)	0.93 (0.93)	0.62 (0.64)
	Biomass storage org.	−830 (− 854)	2854 (2833)	0.94 (0.94)	0.83 (0.83)
	N uptake above gr. [kg N ha^{-1}]	30.1 (14.0)	63.4 (42.1)	0.71 (0.80)	−1.6 (− 0.14)
Fallow	Water content 5 cm [vol%]	10.3 (9.7*)	11.5 (10.7*)	0.44 (0.46*)	−4.0 (− 3.9*)
	45 cm	4.8 (4.6*)	6.8 (6.7*)	0.13 (0.14*)	−181 (− 174*)
	90 cm	3.1	4.9	0.09	−120
	N_{min} 0–20 cm [kg N ha^{-1}]	30.9 (1.6)	34.2 (9.8)	0.65 (0.91)	−2.6 (0.71)
	NO_3 conc. 45 cm [mg NO_3 l^{-1}]	41.3 (− 1.8)	45.3 (35.1)	0.41 (0.24)	−0.73 (− 0.04)
	90 cm	35.7 (21.6)	42.5 (25.9)	0.26 (0.35)	−12.7 (− 4.1)
All plots	Water content 5 cm [vol%]	9.6 (8.0*)	11.0 (9.5*)	0.48 (0.51*)	−3.5 (− 3.2*)
	45 cm	4.6 (4.4*)	8.0 (7.4*)	0.4 (0.41*)	−11.8 (− 11.6*)
	90 cm	0.5	4.8	0.7	−1.1
	N_{min} 0–20 cm [kg N ha^{-1}]	26.1 (3.5)	30.9 (10.0)	0.59 (0.86)	−3.8 (0.49)
	NO_3 conc. 45 cm [mg NO_3 l^{-1}]	22.3 (− 13.6)	46.8 (41.5)	0.38 (0.25)	−1.3 (− 0.78)
	90 cm	30.7 (19.0)	37.6 (23.1)	0.26 (0.34)	−13.5 (− 4.5)

and classical linear regression analysis gave an r^2 of 0.53 and a slope of 0.86 for all plots and crops on a dry matter basis. This indicates that despite the systematic long term trend, the annual fluctuations and differences between the treatments were properly reflected by the model.

Long term organic matter dynamics

The decrease of the total C_{org} was reflected over most of the 99 years for the non fertilized plot 18 by the simulation (Fig. 3). The RMSE of 0.09% (Table 4) was in the order of magnitude (0.12%), which was reported by Smith et al. (1997) in a comparison of nine soil organic matter models for the same experiment over a period between 1954 and 1994. The value was also significantly below the RMSE$_{95\%}$ value for this data (0.345%) indicating that the simulation fell within the 95% confidence interval of the measured data (Powlson et al. 1998).

On the fertilized plot 1 (NPK + FYM), receiving farmyard manure, calculated C contents were obviously lower than observed (Fig. 3). The observed annual fluctuations were much higher than simulated by the model. Körschens (1994) proposed that spatial variability as well as changes of the analysis methods (dry combustion until 1996, later CNS elementary analysis) during the long period might be responsible for the high

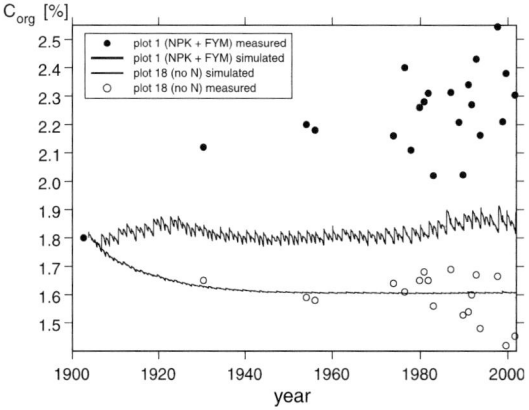

Fig. 3 Calculated organic carbon contents (from organic nitrogen simulation) and measurements on two plots of the Bad Lauchstädt long term experiment (dots: measurements, lines: simulation)

fluctuations of these measurements. The same treatment was included in the model comparison of Smith et al. (1997) who reported RMSE values of about 0.15% on average for different models applied for a shorter period between 1954 and 1994. The $RMSE_{95\%}$ was given by Powlson et al. (1998) for this data set with 0.41%, which is slightly below the achieved value of 0.42% of the HERMES model. The net-mineralization approach of HERMES was obviously insufficient to explain this increase of carbon storage if high amounts of organic matter are supplied with farmyard manure. A large proportion of carbon seems to be stored in organic compounds, which might have led to larger amounts of immobilized nitrogen especially during the first 30 years of the experiment (Körschens 1994). The above mentioned model results were achieved using the given soil texture and the related standard capacity parameters from AG Boden (1994).

Short term experiment at Bad Lauchstädt

N-uptake by crops was mostly well reflected by the "equilibrated" model run (Fig. 4a). Using the "standard pool initialization" of 13% of soil organic nitrogen to derive the slowly decomposable mineralization pool led to a clear overestimation of the soil mineral nitrogen contents in the upper 20 cm depth of the crop rotation plot (Fig. 4b) as well as in the fallow plot (Fig. 4c). Similar results are found for the simulated and observed

Fig. 4 Comparison between simulations and measurements for the short term Bad Lauchstädt crop rotation plot (1999–2002) for a) nitrogen uptake by plants, b) soil mineral nitrogen in 0 – 20 cm of the crop rotation plot and c) soil mineral nitrogen in 0 – 20 cm of the fallow plot. (dots = observations; solid lines = simulation considering history of long term plot 13 (NPK); broken lines = simulation with "standard pool initialization")

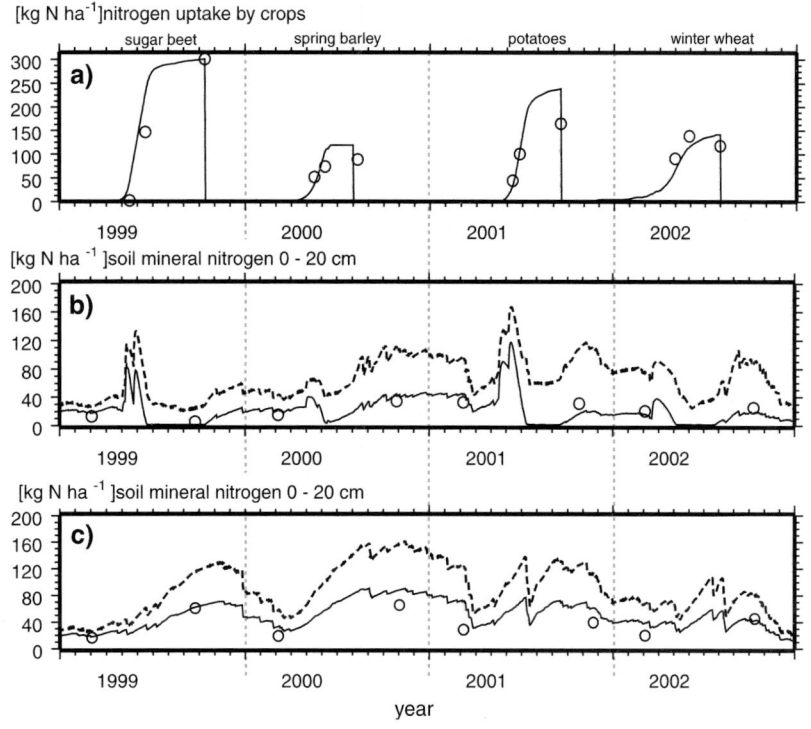

nitrogen concentrations in the deeper soil profile (MBE in 90 cm in Table 4). The results of the "equilibrated" simulation showed a better fit for the soil mineral nitrogen in the 0–20 cm depth without significant changes in crop growth and N uptake. However, ME remains negative because the measurements in spring and at harvest showed only a small variance, which is compared to the deviations of the model results. It is obvious that the temporal resolution of the measurements was not sufficient to reflect the dynamics of soil mineral nitrogen throughout the year. The difference between the cropped and the bare fallow plot was reflected well by the model. The simulated nitrogen balance of the different plots are summarized in Table 3. Nitrogen mineralization of the fallow plot was reduced compared to the cropped plot due to the lack of nitrogen supply from fertilizers and crop residues. The higher mineral nitrogen contents and reduced water depletion of the fallow plot during the summer season resulted in slightly higher denitrification during summer and higher nitrogen leaching during winter periods compared to the cropped plot (Table 3).

The MBE showed an overestimation of water contents near the surface combined with a low index of agreement and negative ME values (Table 4). Part of the overestimation can be explained by the provided FDR data, which indicated low water contents during winter. This decrease in measured soil water contents during the winter was inconsistent with the water balance. Therefore, it can be assumed that these observations represent artefacts caused by the FDR technique, which does not work properly in frozen soil. Eliminating these values for days with air temperatures lower than 0°C improves the performance in 5 and 45 cm depth (see *marked values in brackets of Table 3). Nevertheless, the deviations still indicate an overestimation of the water holding capacity in the upper part of the profile. However, a calibration based on the measured data led to inconsistencies between the simulated water contents and measured values in the lower part of the profile and was therefore rejected. Simulated nitrogen concentrations showed poor agreement with data measured by suction cups (Table 4). The nitrate concentrations in 90 cm were overestimated while those of 45 cm were underestimated. This might be explained by bypass flow in macro-pores, which might not be detected by the upper suction cup and, which was not considered by the capacity approach of the model. Nevertheless, the IA and ME values for the crop biomass and the storage organs indicate a good performance of the model regarding crop growth. Simulation of nitrogen uptake was improved by the "equilibrated" model run (MBE = 14 kg ha^{-1}) compared to the "standard pool initialization" simulation (MBE = 30 kg ha^{-1}). However, the relationship between MBE and RMSE indicates the occurrence of outliers in the measurements, which is reflected by the negative ME coefficient but a good index of agreement (IA).

Conclusions

The simulations of the short term experiment in Bad Lauchstädt and of the Müncheberg plots demonstrated that a generalized model initialization at a certain time often lacks enough information about the site's history, which might have an important bearing on the organic nitrogen pool and the following simulation of N-mineralization. Results from the Müncheberg plots suggest an influence of fairly recent management changes on the fast decomposable pool, while in Bad Lauchstädt long term historical management affected the pool size of the slowly decomposable organic nitrogen fraction. Therefore, to consider the history of a field, it is important to equilibrate the organic pool sizes during a pre-simulation run.

The results of the simulations for the unfertilized plot of the long term experiment were similar to those of several organic matter models applied to this treatment (Smith et al. 1997), although the latter models were tuned and covered only a period of 40 years. The increase of soil organic matter in the manure amended plots was underestimated, because HERMES does not consider explicitly carbon dynamics and related nitrogen immobilization. The "self generation" of crop residues from the HERMES crop growth model increases the risk of failure compared to an optimized or measured input, which is often used in soil organic matter models. Nevertheless, HERMES offers a better opportunity to simulate the interrelation between crop growth and soil organic matter than many organic matter models. However, there is a need to actualize parameters for the crop growth component to consider advances by crop breeding if longer periods have to be simulated.

For all sites the question of the representativeness of point measurements, their spatial variability and uncertainty within the field arises, which should be considered in the evaluation of model performance. Concerning regionalization, the validity of standard soil parameters, which are widely used for regional assessments, have to be discussed. The tabulated values of the AG Boden (1994), which were used here, were critically evaluated and discussed by Vorderbrügge (1997) who found some inconsistencies between the tabulated values.

Acknowledgements This contribution was supported by the German Federal Ministry of Consumer Protection, Food and Agriculture and the Ministry of Agriculture, Environmental Protection and Regional Planning of the Federal State of Brandenburg (Germany). A special thank is given to the data providers from Bad Lauchstädt and Müncheberg Experimental Stations and to the European Science Foundation and COST 718 for sponsoring the model workshop.

References

Addiscott TM, Whitmore AP (1987) Computer simulation of changes in soil mineral nitrogen and crop nitrogen during autumn, winter and spring. J Agric Sci Cambridge 109:141–157

Ångström A (1924) Solar and terrestrial radiation. Quart J Roy Meteor Soc 50:121–131

Asseng S, van Keulen H, Stol W (2000) Performance and application of the APSIM NWHEAT model in the Netherlands. Eur J Agron 12:37–54

Austin RB (1989) Genetic variation in photosynthesis. J Agric Sci Cambridge 112:287–294

Bellocci G, Silvestri N, Mazzoncini M, Menini S (2002) Using the CROPSYST model in continous rainfed maize (*Zea mais* L,) using alternative management options. Ital J Agron 6:43–56

AG Boden (1994) Bodenkundliche Kartieranleitung, 4th edn. Schweizerbart, Stuttgart

Boons-Prins ER, de Koning GHJ, van Diepen CA, Penning de Vries FWT (1993) Crop specific simulation parameters for yield forecasting across the European Community. Simulation reports CABO-TT, No. 32, Wageningen, The Netherlands

Burns IG (1974) A model for predicting the redistribution of salt applied to fallow soils after excess rainfall or evaporation. J Soil Sci 25:165–178

Evans LT (1993) Crop evolution, adaptation and yield. Cambridge University Press, Cambridge, UK

Fox DG (1981) Judging air quality model performance: a summary of the AMS Workshop on dispersion model performance. Bull Am Meteorol Soc 62:599–609

Franko U, Puhlmann M, Kuka K, Böhme F, Merbach I (2007) Dynamics of water, carbon and nitrogen in an agricultural used Chernozem soil in Central Germany. In: Kersebaum KC, Hecker J-M, Mirschel W, Wegehenkel M (eds) Modelling water and nutrient dynamics in soil-crop systems. Springer, Dordrecht, pp 245-258

Greenwood DJ, Lemaire G, Gosse G, Cruz P, Draycott A, Neeteson JJ (1990) Decline of percentage N of C3 and C4 crops with increasing plant mass. Ann Bot 66:425–436

Greenwood DJ, Draycott A (1995) Modelling uptake of nitrogen, phosphate and potassium in relation to crop growth. In: Kabat P, Marshall B, van den Broek BJ, Vos J, van Keulen H (eds) Modelling and parameterization of the soil plant system—a comparison of potato growth models. Wageningen Press, Wageningen The Netherlands, pp 155–175

Haude W (1955) Zur Bestimmung der Verdunstung auf möglichst einfache Weise. Mitt Dt Wetterd 11:1–24

Heger K (1978) Bestimmung der potentiellen Evapotranspiration über unterschiedlichen landwirtschaftlichen Kulturen. Mitt Dtsch Bodenkd Ges 26:21–40

Henning F-W (1988) Landwirtschaft und ländliche Gesellschaft in Deutschland. Bd. 2:1750 bis 1986. UTB 774, 2nd edn. Schöningh, Paderborn, Germany

Holling CS (1978) Adaptive environmental assessment and management. John Wiley and Sons, New York

IIRB (Institut International de Recherches Betteravières) (2003) Sugar beet growth and growth modelling Advances in sugar beet research, Vol. 5. IIRB, Brussels, Belgium

Kabat P, Marshall B, van den Broek BJ, Vos J, van Keulen H (eds) (1995) Modelling and parameterization of the soil plant system—a comparison of potato growth models. Wageningen Press, Wageningen, The Netherlands

Kersebaum KC (1995) Application of a simple management model to simulate water and nitrogen dynamics. Ecol Model 81:145–156

Kersebaum KC, Beblik AJ (2001) Performance of a nitrogen dynamics model applied to evaluate agricultural management practices. In: Shaffer M, Ma L, Hansen S (eds) Modeling carbon and nitrogen dynamics for soil management. Lewis Publishers, Boca Raton, USA, pp 549–569

Kersebaum KC, Hecker J-M, Mirschel W, Wegehenkel M (eds) (2007) Modelling water and nutrient dynamics in soil-crop systems. Springer, Dordrecht

Kersebaum KC, Lorenz K, Reuter HI, Wendroth O (2002) Modeling crop growth and nitrogen dynamics for advisory purposes regarding spatial variability. In: Ahuja LR, Ma L, Howell TA (eds) Agricultural system models in field research and technology transfer. Lewis Publishers, Boca Raton, USA, pp 229–252

Körschens M (ed) (1994) Der statische Düngungsversuch Bad Lauchstädt nach 90 Jahren. B.G. Teubner, Stuttgart, Germany

Legates DR, McCabe GJ (1999) Evaluating the use of, goodness of fit" measures in hydrologic and hydroclimatic model validation. Water Res Res 35:233–241

Mirschel W, Wenkel K-O, Wegehenkel M, Kersebaum KC, Schindler U, Hecker J-M (2007) Müncheberg field trial data set for agro-ecosystem model validation. In: Kersebaum KC, Hecker J-M, Mirschel W, Wegehenkel M (eds) Modelling water and nutrient dynamics in soil-crop systems. Springer, Dordrecht, pp 219-243

Nash JE, Sutcliffe IV (1970) Riverflow forcasting through conceptual model. J Hydrol 273:282–290

Nuske A (1983) Ein Modell für die Stickstoff-Dynamik von Acker-Lößböden im Winterhalbjahr-Messungen und Simulationen. Ph.D. Thesis, University of Hannover, Hannover, Germany

Powlson DS, Smith P, Coleman K, Smith JU, Glendining MJ, Körschens M, Franko U (1998) A European network of long term sites for studies on soil organic matter. Soil Till Res 47:263–274

Richter D (1995) Ergebnisse methodischer Untersuchungen zur Korrektur des systematischen Messfehlers des Hellmann-Niederschlagsmessers. Ber Dtsch Wetterd 159:93

Richter J, Nuske A, Habenicht W, Bauer J (1982) Optimized N-mineralization parameters of loess soils from incubation experiments. Plant and Soil 68:379–388

Smith P, Smith JU, Powlson DS, McGill WB, Arah JRM, Chertov OG, Coleman K, Franko U, Frolking S, Jenkinson DS, Jensen LS, Kelly RH, Klein-Gunnewiek H, Komarov AS, Li C, Molina JAE, Mueller T, Parton WJ, Thornley JHM, Whitmore AP (1997) A comparison of the performance of nine soil organic matter models using data sets from seven long-term experiments. Geoderma 81:153–225

Supit I, Hooijer AA, van Diepen CA (eds) (1994) System description of the WOFOST 6.0 crop simulation model implemented in CGMS. Vol. 1: Theory and Algorithms. EC Publication EUR 15956, Luxemburg

van Dobben WH (1962) Influence of temperature and light conditions on dry matter distribution, development rate and yield of arable crops. Neth J Agric Sci 10:37–389

van Heemst HDJ (1988) Plant data values required for simple crop growth simulation models: review and bibliography. Simulation reports CABO-TT 17, Centre for Agrobiological research and Agricultural University Wageningen, Wageningen, The Netherlands

van Keulen H, Penning de Vries FWT, Drees EM (1982) A summary model for crop growth. In: Penning de Vries FWT, van Laar HH (eds) Simulation of plant growth and crop production. PUDOC, Wageningen, The Netherlands, pp 87–97

Kooman PL, Spitters CJT (1995) Coherent set of models to simulate potato growth. In: Kabat P, Marshall B, van den Broek BJ, Vos J, van Keulen H (eds) Modelling and parameterization of the soil plant system—a comparison of potato growth models. Wageningen Press, Wageningen, The Netherlands, pp 253–274

Vorderbrügge T (1997) Vergleich von bodenphysikalischen Kennwerten der Bodenkundlichen Kartieranleitung mit gemessenen Werten. Mitt Dtsch Bodenk Ges 85/II: 1267–1270

Wegehenkel M, Mirschel W (2006) Crop growth, soil water and nitrogen balance simulation on three experimental field plots using the OPUS model—a case study. Ecol Mod 190:116–132

Wendling U, Schellin H-G, Thomä M (1991) Bereitstellung von täglichen Informationen zum Wasserhaushalt des Bodens für die Zwecke der agrarmeteorologischen Beratung. Z Meteorol 41:468–475

Willmott CJ (1982) Some comments on the evaluation of model performance. Bull Am Meteorol Soc 64:1309–1313

CHAPTER TWELVE

Calibration and validation of CERES model for simulating water and nutrients in Germany

Ajeet Singh Nain[1] and Kurt Christian Kersebaum[2]

Abstract Crop simulation models (CSM) were built to simulate the crop response to environment and soil on plot levels and were gradually applied for simulating regional behaviours of crop. Most of the models applied for simulating regional behaviour are general crop simulation models or crop-specific simulation models. The models, which use cultivar specific information, are difficult to apply on regional scale. The reason is that more than one crop cultivar are grown in even very small geographical regions. The present study is one step ahead in the direction of regional application of a crop simulation model, in which we calibrated and validated the Crop Environment Resource Synthesis (CERES) model to simulate the averaged conditions for wheat and barley. The model was calibrated to the data of two plots for the years 1993–1994 for winter wheat and for the year 1994–1995 for winter barely. The performance of the calibrated model was verified on an independent data set (one plot of 1993–1994 and three plots of 1997–1998 for winter wheat and one plots of 1994–1995 for winter barley) and the data set, which was not used

Ajeet Singh Nain and Kurt Christian Kersebaum
[1]Corresponding Author
Department of Agrometeorology, Indira Gandhi Agricultural University, Krishaknagar, Raipur-492006, Chhattisgarh, India.
Telephone: +91 771 2442557, fax: +91 771 2442557
e-mail: nain_ajeet@rediffmail.com
[2]Institute for Landscape Systems Analysis, Leibniz-Centre for Agricultural Landscape Research (ZALF) e. V. Müncheberg, D-15374, Germany.

for the calibration such as water and nitrogen dynamics in soil. The results showed that the model could very well simulate the crop variables such as crop phenology, above- and below-ground biomass and grain yield (The root mean square error (RMSE) was below 20% and the coefficient of determination (R^2) value more than 0.85, which was significant at 90% probability level). The performance of the model for simulating water dynamics in different soil layers was also good with RMSE less than 20% and a correlation coefficient more than 0.69 ($t \geq 48.10$; $p < 0.001$). However, the model showed limited accuracy in simulating nitrogen dynamics (nitrate and ammonium). The model performance for simulating nitrate content was satisfactory to some extent as the model could at least capture the trends of nitrate dynamics in different soil layers (correlation coefficient more than 0.61, which is significant on 99% probability level, $t \geq 4.41$; $p < 0.001$), but the ability of the model to simulate ammonium dynamics in soil is poor (RMSE more than 20% and correlation coefficient less than 0.23).

Keywords CERES model, model calibration, model validation, winter barley, winter wheat, yield forecast

Introduction

The use of a crop simulation (CSM) model in combination with remote sensing and geographic information systems (GIS) is increasingly getting attention for the simulation of the regional behaviour of crops. The large area simulation of crop behaviour and to forecast the crop yield requires that CSM has to be calibrated and validated so that the model could mimic the regional behaviour of the crop. Several CSMs have been developed so far on the basis of physical relationship between crop growth and weather input interaction (Whisler et al. 1986). These require crop specific parameters, soil characteristics, agricultural practices and daily weather inputs to simulate crop growth and yield. These models are generally calibrated and validated using observations from experimental plots and can mimic crop response to various inputs at plot level (Kenneth et al. 1996). With the development of the latest technology, such as remote sensing and GIS, the CSMs have been used for regional crop yield forecast (Nain et al. 2004a). The CSMs (validated for single crop cultivars) were used in different ways such as: (1) the CSM was used for simulating probable weather response of crop and then weather response was incorporated with the regional technological trend yields (Nain et al. 2000, Nain et al. 2002), (2) it was assumed that the region is subject to the yield forecast has only one single cultivar and the genetic coefficients of that particular cultivar were used to run CSM (Nain et al. 2000). The possible uses of CSM for simulating the regional behaviour of crops are prone to errors and may lead to wrong estimates of crop yield. For the precise estimation of regional crop yield, it is warranted to have the accurate information on crop cultivars and sowing dates of each field. Getting information on every field is not only exhaustive and time-consuming, but also involves huge cost. The highly advanced technique, such as remote sensing, has not been able to address this problem until now, except in few cases, where species of crops highly differ from each other (Jacobsen et al. 1995). There are some other models available, which do not require the cultivar specific information, such as WOFOST and SWAP, but their performance is limited (Eitzinger et al. 2004). Jamieson et al. (1998) compared the models AFRCWHEAT2, Crop Environment REsource Synthesis (CERES)-Wheat, Sirius, SUCROS2 and SWHEAT with measurements from wheat grown under drought and concluded that CERES-Wheat has proved to be useful for evaluating drought effects on crops at specific locations (the uncertainty of yield prediction was found to be below 10%).

In the present study, we have attempted to calibrate the model so that it could be applied to all the cultivars in the region for the simulation of regional behaviour of the crop. We have considered the CERES model, part of DSSAT 3.5 shell, for simulating wheat and barley crop. The CERES model was originally developed by Ritchie (1985) for the estimation of wheat yield in USA, but later on the model was validated and calibrated by different researchers all over the world (Hodges et al. 1987; Moulin and Beckie 1993; Toure et al. 1995; Chipanshi et al. 1997; Hundal and Kaur 1997; Timsina et al. 1998). The CERES models were successfully applied for different uses, such as crop productivity in response to nitrogen fertilizers (Kovacs et al. 1995), to derive the optimum transplanting dates under rain-fed condition (Saseendran et al.1998), for devising agronomic management strategies (Singh and Thornton 1992; Hundal and Kaur 1999; Heng et al. 2000), for the evaluation of the performance of medium range weather forecast (Singh et al. 1997), for regional crop yield forecast (Rosenthal et al. 1998, Bannayan et al. 2003; Nain et al. 2004a), for yield gap analysis (Jintrawat 1995), to predict yield trends of rice (Aggarwal et al. 2000), for pest and disease management (Pinnschmidt et al. 1995), for climate change analysis (Bachelet and Gray 1993; Singh and Ritchie 1993; Timsina et al. 1997; Jamieson et al. 2000), and aiding government policy and strategic planning (Bowen and Papajorgji 1992).

Though the CERES models have a wide applicability and were used for different purposes, yet the model has not been evaluated extensively under German conditions. The performance of a crop simulation model depends upon the ability of the model to simulate the water and nitrogen dynamics. The water balance components are indispensable for most physical and physiological processes within the soil–crop–climate system. It is important, therefore, to calculate the water budget parameters as accurately as possible to reduce uncertainties in the simulated outputs of crop and ecosystem models (Aggarwal 1995; Addiscott et al. 1995). On the other hand, nitrogen is the main nutrient, which regulate the growth and development of wheat and other cereal crops (Miller 1939). If models are to be effective for evaluating alternative management strategies, they need to reproduce accurately the seasonal variations of nitrogen found in the soil and the long-term changes in

soil properties. Nitrogen is an important component of many important structural, genetic and metabolic compounds in plant cells. It is a major component of chlorophyll, the compound by which plants use sunlight energy to produce sugars from water and carbon dioxide (i.e. photosynthesis). It is also a major component of amino acids, the building blocks of proteins. Some proteins act as structural units in plant cells, while others act as enzymes, making possible many of the biochemical reactions on which life is based. Nitrogen is a component of energy-transfer compounds, such as adenosine triphosphate (ATP), which allows the cells to conserve and use the energy released in metabolism. Finally, nitrogen is a significant component of nucleic acids such as DNA, the genetic material that allows cells (and eventually whole plants) to grow and reproduce (Eckert 2004).

Materials and methods

Wheat (here: winter wheat *Triticum aestivum* Linn.) is sown in mid October and harvested in end of July in the major wheat grown area of Germany. This time corresponds to winter and wheat is subjected to vernalization. The active growth of wheat is started after extreme winter period is over, i.e. during early March. The sowing of barley (here: winter barley, *Hordeum vulgare*) is done before the sowing of wheat and takes place during late September. Barley attains active vegetative growth stage in March and is harvested during late July. The crop calendar for wheat and barley in Germany has been given in Table 1.

Study area and data

The study was carried out at the Experimental Sration of the Leibniz-Centre for Agricultural Landscape Research (ZALF) e. V. Müncheberg, which is located at 52° 515′ E and 14° 07′ N at the elevation of 62 m from mean sea level. This region has evident cool temperate climate and sandy soils with low groundwater table (>1200 cm). More information about the study region can be obtained from Mirschel et al. (2007). The data for the present study was extracted from the data set of three plots from 1992 to 1997, which were especially designed for the crop modelling studies. In this duration, various crops were grown on all three plots ranging from oil radish (green manure) to sugar beet to winter wheat to winter barley to winter rye. As CERES can only simulate grain crops, we selected winter wheat and winter barley for our study. Winter wheat was grown twice (1993–1994 and 1997–1998) during the above said period, while winter barley was grown only once (1994–1995) on all three plots (plot 1, plot 2 and plot 3). Keeping in mind the availability of the data, we had very limited choice to use data for the calibration and validation of the CERES model. We decided to use plot 2 and plot 3 data for the year 1993–1994 for model calibration, data of plot 1 for the year 1993–1994 and data of all three plots for the year 1997–1998 for model validation for the wheat crop, while for barley we used data of plot 2 and plot 3 for the model calibration and plot 1 for the model validation for the year 1994–1995. The three wheat cultivars (Bussard, Ramiro and Greif) and three barley cultivars (Grete, Berit and Noveta) were grown during the observation period, which fulfill the basic requirement of our study (the regional application of the CERES model), as well as the basic theme (to test the performance of un-calibrated model or once calibrated) of International workshop on "Modelling water and nutrient dynamics in soil-crop systems" in which the paper was presented.

Table 1 The principal growth stages of winter wheat and winter barley in Germany

Crop stages	Winter wheat	Winter barley
Sowing	Mid October	Late September
Emergence	Early December	Mid October
Maximum growth Stage/tillering	Early March	March
Stem elongation	Mid April	Mid April
Booting	Mid May	Mid May
Flowering	Early July	Early June
Ripening/harvest	Early August	Late July

Model description

Many crop simulation models are now available to users; these range from multispecies models to suites of models with shared characteristics (Jones et al. 1995). A set of crop models that share a common input–output data format has been developed and embedded in a software package called the decision support system for agrotechnology transfer (DSSAT). DSSAT itself (IBSNAT 1989; Jones 1993; Tsuji et al. 1994) is a shell that allows the user to organize and manipulate crop, soils and weather data and to run crop models in various ways and analyse their outputs. The models running under DSSAT include the CERES models for rice, wheat, maize, sorghum, pearl millet and barley (Ritchie 1985; Ritchie and Otter 1985; Ritchie 1986); the CROPGRO model for peanut, soybean, and phaseolus bean; and a model for cassava and potato (Tsuji et al. 1994). In this study, two CERES models were used, namely CERES-Barley (Otter-Nacke et al. 1991) and CERES-Wheat (Godwin et al. 1989).

Growth and development

The CSMs simulate crop growth with a daily time step from sowing to maturity, based on physiological processes that describe the crop's response to soil and aerial environment conditions. The phase development is quantified according to the plant's physiological age. Phenological development and growth in the CERES models are specified by cultivar-specific genotype coefficients depending on the photoperiod, thermal time, temperature response, and dry matter partitioning. The CERES model calculates dry matter accumulation as a linear function based on intercepted photosynthetically active radiation. Potential dry matter accumulation depends on the amount of biomass already produced and the actual leaf area index. The actual biomass production on any day is constrained by suboptimal temperatures, soil water deficits and nitrogen and phosphorous deficiencies. The dry matter production is partitioned into different parts of the plant, on the basis of temperature and phenological stage of the crop. Rooting depth is assumed to increase each day as a function of the crop development rate and is thus predicted as the product of a constant (set for each crop) and the thermal time for the day plus the rooting depth from the previous day. Another parameter taken into account is the root-weighting factor (ranging from 0 to 1), which characterizes the potential suitability of each soil layer for root growth. It is multiplied by a soil water factor to obtain the actual root distribution. In this way, vertical distribution of soil water content is taken into account and root growth is restricted to zones with higher soil water availability.

Water dynamics

The soil water balance, nitrogen balance and phosphorus balance submodels operate on the basis of soil layers. The soil water balance component simulates surface run-off, evaporation, drainage, irrigation, and water extraction by the plant. The model utilizes the lower limit (LL-corresponding to the wilting point) and drained upper limit (DUL-corresponding to the field capacity) of plant-extractable water. The soil water content of vertical soil layers is calculated by distributing total infiltrated water among various soil layers by a simple cascading principle. Run-off is calculated by modification of the USDA soil conservation service curve number method (Williams 1991), which is suitable when only daily sums of rainfall are available. The slowest draining layer controls drainage of the water from the soil profile. If precipitation occurs on a given day, run-off is calculated by the above-mentioned method and the excess water remaining after run-off (PINF) is moved into the first layer of the soil profile. The amount of water that the soil within a particular layer can hold (HOLD) lies between the actual soil water and saturation in the particular layer. If PINF is less than or equal to HOLD, a new actual soil water content value prior to drainage is calculated. If this new actual water content value is less than the drained upper limit of water content (DUL_L), no drainage occurs into the lower layers. By contrast, if the new actual water content in the layer is greater than DUL_L, drainage into lower layers occurs according to a constant drainage rate, which is part of the input data set. Soil resistance, root resistance, and atmospheric demand control the rate of water flow into roots. The distribution of the roots within a soil layer is assumed to be uniform, and maximum daily water uptake by roots in a layer has been determined by trial and error to be 0.03 cm^3 of water per cm of root. It is also assumed that the water potential gradient between the root and the soil remains constant during the day, even when the soil dries out. The model defines the atmospherically determined demand for water (Etp) as the potential evaporation rate (Ep) determined from a version of Penman's (1948) equation modified by the current value of leaf area index (LAI). The equation used in the CERES models is as follows:

$$E_{tp} = E_p [1 - \exp * k * \text{LAI}] \qquad (1)$$

The Priestley–Taylor model (Priestley and Taylor 1972) is used to calculate E_p as the product of the equilibrium evaporation rate, calculated from solar radiation and ambient air temperature, and a factor "a". The latter was set at 1.1 for temperatures in the range 5–32°C for barley and 5–24°C for wheat. The value of factor "a" increases by 0.05 per 1°C above the upper limit of the interval, i.e. 32°C (24°C for wheat). Below 5°C the value of factor "a" decreases exponentially to a value 0.7 at 0°C. The value of k is 1 (Ritchie 1998). The evaporation from the soil surface is based on a two-stage evaporation model, which is discussed in detail by Ritchie (1972). In order to reduce the potential water uptake from the entire root zone (TWRU), a water uptake fraction (WUF) is calculated as follows:

$$\text{WUF} = E_{tp}/\text{TWRU} \qquad (2)$$

If WUF is greater than 1, the actual plant evaporation is TWRU. If WUF is less than or equal to 1, the plant is considered to be free of a water deficit for water uptake calculations. The potential biomass production rate is assumed to decrease in the same proportion as the transpiration. The potential transpiration and biomass production rates are reduced by multiplying their potential rates by the soil water deficit factor (SWDF1), which is:

$$\text{SWDF1} = \text{TWRU}/E_{tp} \qquad (3)$$

The value of SWDF1 is set to equal 1 when the ratio exceeds 1. The CERES model also calculates the second water deficit factor SWDF2 to account for water deficit effects on plant physiological processes that are more sensitive than transpiration and biomass production, such as leaf expansion or tillering. Values of SWDF2 are assumed to fall below 1 when SWDF1 is 1.5, and SWDF2 is assumed to decrease linear from 1.0 to 0.0 in proportion to this ratio.

Nutrient dynamics

The nitrogen dynamics routines of the CERES models were designed to simulate each of the major N loss processes and the contributions to the N balance made by mineralization. The routines also describe the uptake of N by the crop and the effects of N deficiency on crop growth processes. The transformations simulated are mineralization and/or immobilization, nitrification, denitrification, and urea hydrolysis. Nitrate movement, associated with water movement in both directions upward and downward direction, is also simulated. Since the rates of transformation of nitrogen are very much influenced by soil water status, the simulation of nitrogen dynamics requires that the water balance is also simulated. Soil temperature greatly influences many of the transformation rates. Therefore, a procedure to calculate soil temperature at various depths, based on the soil temperature routine (Williams et al. 1984), is also invoked in the nitrogen component of the model. The model does not simulate losses by ammonia volatilization or ammonium exchange equilibria and fixation. Under conditions of good fertilizer practice, where the fertilizer is either incorporated or placed beneath the soil surface, volatile ammonia losses should be small. The nitrogen components of the model have been constructed with the philosophy of maintaining simplicity and a state of balance with the remainder of the model (Ritchie and Otter 1985), but in the latest version of DSSAT (Jones et al. 2004), the advanced soil module based on the CENTURY model (Parton et al. 1988, 1994) have been used

The model requires the organic carbon concentration in each layer (OC_L) as an input using an assumed soil C:N ratio of 10:1 to calculate the amount of organic N associated with the organic matter (HUM_L). The contribution of recent crop residues to the supply of nitrogen in the soil is also estimated on the basis of the depth of incorporation (SDEP) of the crop residue.

The leaching of N is simulated using a simple approach based on the cascading system for drainage. Nitrate N may move between layers of the soil profile in the CERES models, but the movement of ammonium is not considered. Nitrate movement in the soil profile is highly dependent upon water movement. Therefore, the volume of water present in each layer (SW_L * DLAYR_L; where SW_L is the water content at the depth increment (L), and DLAYR_L is the thickness of the layer), and the water draining from each layer (FLUX_L) in the profile is used to calculate the nitrate lost from each layer (NOUT) based on SNO3_L (fraction of the mass of nitrate present in each layer) as follows:

$$NOUT = (SNO3\_L * FLUX\_L) / (SW\_L * DLAYR\_L + FLUX\_L) \qquad (4)$$

When the concentration of nitrate in a layer falls to 0.025 mg NO3 per 100g of soil no further leaching from that layer is deemed to occur. Most of the difference in the simulated leaching rate between soils of different texture is explained by this difference in proportion of water, which is mobile.

The CERES model simulates the decay of organic matter and the subsequent mineralization and/or immobilization of N, the nitrification of ammonium and denitrification in subroutine NTRANS. Fertilizer addition and transformations (assumed to be instantaneous) are also performed in this subroutine. Fertilizer N is partitioned in the model between nitrate and ammonium pools according to the nature of the fertilizer used.

The approach used in the CERES model for mineralization and immobilization is based on a modified version of the PAPRAN model (Seligman and van Keulen 1981). C:N ratio, water dynamics and soil temperature play a great role in mineralization and immobilization. The mineralization and immobilization routine simulates the decay of two types of organic matter: fresh organic matter (FOM), which comprises crop residues or green manure and a stable organic or humic pool (HUM). The FOM pool in each layer (L), is the addition of three sub-pools (FPOOL_L,1 = carbohydrate; FPOOL_L,2 = cellulose; FPOOL_L,3 = lignin).

This potential nitrification (process of oxidation of ammonium to nitrate) rate is a Michaelis-Menten kinetic function dependent only on the ammonium concentration and thus independent on soil type. The main factors, which limit nitrification, are: substrate NH4, oxygen, soil pH and temperature. Actual nitrification capacity is calculated by reducing the potential rate by the most limiting of the environmental indices and the capacity index. A temperature factor (TF) and a soil water factor for nitrification (WFD) are used together with the ammonium concentration factor (SANC) to determine an environmental limit of nitrification capacity (ELNC). Denitrification (opposite to nitrification) is a microbial process, which occurs under anaerobic conditions and is influenced by organic carbon content, soil aeration, temperature, and soil pH. The approach for denitrification was adopted from the functions described by Rolston et al. (1980). N uptake by plant is determined by the calculation of the components of demand and supply and then use the lesser of these two to determine the actual rate of uptake.

The phosphorus components, still under development, simulate the processes of absorption and desorption for phosphorus, organic phosphorus turnover and the dissolution of rock and fertilizer phosphate. The phosphorous components have been incorporated in the new version of DSSAT (Jones et al. 2004).

Inputs

The data encompass data on the site, where the model is to be operated, on the daily weather during the growing cycle, on the characteristics of the soil at the start of the growing cycle or crop sequence, and on the management of the crop. Required weather data encompass daily records of total solar radiation incident on the top of the crop canopy, maximum and minimum air temperature above the crop and rainfall. The CERES model requires information on the water-holding characteristics of different soil layers. It needs a root-weighting factor that accommodates the impact of several adverse soil factors on root growth in different soil layers, such as soil pH, soil impedance, and salinity. Additional soil parameters are needed for computing surface run-off, evaporation from the soil surface, and drainage (Ritchie 1972). Initial values of soil water, nitrate and ammonium are needed, as well as an estimation of the above- and below-ground residues from the previous crop. All aspects of crop management including modifications to the environment (e.g. photoperiod extension) as imposed in some crop physiology studies, also are needed. Typical crop management factors include planting date, planting depth, row spacing, plant population, fertilization, irrigation, and inoculation. Plant bed configuration and bund height are also necessary for some crops. The model also requires coefficients for the genotypes involved (Hunt 1993; Ritchie 1993). The input data required to run the CERES model have been tabulated in Table 2.

Model calibration

The major part of present study was the calibration of the CERES model so that the model could be used for the regional simulation of crop behaviour and for the yield forecast. The water module and nitrogen module have been tested successfully worldwide with different sets of soil properties and soil textures (Timsina and Humphreys 2003). The model use at a new location requires that the growth and development modules have to be validated or calibrated (if required) with cultivars of the location. The growth and development modules of the CERES model use different sets of genotype coefficients (P1V, P1D, P5, G1, G2, G3 and PHINT), which define the phenology and crop growth in time domain. We used the Hunt et al. (1993) approach for the development of a new general set of cultivar coefficients. These will not be cultivar-specific, but generalized, so that these represent all the major cultivars being grown in the region. The genetic

Table 2 Contents of minimum data sets for operation and evaluation of the CERES model

(a) For operation of model	
Site	Latitude and longitude, elevation; average annual temperature; average annual amplitude in temperature Slope and aspect; major obstruction to the sun (e.g. nearby mountain); drainage (type, spacing and depth); surface stones (coverage and size)
Weather	Daily global solar radiation, maximum and minimum air temperatures, precipitation
Soil	Classification using the local system and (to family level) the USDA-NRCS taxonomic system
Initial conditions	Basic profile characteristics for each soil layer: in situ water release curve characteristics (saturated drained upper limit, lower limit); bulk density, organic carbon; pH; root growth factor; drainage coefficient
Management	Previous crop, root, and nodule amounts; numbers and effectiveness of rhizobia (nodulating crop)
	Water, ammonium and nitrate for each soil layer
	Cultivar name and type
	Planting date, depth and method; row spacing and direction; plant population
	Irrigation and water management, dates, methods and amounts or depths
	Fertilizer (inorganic) and inoculants applications
	Residue (organic fertilizer) applications (material, depth of incorporation, amount and nutrient concentrations), tillage
	Environment (aerial) adjustments
	Harvest schedule
(b) For evaluation of models	
	Date of emergence
	Date of flowering or pollination (where appropriate)
	Date of onset of bulking in vegetative storage organ (where appropriate)
	Date of physiological maturity
	LAI and canopy dry weight at three stages during the life cycle
	Canopy height and breadth at maturity
	Yield of appropriate economic unit (e.g. kernels) in dry weight terms
	Canopy (above-ground) dry weight to harvest index (plus shelling percentage for legumes)
	Harvest product individual dry weight (e.g. weight per grain, weight per tuber)
	Harvest product number per unit at maturity (e.g. seeds per spike, seeds per pod)
	Soil water measurements vs. time at selected depth intervals
	Soil nitrogen measurements vs. time
	Soil C measurements vs. time, for long-term experiments
	Damage level of pest (disease, weeds, etc.) infestation (recorded when infestation first noted, and at maximum)
	Number of leaves produced on the main stem
	N percentage of economic unit
	N percentage of non-economic crop parts

coefficients that influence the occurrence of developmental stages in the CERES models can be derived iteratively, by manipulating the relevant coefficients to achieve the best possible match between the simulated and observed number of days to the phenological events. Other coefficients can be derived from determinations of non-limited grain weight, number of grains per panicle, and rate of grain filling. Alternatively, genetic coefficients can be determined using the GENCALC software that uses the observations of phenological events from one or several experiments from a range of environments (Hunt and Pararajasingham 1994). The DSSAT shell includes default genetic coefficients for a range of species and cultivars, and the model user can also develop estimated coefficients for local conditions from the default coefficients. Many commonly grown cultivars are not included in the DSSAT shell, many of the cultivars that are included are not commonly grown, and the source of the genetic coefficients is not provided. The DSSAT shell contains genetic coefficients for very few varieties of wheat and barley grown under German conditions. For the initialization of the model we used genotype coefficients developed by Mirschel et al. (1993) for wheat (cultivar Faktor) and default barley coefficients given in DSSAT shell for high latitude winter barley cultivar (Tsuji et al. 1994). The values of genetic coefficients for initializing the mode were: P1V = 5.0, P1D = 5.0, P5 = 8.0, G1 = 3.9, G2 = 3.0 and G3 = 3.0 for wheat, P1V = 6.0, P1D = 0.0, P5 = 5.0, G1 = 7.0, G2 = 10.0, G3 = 3.0 and PHINT = 65 for barley. The crop variables used for the calibration include phenological stages, above- and below-ground biomass and grain yield. The data of plot 2 and plot 3 for the years 1993–1994 (for wheat) and for the years 1994–1995 (for barley) were used for the calibration of the CERES model.

Model validation

Validation is an important step in model verification (Addiscott et al. 1995; Power 1993). It involves a comparison between independent field measurements (data) and outputs created by the model. Water and nitrogen contents in different soil layers, grain yield, above-ground biomass, below-ground biomass and phenological stages were considered as the evaluation parameters for the CERES model. The data of water and nitrogen content in soil for all three growing periods (1993–1994, 1994–1995 and 1997–1998) and all three plots (plot 1, 2 and 3) and data of crop phenological stages, above- and below-ground biomass and crop yield for the years 1993–1994 (plot 1) and 1997–1998 (all three plots 1, 2 and 3) for wheat and for the years 1994–1995 (plot 1) for barley were used for model validation. The model validation is based on successful calibration and validation (deviation in observed and simulated variable below 20%) on independent data sets. We used two standard criteria: the mean deviation (MD) and RMSE. Here, they are defined as: MD = $[\Sigma(S_i - O_i)]/n$ and RMSE = $[(\Sigma(S_i - O_i)^2)/n]^{1/2}$, where S_i and O_i are the time series of the simulated and observed data and n denotes the number of observations. The RMSE was then converted into % errors by dividing mean of observation and multiplying by 100 (% Error = $(RMSE/Mean_{obs})*100$). MD indicates an overall bias with the predicted variable, while RMSE quantifies the scatter between observed and predicted data, which is readily comparable with the error on the observed data. To test the significance of relationship between observed and simulated values, the $t = \hat{\beta}_j / S_{\hat{\beta}_i}$ statistics, which is based on ratio between estimated parameter value ($\hat{\beta}_j$) to the estimated parameter standard deviation ($S_{\hat{\beta}_i}$), was also determined, which is based on correlation coefficient (r) and the coefficient of determination (R^2).

Simulation run

The model was run in two environments: (1) for the calibration purpose and (2) validation purpose. The initial values of genotype coefficients for calibration purpose were adopted from Mirschel et al. (1993) for wheat and Tsuji et al. (1994) for barley, while for validation purpose new genotype coefficients (after calibration) were used. The actual hydrophysical properties of soil furnished by Mirschel et al. (2007) were prescribed for the model run. The soil observations were available up to 210 cm and the representative soil values for plot 1 have been given in Table 3. The initial conditions to start the model were derived from the observed water and nitrogen content in soil. The data on rooting factor in various soil layers were not available and were defined as 0.75 (0–30 cm), 0.50 (30–60 cm), 0.20 (60–90 cm), 0.10 (90–110 cm) and 0.0 (110–160 and 160–210 cm) on the basis of general

Table 3 Description of soil data used for the study. The soil properties of all three plots are almost similar and only data of plot 1 have been given here for representation

Horizon	Depth (cm)	Sand (%)	Silt (%)	Clay (%)	Organic carbon (%)	Total nitrogen (%)	C:N	pH	Bulk density (g/cm^3)
Ap	0–30	83	9	8	0.66	0.054	12.1	6.1	1.45
Ael	30–60	86	8	6	0.16	0.015	11.0	6.1	1.50
Bt	60–90	83	10	7	0.08	0.007	11.1	6.3	1.55
C1	90–110	72	14	14	n.m	n.m	n.m	n.m	n.m
C2	110–160	92	7	1	n.m	n.m	n.m	n.m	n.m
C3	160–210	98	1	1	n.m	n.m	n.m	n.m	n.m

n.m.: not measured

understanding of the region. The upper and lower drained limits were worked out for each soil layer from the long-term observations on water dynamics in different soil layers. The data of hydraulic conductivity and electrical conductivity were not available, so default values calculated or set by DSSAT were adopted. The daily meteorological data, which include maximum and minimum air temperature, solar radiation and rainfall, were used from the automatic weather station installed in the field. Relative humidity, which is optional for model run, was also used from the automatic weather station.

Results and discussion

In the present study we calibrated the CERES model to adopt the model for simulation of regional behaviour of wheat and barley. We used a part of the data for the model calibration and remaining independent data for the validation of model. Results have been discussed under two aspects: (a) model calibration and (b) model validation.

Model calibration

The genotype coefficients for wheat and barley were developed for the German conditions to adopt the model to simulate the regional behaviour of crops. It is important to state here that we have not changed/altered the original source code of the model. The data of plot 2 and plot 3 (1993–1994 for wheat and 1994–1995 for barley) were used to calibrate the model. The crop variables, which we considered for the model calibration, were crop phenological stages, crop biomass (above- and below-ground), and crop yield. The observed and predicted phenological stages of wheat and barley have been depicted in Fig. 1. As the experiment was not especially designed for the CERES model, only some phenological stages could be directly examined and have been given in Fig. 2. Phenological stages were converted in crop development code (DC) on the basis of Witzenberger et al. (1989) and Lancashire et al. (1991) approach. The visual interpretation of the results shows that there was a good agreement between simulated and observed phenological stages. The comparable phenological stages of wheat (Fig. 2a) with 1:1 line justify the calibration process with a RMSE of 9.3%. The value of coefficient of determination was 0.99, which is significant on 99% probability level (t = 27.19; $p < 0.001$). The results for barley were also on the same trend-line as of wheat with RMSE of 4.05% and R^2 = 0.95 (t = 10.71; $p < 0.001$). The model could not exactly simulate above- and below-ground biomass (RMSE > 20%) but could capture the temporal pattern of biomass with a coefficient of determination (R^2) of 0.94 and 0.77 for above- and below-ground biomass for wheat and of 0.98 and 0.80 for above- and below-ground biomass for barley, respectively. The pattern relationships are found to be significant at 95% probability level (t ≥ 4.5; $p < 0.05$). The temporal behaviour of simulated above- and below-ground biomass and its scatter graph shown in Fig. 3. Statistical analysis could not be done for the grain yield due to insufficient number of observations, but results show that there is deviation between observed and simulated mean grain yields of wheat and barley of about 3% (Table 4). The genotype coefficients resulting from the calibration process have been depicted in Table 5. The only difference in genotype coefficients in the case of wheat is the latter used PHINT values, which

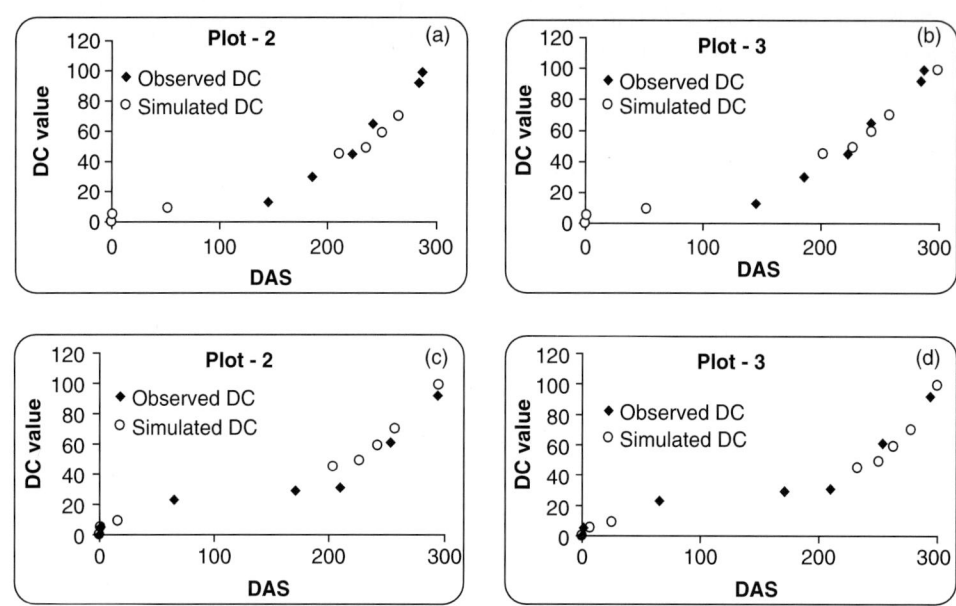

Fig. 1 Comparison between observed phenological stages and simulated phenological stages of wheat (a,b) and barley (c,d) for the calibration period. The phenological stages have been converted into crop development code (1–100 scale) according to Witzenberger et al. (1989) and Lancashire et al. (1991) from days after sowing (DAS)

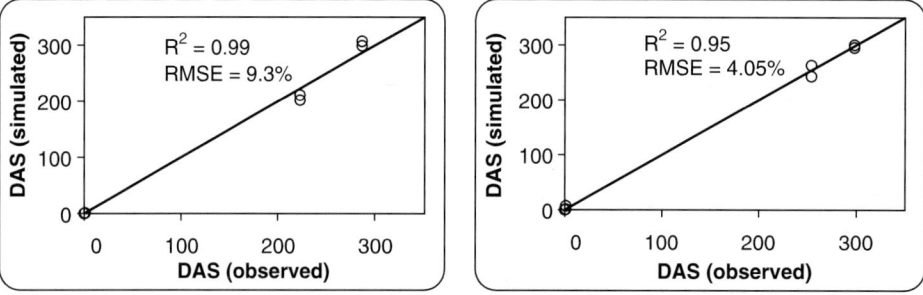

Fig. 2 Scatter plot between observed phenological stages and simulated phenological stages from days after sowing (DAS) for wheat (left) and barley (right). RMSE is the root mean square error, while R^2 is the coefficient of determination

was not used in previous studies (Mirschel et al. 1993). The coefficients of barley deviated largely from their original values as the initial coefficients were generalized and were given to represent all kinds of cultivars grown on high latitude (Tsuji et al. 1994).

Model validation

The performance of the calibrated model was evaluated against the independent data set of plot 1 (for 1993–1994) and data set of all three plots (1997–1998) for wheat and data set of plot 1 (1994–1995) for barley. Due to unavailability of large data sets the data sets of plot 2 and plot 3, which were not used for model calibration (such as nitrogen and water dynamics in soil), were also used for model validation. The crop and soil parameters, which were considered for the model evaluation are: phenological stages, grain weight, crop biomass (above- and below-ground), soil water content of different layers, and soil nitrogen content of different layers.

Crop parameters

The model could very well simulate the crop phenological stages as evidenced by Fig. 4. The curves of

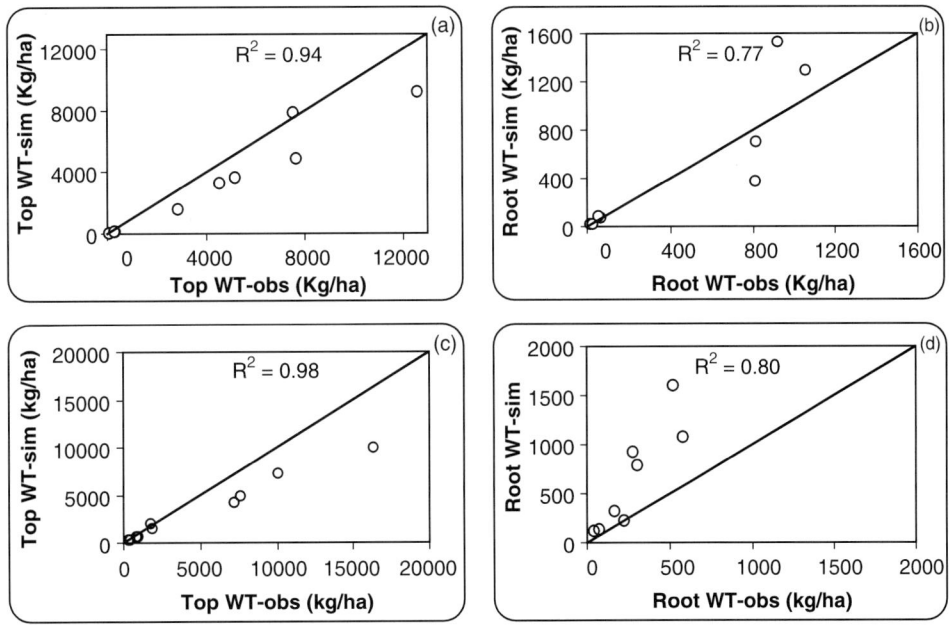

Fig. 3 The relationship between simulated above- (a) and below-ground (b) biomass for wheat and for barley (c,d), respectively. The data were used for calibration

Table 4 Comparison between observed and simulated grain yield (kg/ha) of wheat and barley for calibration data

Crop	Plot	Observed yield	Simulated yield	Deviation (%)
Wheat	Plot 2 (1993–1994)	3245	3736	−15.13
Wheat	Plot 3 (1993–1994)	4797	3675	23.39
Barley	Plot 2 (1994–1995)	3020	3248	−7.55
Barley	Plot 3 (1994–1995)	5680	5579	1.78
Mean		4185.5	4059.5	3.01

Table 5 Modified genotype coefficients for wheat and barley

Parameter	Description of parameter	Coefficient Wheat	Barley
P1V	Relative amount that development is slowed for, each day of unfulfilled vernalization, assuming that 50 days of vernalization is sufficient for all cultivars.	5.0	6.0
P1D	Relative amount that development is slowed when plants are grown in a photoperiod 1 h shorter than optimum (which is considered to be 20 h).	5.0	2.5
P5	Relative grain filling duration based on thermal time (degree days above a base temperature of 1°C), where each unit increased above zero adds 20 degree days to an initial value of 430 degree days.	8.0	5.0
G1	Kernel numbers per unit weight of stem (less leaf blades and sheaths) plus spike at anthesis (1/g).	3.9	5.0
G2	Kernel filling rate under optimum conditions (mg/day).	3.0	5.0
G3	Non-stressed dry weight of a single stem (excluding leaf blades and sheaths) and spike when elongation is ceased (g).	3.0	3.0
PHINT	Phylochron interval; the interval in thermal time between successive leaf tip appearances.	95	65

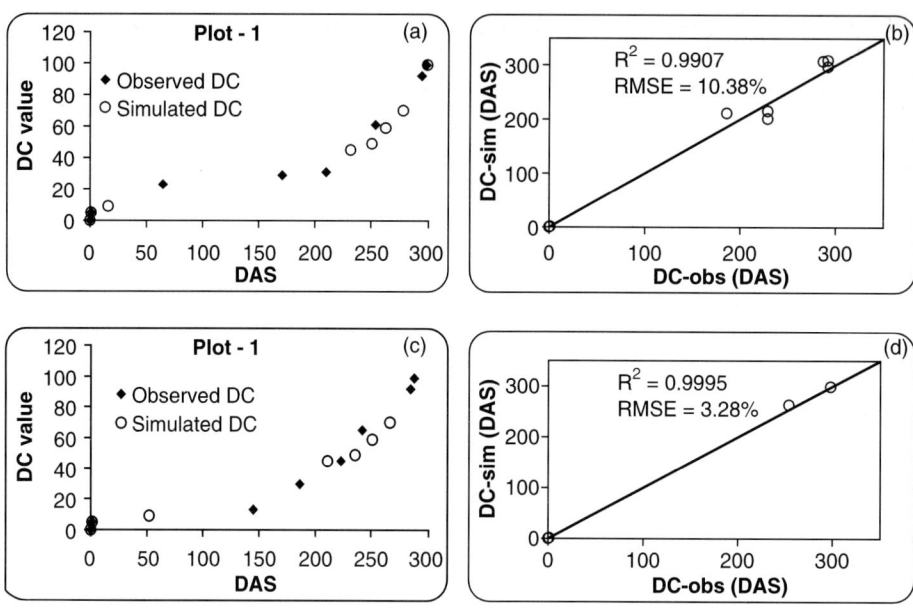

Fig. 4 Comparison between observed phenological stages and simulated phenological stages of wheat for 1993–1994 (a,b) and barley for 1994–1995 (c,d) for the validation data. The scatter plot (b) contain data from one plot of 1993–1994 and three plots of 1997–1998

phenological stages suggest that the simulated phenological stages of wheat and barley were in close agreement with the observed phenological stages. As discussed above, the predicted simulated stages of wheat and barley were not exact to that observed in phenological stages. Only few phenological stages of plot 1 (1993–1994) and plots 1, 2 and 3 (1997–1998) could be directly compared, such as sowing, germination, terminal spikelet, and maturity for wheat, and sowing, germination, end of ear growth, and maturity for barley. The difference between simulated and observed phenological stages varied from 0 to 28 days for wheat and from 0 to 11 days for barley with an average difference of 9 and 3 days, respectively. The RMSE for wheat was 10.38%, while that for barley was 3.28%. The coefficient of determination was more than 0.99 ($t \geq 6.18$; $p < 0.001$) in both cases. The observed and simulated phenological stages with scatter plots are shown in Fig. 4. The total duration of wheat was over predicted on average by 14 days, which is due to the insufficient tuning of the parameter responsible for grain filling duration (P5). With a low value of P5 the simulated yields were not matching exactly, so compensatory adjustment was done by increasing the value of P5. The average wheat yield is relatively high under European conditions (FAO 2004) and requires the additional tuning of parameters responsible for grain filling. Timsina et al. (1995) and Timsina et al. (1998) also observed the overestimation of maturity dates in Pantnagar, India by 4–9 days and by 3–6 days in subtropical northern Bangladesh.

The simulated above-ground biomass of wheat at different stages agreed with the observed biomass (Fig. 5a). A strong relationship (Fig. 5b) was found between observed and simulated biomass ($R^2 = 0.97$), which is significant at 99% probability level ($t = 10.54$; $p < 0.01$). The model could also simulate the pattern of below-ground biomass (Figure 5c,d) of wheat with $R^2 = 0.99$ ($t = 17.39$; $p < 0.05$). For simulations in Indian Punjab, Hundal and Kaur similar results were found and the simulated biomass of wheat varied from 93% to 128% of the measured biomass. Heng et al. (2000) also reported a good agreement between observed and simulated biomass in many locations across the world including India, Bangladesh and China. The simulated grain yields of wheat also showed a good agreement with observed wheat yields. The simulated wheat yield varied from 1640 to 4285 kg/ha with an average yield of 3275 kg/ha, while the observed yield varied from 1429 to 4516 kg/ha with

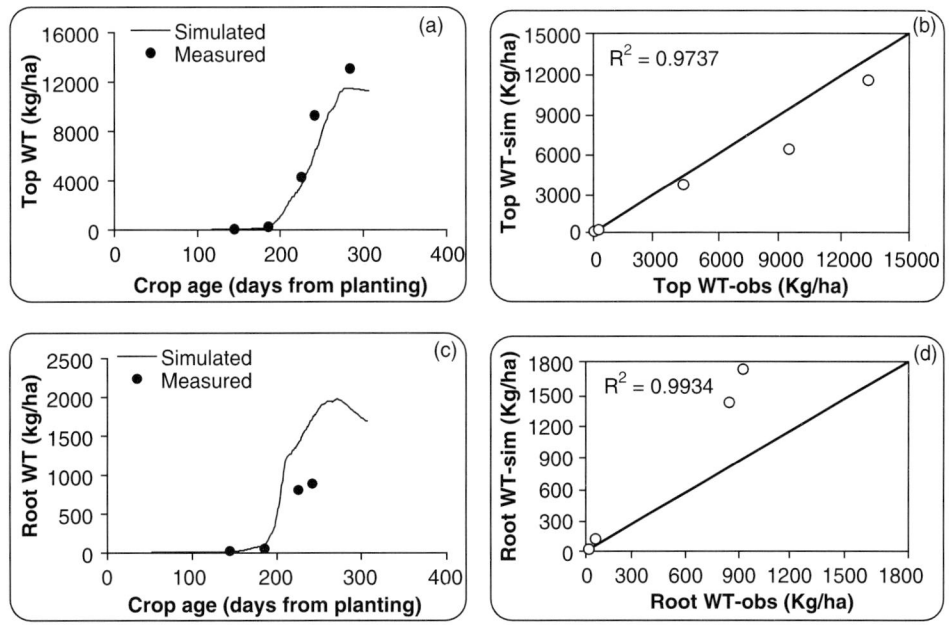

Fig. 5 Temporal pattern of simulation and measured above-ground (a) and below-ground (c) biomass of wheat and their relationships (b,d)

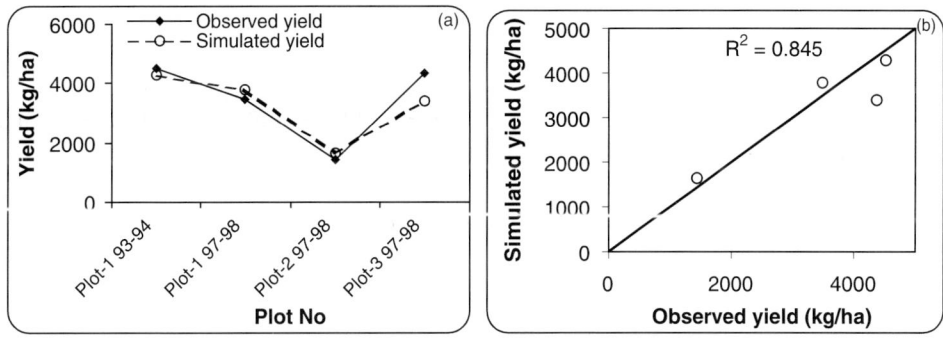

Fig. 6 Observed and simulated wheat yields (a) with their scatter plot (b)

an average yield of 3450.75 kg/ha. The RMSE between observed and simulated wheat yield was 15.45% with $R^2 = 0.85$, which was significant at 90% probability level ($t = 3.31$). The simulated and observed wheat yields and their relationship are given in Fig. 6. Hundal and Kaur (1997) found generally close agreement between observed and simulated wheat yield (cv. HD-2329) over 8 years (1985–1986 to 1992–1993) in the subtropical environment of Ludhiana, north-west India. Simulated grain yields were within 80% to 115% of measured yields. Nain et al. (2000, 2002, 2004a) applying CERES wheat model on district and regional yield forecast also found close agreement between observed and predicted wheat yield (RMSE less than 10%). The grain yield of barley could not be compared due to unavailability of observed data.

Water dynamics

The CERES model simulates water in 10 soil layers (0–5, 5–15, 15–30, 30–45, 45–60, 60–90, 90–120, 120–150, 150–180 and 180–210 cm) and have common module for simulating water dynamics for wheat, as well as barley. The observations were available only for five soil layers, i.e. 0–30 cm, 30–60 cm,

60–90 cm, 90–120 cm and 120–150 cm. Thus, for comparison, we aggregated the simulated output of soil layers 0–5, 5–15 and 15–30 (representing soil layer 0–30 cm), and 30–45 and 45–60 cm (representing soil layer 30–60 cm). From here onwards the layer 1 represents 0–30 cm soil depth, layer 2: 30–60 cm, layer 3: 60–90 cm, layer 4: 90–120 cm and layer 5: 120–150 cm of soil depth. The data from wheat and barley were analysed together. The represented graphs for all five soil layers are given in Fig. 7. The results show that the model performed very well for simulating water dynamics. The RMSE ranged from 10.12% (layer 5) to 20.59% (layer 1) with a mean error of 16.21%. The correlation coefficients for the soil layers 1, 2, 3, 4 and 5 were 0.71, 0.85, 0.69, 0.93 and 0.90, respectively (Fig. 8). The relationship between observed and simulated water content in soil layer is significant at 99% probability level ($t \geq 48.10$; $p < 0.001$). The analysis shows that RMSE for layer 1 is highest and that of layer 5 is lowest, which may be due to the fact that variation in layer 1 is highest and variation in layer 5 is lowest. The amount of rain per rainfall event at in Müncheberg site is not high and water only enters up to first or sometime second and third soil layer. The correlation coefficient of layer 5 is high and RMSE is low, which explains that the model could simulate the amount of water content and its temporal trend precisely. This also suggests that if precise input data is used without any change/modification, the model can give accurate results up to 150 cm or even more soil depth. Figure 7 shows that the model has some bias for simulating water after some rainfall events as the model showed higher amounts of water content in the upper two soil layers, which might be due to the wrong calculation of run-off amount. The run-off amount is calculated on the basis of soil texture and slope of the soil (Ritchie and Otter 1985); but rainfall intensity is not considered. The model also showed a low water content in the later phase of crop growth, which is due to the longer crop growth cycle defined through genotype coefficients. The crop gets matured early in actual condition and transpiration is stopped, while in simulation due the longer period of growth cycle the transpiration continues. The results of present study are comparable with other similar studies conducted by Xevi et al. (1996) on CERES-Maize, in which RMSE values ranged from 9.8 vol.% to 16.6 vol.% depending on the soil depth. Garrison et al. (1999) applying the same model on loamy soils reported RMSE in the range of 3.0 vol.% to 5.4 vol%.

Nitrogen dynamics

The CERES model also simulates nitrogen dynamics in 10 soil layers (e.g. water dynamics as mentioned above). The observations of nitrogen content were only available for three soil depths (0–30, 30–60 and 60–90 cm). The, simulated nitrogen contents of upper three layers have been combined to represent the soil layer 0–30 cm, and 4th and 5th simulation layers were combined to represent the soil layer 30–60 cm. The representative outputs of nitrogen dynamics in soil are shown in Fig. 9 (nitrate) and Fig. 10 (ammonium). The results suggest that the ability of the CERES model for simulating nitrate content is satisfactory to some extent, but that for ammonium is poor. The correlation coefficients for combined data set of wheat and barley for nitrate simulation in soil were 0.61, 0.74 and 0.89 ($t \geq 4.41$; $p < 0.001$) for soil layers 0–30, 30–60 and 60–90 cm, respectively. The RMSE was found to be higher (more than 20%), which suggest that the model could capture the pattern of nitrate content but could not capture the amount of nitrate in soil precisely. The nitrate content in soil was underestimated by the model in most of the cases as suggested in Fig. 9 (scatter plots). Most of comparative nitrate values are falling on the right hand side (towards observed values) from 1:1 scatter line. Gabrielle et al. (2002) found that CERES could not simulate the nitrate concentration peaks measured after fertilizer application in Villamblain. They attributed these discrepancies to a failure in some of the routines rather than to a wrong setting of their parameters.

The correlation coefficients for ammonium content in soil were 0.04, 0.09 and 0.23, for soil layer 0–30, 30–60 and 60–90 cm, respectively, which are not significant at 95% probability level. The RMSE of simulation was also high (more than 50%). This shows that the CERES model could not perform well for simulating ammonium content in different layers of soil. In most cases the amount of ammonium is stable in the 2nd and the 3rd soil layers (Fig. 10), which is due the fact that the movement of ammonium in different soil layers is not considered in the model (Ritchie and Otter 1985). The reason for inefficiency of the model

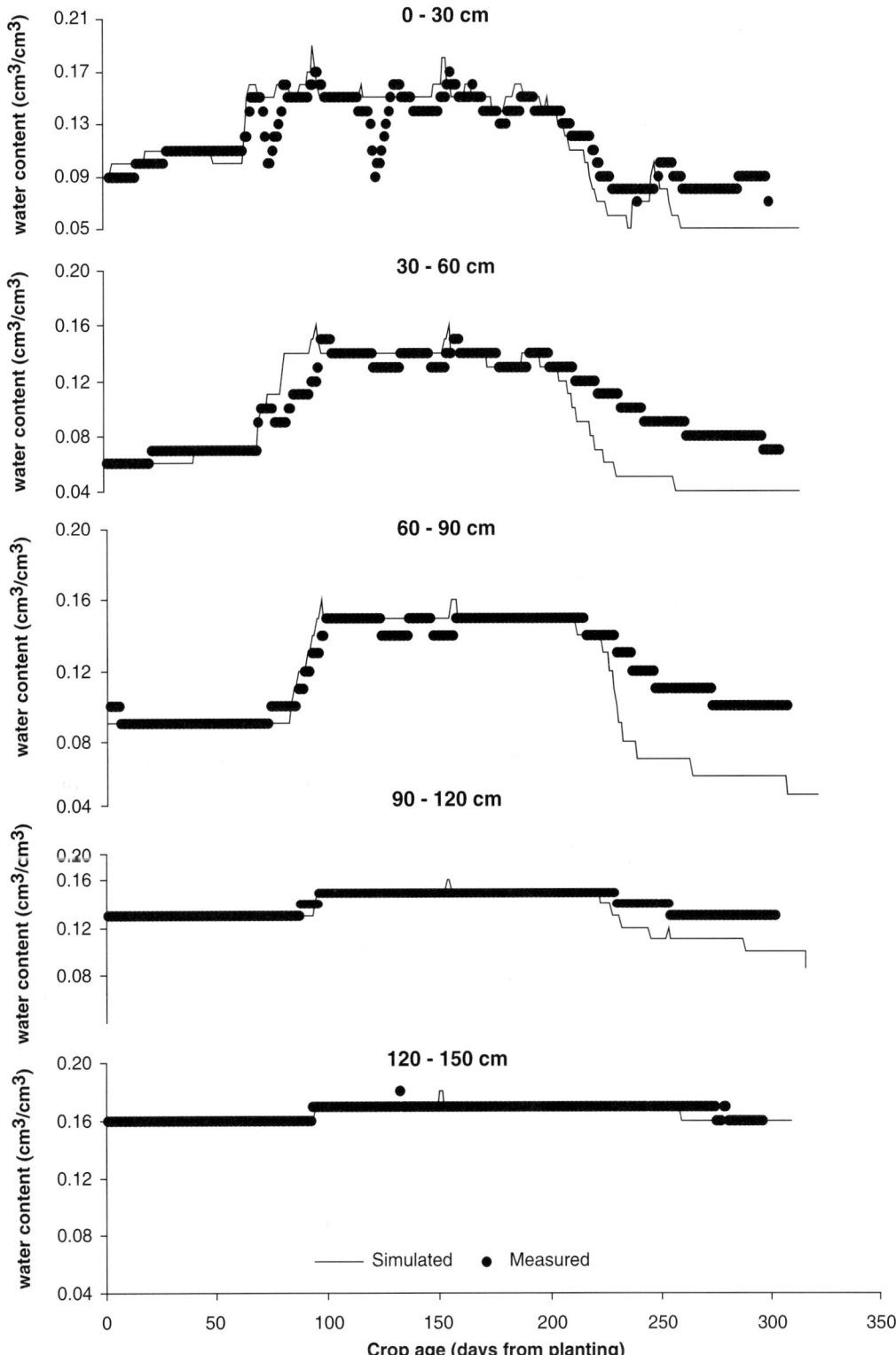

Fig. 7 Comparison of simulated and measured dynamics of water in different soil layers

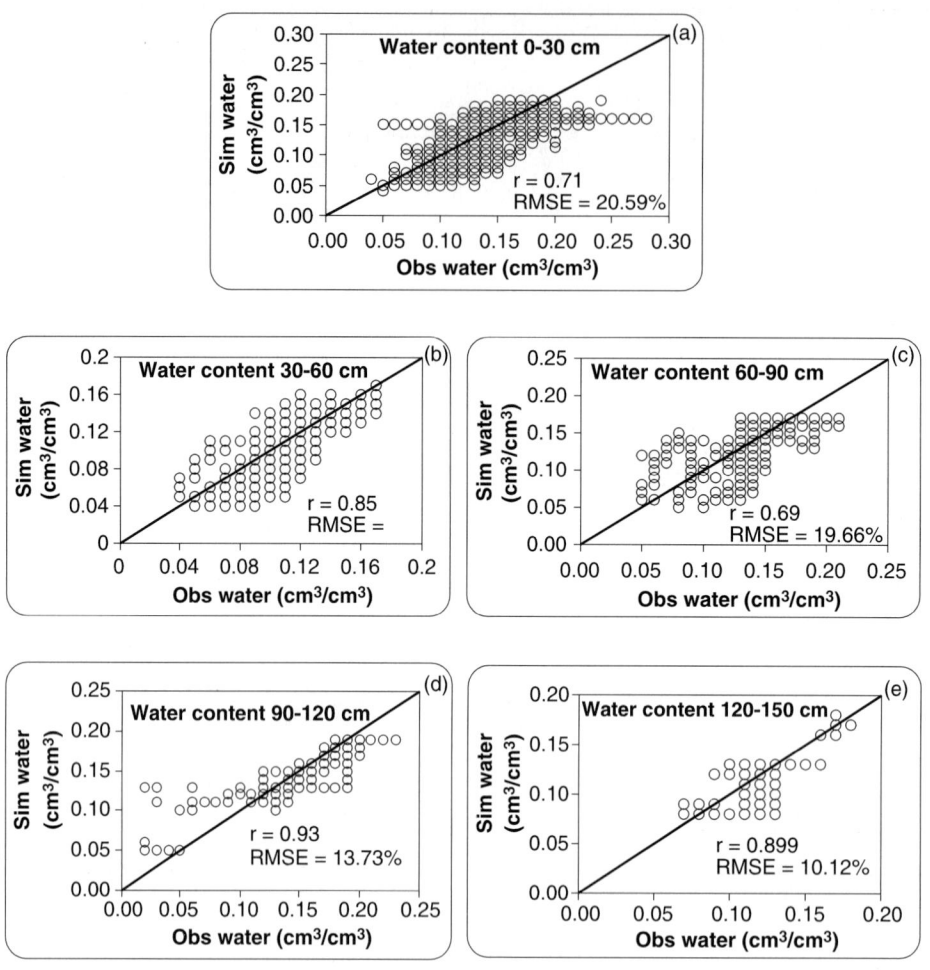

Fig. 8 Relationship between measured and simulated water content in different soil layers with correlation coefficients and RMSE (%)

to simulate the ammonium content may be attributed to: (a) the model does not simulate the loss of ammonium through volatilization (Ritchie and Otter 1985), (b) the fertilizer applied on the field contained more ammonium, which is highly volatile (Vitosh et al. 1995) and the content of ammonium in soil depends on depth of fertilizers application (Anonymous 2001) and (c) the soils of the study region are sandy (Mirschel et al. 2007) in which losses of ammonium are prominent (Vitosh et al. 1995).

Conclusions

It can be concluded from the obtained results that the CERES model can be applied for the simulation of regional behaviour of wheat and barley after a calibration process. The model showed reasonable accuracy for simulating crop variables, such as phenological stages, crop above- and below-ground biomass, and grain yield. The CERES model also showed a considerable accuracy in simulating water dynamics in different soil layers. Though the model was tested on a sandy soil, where variation of water content in different soil layers is very prominent. The model could simulate nitrate content in different soil layers satisfactorily, but the ability of the model to simulate ammonium content is poor. The next generation of the CERES model will overcome this problem because the CENTURY module has been incorporated to simulate the soil process behaviours (Jones et al. 2004). In view of the good model performance for simulating phenological stages, and grain yield and satisfactory model performance for simulating

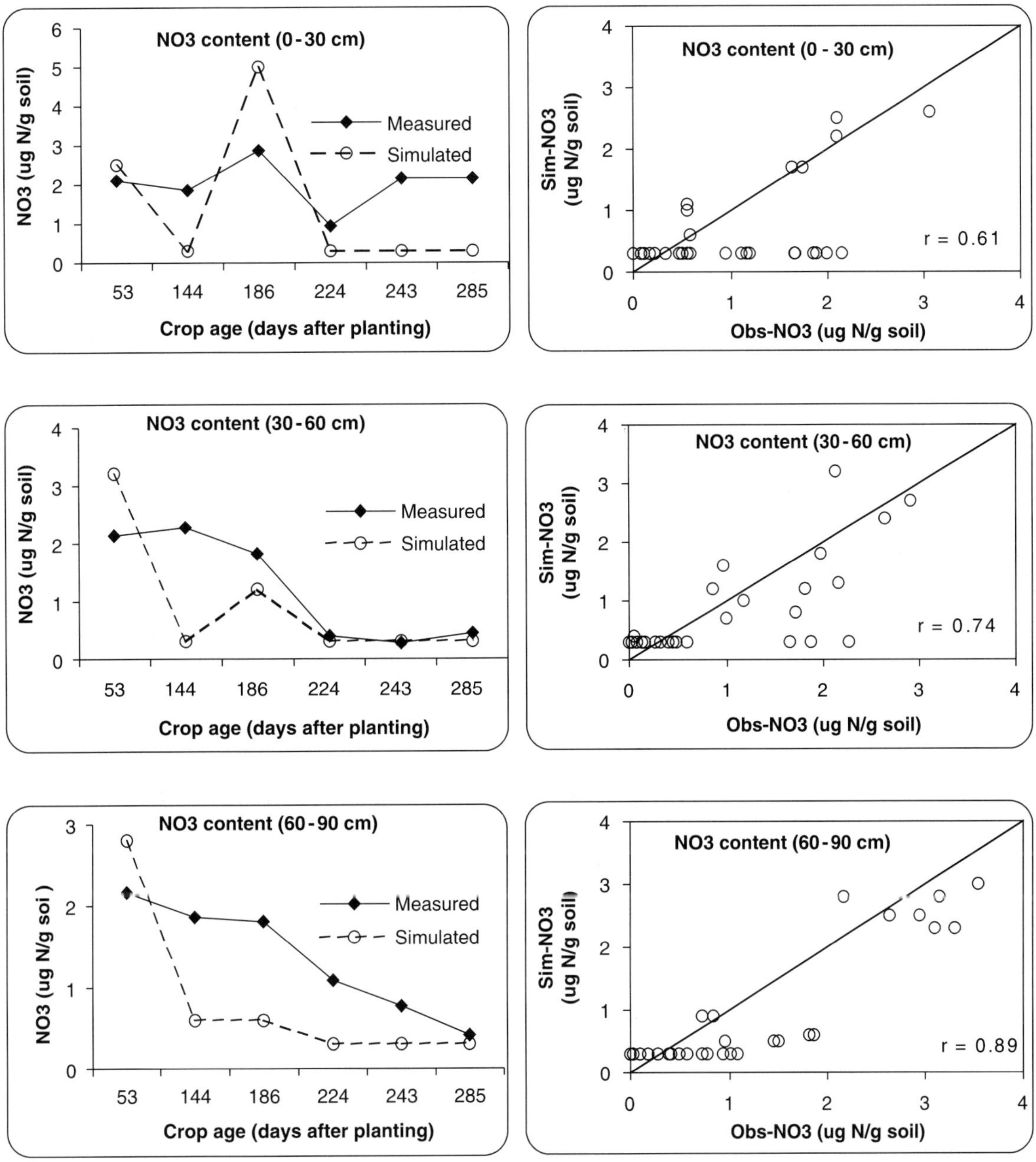

Fig. 9 The left hand graphs show observed and simulated nitrate (NO3) dynamics in different soil layers of plot 1, while right hand side graphs show their relationship (data of all three plots for the years 1993–1994, 1994–1995 and 1997–1998) on 1:1 scatter line with the correlation coefficient (r)

water content after a considered validation the model can be used for a regional simulation of crop behaviour. With the prime aim of yield forecast, the model performance towards simulation of ammonium content in soil can be tolerated. The present study was the first step of a regional application of the CERES model. Future study would be focused on the integration of the CERES model with remote sensing in a GIS environment, so that the whole system could simulate the regional behaviour of the crop precisely. Remote sensing is becoming a promising tool to generate the vital information for model inputs, such as sowing dates (Nain et al. 2004b), LAI, and crop biomass (Colwell 1983).

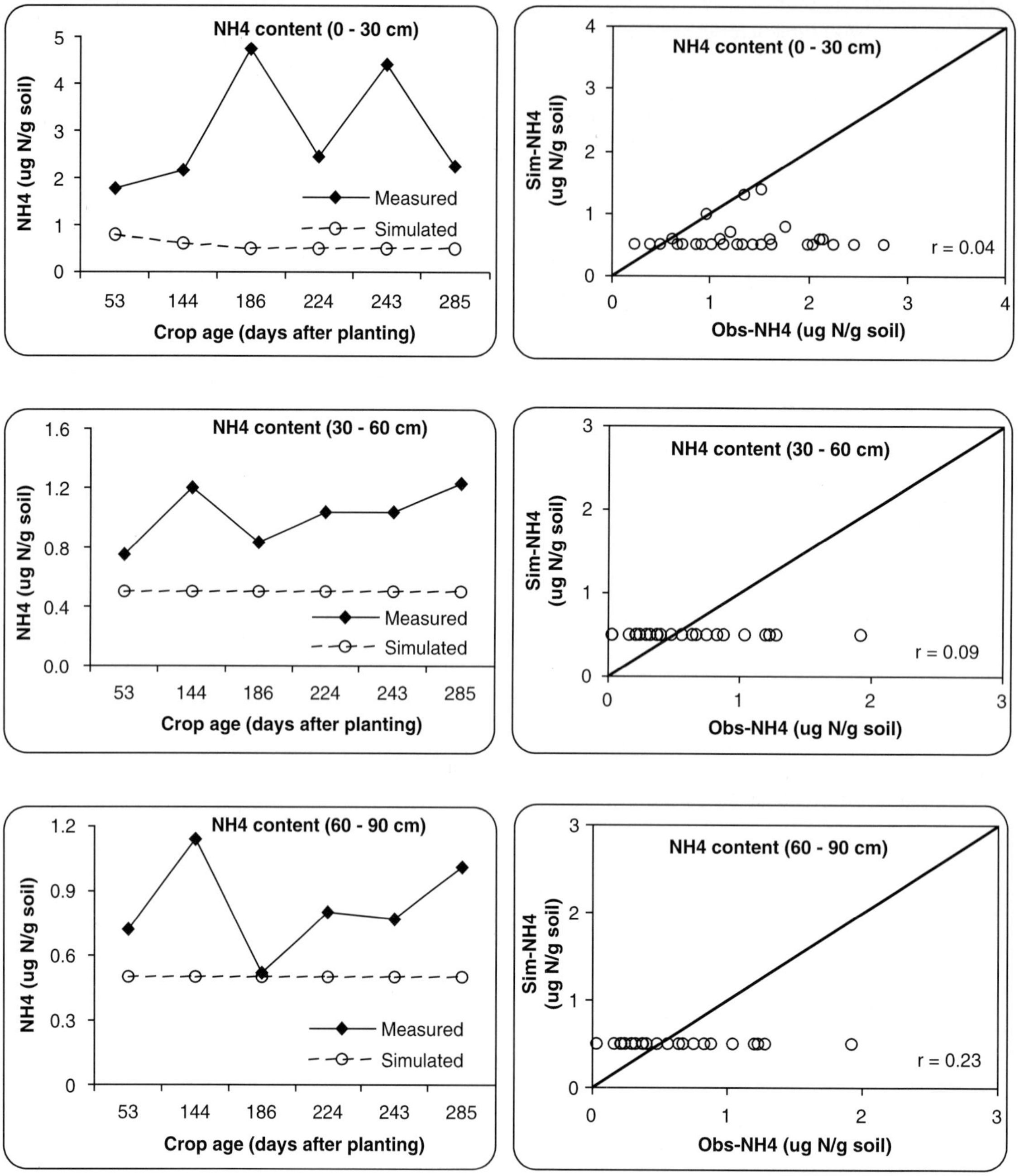

Fig. 10 The left hand graphs show observed and simulated ammonium (NH4) dynamics in different soil layers of plot 1, while right hand graphs show their relationship (data of all three plots for the years 1993–1994, 1994–1995 and 1997–1998) on 1:1 scatter line with the correlation coefficient (r)

Acknowledgements Authors acknowledge the organizer committee for organizing workshop on "Modelling water and nutrient dynamics in soil-crop systems" at ZALF Müncheberg, Germany and endowing with data for the study. Ajeet Singh Nain also acknowledges DAAD, Bonn, Germany for the grant of Advanced Research Fellowship and Leibniz-Centre for Agricultural Landscape Research (ZALF) e. V. Müncheberg, Germany for making available the facilities required for analysis.

References

Addiscott T, Smith J, Bradbury N (1995) Critical evaluation of models and their parameters. J Environ Qual 24:803–807

Aggarwal PK (1995) Uncertainties in crop, soil and weather inputs used in growth models implications for simulated outputs and their applications. Agric Syst 48(3):361–384

Aggarwal PK, Bandyopadhyay S.K, Pathak H, Kalra N, Chander S, and Kumar S (2000) Analysis of yield trends of the rice-wheat system in north-western India. Outlook on Agric 29:259–268

Anonymous (2001) Soil fertility guide. Manitoba Agriculture, Food and Rural Initiatives Publications Office 204-945-3893, USA, p 46 (available for download on http://www.gov.mb.ca/agriculture/crops/cropproduction/gaa01d25.html)

Bachelet D, Gray CA (1993) The impacts of climate change on rice yield: a comparison of four model performances. Ecol Modell 65:71–93

Bannayan M, Crout NM Jr, Hoogenboom G (2003) Application of the CERES-Wheat model for within-season prediction of winter wheat yield in the United Kingdom. Agronomy J 95:114–125

Bowen WT, Papajorghi P (1992) DSSAT estimated wheat productivity following late-season nitrogen application in Albania. Agrotechnol Transf 16:9–12

Chipanshi AC, Ripley EA, Lawford R.G (1997) Early prediction of spring wheat yields in Saskatchewan from current and historical weather data using the CERES-Wheat model. Agric For Meteorol 84:223–232

Colwell RN (1983) Manual of remote sensing, 2nd edn. American Society of Photogrammetry, Falls Church, Virginia, USA

Eckert DJ (2004) Role of nitrogen in plant, Manual of Efficient Fertilizer Use, downloaded from: http://www.imcglobal.com/general/efumanual/pdf/nitrogen.pdf on 22 September, 2004

Eitzinger J, Trnka M, Hösch J, Žalud Z, Dubrovský M (2004) Comparison of CERES, WOFOST and SWAP models in simulating soil water content during growing season under different soil conditions. Ecol Modell 171:223–246

FAO (2004) Statistical database, FAOSTAT-Agriculture, Food and Agriculture Organization of United Nations (http://faostat.fao.org/faostat/collections?subset=agriculture)

Gabrielle B, Roche R, Angas P, Cantero-Martinez C, Cosentino L, Mantineo M, Langensiepen M, Hénault C, Laville P, Nicoullaud B, Gosse G (2002) A priori parameterisation of the CERES soil-crop models and tests against several European data sets. Agronomie 22:119–132

Garrison MV, Batchelor WD, Kanwar RS., Ritchie JT (1999) Evaluation of the CERES-Maize water and nitrogen balances under tile-drained conditions. Agric Syst 62:189–200

Godwin DC, Ritchie JT, Singh U, Hunt L (1989) A user's guide to CERES Wheat-V2.1. International Fertilizer Development Center, Muscle Shoals, AL, USA

Heng LK, Baethgen WE, Moutoonnet P (2000) The collection of a minimum dataset and the application of DSSAT for optimizing wheat yield in irrigated cropping systems. In: Optimizing nitrogen fertilizer application to irrigated wheat, IAEA TECDOC-1164, pp 7–17

Hodges T, Botner D, Sakamoto C, and HaysHaug J (1987) Using the CERES-Maize model to estimate production for the U.S. Corn-belt Agric For Meteorol 40:293–303

Hundal SS, Kaur P (1997) Application of the CERES-Wheat model to yield predictions in the irrigated plains of Indian Punjab. J Agric Science (Cambridge) 129:13–18

Hundal S.S, Kaur P (1999) Evaluation of agronomic practices for rice using computer simulation model, CERES-rice. Oryza 36:63–65

Hunt LA (1993) Designing improved plant types: a breeder's viewpoint. In: Penning de Vries FWT., Teng PS. and Metselaar K. (eds.) Proceedings, Systems Approach for Agricultural Development. Kluwer Academic Press, Dordrecht, The Netherlands, pp 3–17

Hunt LA, Pararajasingham S (1993) GenCalc: Genotype coefficient calculator, user's guide, version 2.0. Crop Science Publication No. LAH-01-93, University of Guelph.

Hunt LA, Pararajasingham S, Jones JW, Hoogenboom G, Imamura DT, Goshi R.M (1993) GENCALC -Software to facilitate the use of crop models for analyzing field experiments. Agronomy J 85:1090–1094

IBSNAT 1989. Decision Support System for Agrotechnology Transfer (DSSAT) v2.1. Department of Agronomy and Soil Science, University of Hawaii, Honolulu, Hawaii

Jacobsen A, Broge NH, Hansen BU (1995) Monitoring wheat fields and grasslands using spectral reflectance data, In: Proc. of International Symposium on Spectral Sensing Research (ISSSR), Melbourne, Australia, (http://www.gsfc.nasa.gov/ ISSSR-95/monitori.html)

Jamieson PD, Bernsten J, Ewert F, Kimball BA, Olesen JE, Pinter Jr. PJ, Porter JR, and Semenov MA (2000) Modelling CO_2 effects on wheat with varying nitrogen supplies. Agric Ecosyst Environ 82:27–37

Jamieson PD, Porter JR, Goudriaan J, Ritchie JT, Keulen H. van, Stol W (1998) A comparison of the models AFR-CWHEAT2, CERES-Wheat, Sirius, SUCROS2 and SWHEAT with measurements from wheat grown under drought. Field Crops Res 55:23–44

Jintrawat A (1995) A decision support system for rapid assessment of lowland rice-based cropping alternatives in Thailand. Agric Systems 47:245–258

Jones JW (1993) Decision support systems for agricultural development. In: Penning de Vries FWT, Teng PS, Metselaar K (eds) Proceedings, Systems Approach for Agricultural Development. Kluwer Academic Press, Dordrecht, The Netherlands, pp 459–472

Jones JW, Hoogenboom G, Porter CH, Boote KJ, Batchelor WD, Hunt LA, Wilkens PW, Singh U, Gijsman AJ, Ritchie JT (2004) The DSSAT cropping system model. Europ J Agronomy 18:235–265

Jones PG, Thornton PK, Hill P (1995) Agrometeorological models: crop yield and stress indices. In: Proceedings, EU/FAO Expert Consultation on Crop Yield Forecasting Methods, Villefranche-sur-Meret Ministere de La Cooperation, FAO, Rome, pp 59–74

Kenneth JB, James WJ, Nigel BP (1996) Potential uses and limitations of crop models. Agronomy J 88:704–716

Kovacs GJ, Nemeth T, Ritchie JT (1995) Testing simulation models for the assessment of crop production and nitrate leaching in Hungary. Agric Sys 49:385–397

Lancashire PD, Bleiholder H, Langelüddecke P, Stauss R, Van den boom T, Weber E, Witzenberger A (1991) An uniform decimal code for growth stages of crops and weeds. Ann Applied Biol 119:561–601

Miller EC (1939) A physiological study of the winter wheat plant at different stages of its development. Kansas Agric Exp Stn Tech Bull 47

Mirschel W, Wenkel K-O, Wegehenkel M, Kersebaum KC, Schindler U, Hecker J-M (2007) Müncheberg field trial data set for agro-ecosystem model validation. In: Kersebaum KC, Hecker JM, Mirschel W, Wegehenkel M (eds) Modelling water and nutrient dynamics in soil-crop systems. Springer, Dordrecht, pp 219–243

Mirschel W, Schultz A, Wenkel KO (1993) Vergleich der Winterweizenmodelle AGROSIM-Wheat und CERES-Wheat. In: Schulze E, Petersen B, Geidel H (eds) Referate der 14. GIL-Jahrestagung, Ökologie und Informatik-Neu Impulse für die Landwirtschaft, Leipzig, 04-06 Oktober, Bericthte der Gesellschaft für Informatik in der Land, Forst und Ernährungswirtschaft, Bd. 5, pp 29–34

Moulin AP, Beckie HJ (1993) Evaluation of the CERES and EPIC models for predicting spring wheat grain yield over time. Can J Plant Sci 73:713–719

Nain AS, Dadhwal VK, Singh TP (2000) Use of CERES-Wheat model for predicting wheat yields of Nainital district (UP), India J Agrometeorology 2:113–122

Nain AS, Dadhwal VK, Singh TP (2002) Real time wheat yields assessment using technology trend and crop simulation model with minimal data set. Curr Sci 82:1255–1258

Nain AS, Dadhwal VK, Singh TP, (2004a) Use of CERES-Wheat model for wheat yield forecast in central Indo-Gangetic Plains of India, J Agric Science (Cambridge) 142:59–70

Nain AS, Kersebaum KC, Dadhwal VK (2004b) Use of satellite imagery for estimating sowing dates of wheat in Uttaranchal, India. In: Ubertini L (ed) Environmental modelling and simulation. IASTED Intern. Conference, St. Thomas, Virgin Islands, USA. Acta Press, Anaheim, pp 139–144

Otter-Nacke S, Ritchie JT, Godwin DC, Singh U (1991) A user's guide to CERES Barley-V2.1. International Fertilizer Development Center, IFDC-SM-3, USA

Parton WJ, Ojima DS, Cole CV, Schimel DS (1994) A general model for soil organic matter dynamics: sensitivity to litter chemistry, texture and management. In: Bryant R.B, Arnold (eds) Quantitative modelling of soil forming processes. Special Publication 39, SSSA, Madison, WI, pp 147–167

Parton WJ, Stewart JWB, Cole CV (1988) Dynamics of C, N, P and S in grassland soils: a model. Biogeochemistry 5:109–131

Penman H.L (1948) Natural evaporation from open water, bare soil and grass. In: Proc R Soc Ser A 193:120–146

Pinnschmidt HO, Batchelor WD, Teng PS (1995) Simulation of multiple species pest damage in rice using CERES-rice. Agri Syst 48:193–222

Power M (1993) The predictive validation of ecological and environmental models. Ecol Model 68:33–50

Priestley CHB, Taylor RJ (1972) On the assessment of surface heat flux and evaporation large scale parameters. Monthly Weather Rev 100(2):81–82

Ritchie JT (1972) Model for predicting evaporation from a row crop with incomplete cover. Water Resour Res 8:1024–1213

Ritchie JT (1985) A user-oriented model of the soil water balance in Wheat. In: Day W, Atkins RK (eds) Wheat growth and modelling, Plenum Publishing Corporation, NATO-ASI Series, pp 293–305

Ritchie JT (1986) CERES-Wheat: a general documentation. USDA–ARS. Grassland, Soil and Water Resource Laboratory, Temple, Texas

Ritchie JT (1993) Genetic specific data for crop modelling. In: Penning de Vries FWT, Teng PS, Metselaar K (eds) Proceedings, Systems Approach for Agricultural Development. Kluwer Academic Press, Dordrecht, The Netherlands, pp 77–93

Ritchie JT (1998) Soil water balance and plant water stress. In: Tsuji GY, Hoogenboom G, Thorton PK (eds) Understanding options for agricultural production. Kluwer Academic Publishers, Dordrecht, The Netherlands, pp 41–55

Ritchie JT, Otter S (1985) Description and performance of CERES-Wheat: a user oriented wheat yield model. In: ARS Wheat Yield Project, ARS-38. National Technical Information Service, Springfield, VA, pp 159–175

Rolston DE (1981) Nitrous oxide and nitrogen gas production in fertilizer loss. In: Delwiche C.C (ed) Denitrification, nitrification and atmospheric nitrous oxide, J Wiley, New York, pp 127–149

Rosenthal WD, Hammer GL, Butler D (1998) Predicting regional grain sorghum production in Australia using spatial data and crop simulation modelling. Agric For Meteorol 91:263–274

Saseendran SA, Hubbard KG, Singh KK, Mendiratta N, Rathore LS, Singh SV (1998) Optimum transplanting dates for rice in Kerala, India, Determined using both CERES v3.0 and Climprob. Agronomy J 90:185–190

Seligman NC, Keulen H van (1981) PAPRAN: a simulation model of annual pasture production limited by rainfall and nitrogen. In: Frissel MJ, Veen JA van (eds) Simulation of nitrogen behaviour of soil–plant systems. Centrum voor Land-bouwpublikaties en Landbouwdocumentatie (PUDOC), Wageningen, The Netherlands, pp 192–221

Singh KK, Kalra N, Mohanty UC, Rathore LS (1997) Performance evaluation of medium-range weather forecast using crop growth simulator. J Environ Sys 25(4): 397–408

Singh U, Ritchie JT (1993) Simulating the impact of climate change on crop growth and nutrient dynamics using the CERES-rice model. J Agr Met 48(5):819–822

Singh U, Thornton PK (1992) Using crop models for sustainability and environmental quality assessment. Outlook Agric 21:209–218

Timsina J, Humphreys E (2003) Performance and application of CERES and SWAGMAN Destiny models for rice-wheat cropping systems in Asia and Australia: a review. CSIRO Land and Water Technical Report 16/03. CSIRO Land and Water, Griffith, NSW 2680, Australia

Timsina J, Adhikari B, Ganesh KC (1997) Modelling and simulation of rice, wheat, and maize crops for selected sites and the potential effects of climate change on their productivity in Nepal. Consultancy Report submitted to Ministry of Agriculture, Harihar Bhawan, Kathmandu, Nepal

Timsina J, Singh U, Badaruddin M, Meisnar C (1998) Cultivar, nitrogen and moisture effects on rice-wheat sequence: experimentation and simulation. Agronomy J 90:119–130

Timsina J, Singh U, Singh Y, Lansigan FP (1995) Addressing sustainability of rice-wheat systems: testing and applications of CERES and SUCROS models. In: Proc Int Rice Res Conf 13–17 February IRRI, Los Banos, Philippines, pp 633–656

Toure A, Major DJ, Lindwall CW (1995) Comparison of five wheat simulation models in southern Alberta. Canadian J Plant Sci 75(1):61–68

Tsuji GY, Uhera A, Balas S (1994) DSSAT v3. Department of Agronomy and Soil Science, University of Hawaii, Honolulu, Hawaii

Vitosh ML, Johnson JW, Mengel DB (1995) Tri-state fertilizer recommendations for corn, soybeans, wheat and alfalfa. Extension Bulletin E-2567 (New), July 1995

Whisler FD, Acock B, Baker D.N, Fye RE, Hodges HF, Lambert JR, Lemmon HE, McKinion JM, Reddy VR (1986) Crop simulation models in agronomic systems. Advances in Agronomy 40:141–208

Wiliams JR (1991) Runoff and water erosion. In: Hanks RJ, Ritchie JT (eds) Modelling plant and soil systems. Agronomy Monograph: 31, American Society of Agronomy, Madison, WI, USA, pp 439–455

Williams JR, Jones CA, Dyke PT (1984) A modelling approach to determining the relationship between erosion and soil productivity. Trans ASAE 27:129–144

Witzenberger A, Hack H, van den boom T (1989) Erläuterungen zum BBCH-Dezimal-Code für die Entwicklungsstadien des Getreides - mit Abbildungen. Gesunde Pflanzen 41:384–388

Xevi E, Gilley J, Feyen J (1996) Comparative study of two crop yield simulation models. Agric Water Manage 30:155–173

CHAPTER THIRTEEN

The impact of crop growth sub-model choice on simulated water and nitrogen balances

Eckart Priesack*, Sebastian Gayler and Hans P. Hartmann
*GSF National Research Center for Environment and Health, Institute of Soil Ecology, PF1129 Neuherberg, Oberschleißheim, D-85758, Germany; *Author for correspondence (e-mail: priesack@gsf.de)*

Received 26 August 2005; accepted in revised form 23 January 2006

Key words: Crop model, Modular simulation, Nitrogen balance, Water dynamics

Abstract

It is the aim of this study to analyse how different crop growth model routines affect the simulation of water flow and nitrogen transport of a crop rotation in agricultural fields. The model system Expert-N is briefly described and used to test the crop growth sub-models against data of a six-year field experiment on sandy soils. Expert-N is a modular soil–plant–atmosphere model system, which comprises different sub-models to simulate one-dimensional vertical transport of water, solute and heat in the unsaturated zone. It includes several sub-models to describe organic matter turnover and has three generic crop growth sub-models. The latter are derived from the crop models CERES, SPASS and SUCROS. Simulations were performed using the different sub-models for each of the cereal crops in the sugar beet, winter wheat, winter barley, winter rye crop rotation. Results show the impact of crop model choice on simulated water balances and turnover of C and N. It is concluded that the simulation of root growth and plant residue mineralisation needs some improvement.

Introduction

Due to their complexity and close interrelationship it is often difficult to describe the fluxes of energy, water, carbon and nitrogen through soil–plant–atmosphere systems. The need to understand the dynamics of these fluxes led to the development of numerous soil–plant–atmosphere system models that can simulate these flow and transport processes and their coupling (De Willigen 1991; Engel et al. 1993; Diekkrüger et al. 1995; Smith et al. 1997; Shaffer et al. 2001; van Ittersum and Donatelli 2003). In particular, soil–plant–atmosphere models are used to predict plant growth and mineralisation of N from soil organic matter aiming at a better adaptation of fertiliser applications to plant demand and at a reduction of nitrate leaching to groundwater. These models may also be applied to describe spatial processes for optimising agricultural management of spatially variable fields (Booltink et al. 2001; Munch et al. 2001; Kersebaum et al. 2002).

The models vary widely in their scope and complexity, and in the way the processes are modelled: more by empirical relationships or more by using fundamental scientific principles (Engel et al. 1993; Grant 1995). The main reason is, that they were developed for different purposes, e.g. for management or research, but also because they are "strongly influenced by the environment, training and preoccupations of their developers" (Addiscott and Wagenet 1985). As a consequence, simulation results from one model usually cannot be reproduced exactly by a somewhat different model when considering the same modelling object, i.e. the same field site for the same time period. This indicates a lack of reproducibility, which reflects the uncertainty due to the differences in model structures and related model parameterisation.

The structural uncertainty of soil–plant–atmosphere models can be analysed using a highly modular modelling system in which different exchangeable sub-models are available to describe one basic process, for example the process of water uptake by roots. Then, by generating a manifold of different whole soil–plant–atmosphere models due to different choices of the exchangeable sub-models a sub-space of the 'model space' in the sense of Beven (2002) can be defined. Furthermore, using the given input data different ways of model parameterisation and initialisation may be applied to define the relation between a model of the model sub-space and the modelling object, i.e. the field or field site which is considered.

In this paper we compare models that are generated by the model system Expert-N (Engel and Priesack 1993; Stenger et al. 1999; Priesack et al. 2001) and that differ by the use of different sub-models for the simulation of crop growth, whereas all other sub-models making up the rest of the soil–plant–atmosphere model remain the same. The comparison is based on a dataset of a six-year field trial with a sugar beet, winter wheat, winter barley, winter rye crop rotation. The models are calibrated using datasets from a field site in Southern Germany and applied to a dataset from North-East Germany provided by Wegehenkel et al. (2004) mainly to get insight about simulation accuracy and how this accuracy is affected by using different sub-models to calculate crop growth.

The Expert-N model system

Expert-N is a model system for the simulation of water, heat, C and N dynamics in soil–plant–atmosphere systems. It comprises a number of modules that provide different approaches to simulate vertical one-dimensional soil water flow, soil heat transfer and solute transport, soil C and N turnover, crop processes and soil management. Each module is made up of different sub-routines that can be selected to simulate each of the important basic processes. The sub-routines cur-

rently available in Expert-N have either been taken from published models such as LEACHN 3.0 (Hutson and Wagenet 1992), CERES-Wheat 2.0 (Jones and Kiniry 1986; Ritchie 1991), HYDRUS 6.0 (Šimunek et al. 1998), SUCROS2 (van Laar et al. 1992), NCSOIL (Nicolardot and Molina 1994), SOILN (Johnsson et al. 1987), and DAISY (Hansen et al. 1991; Svendsen et al. 1995), or have been developed by the Expert-N team (Stenger et al. 1999; Berkenkamp et al. 2002) including the N model N-SIM (Schaaf et al. 1995) and the crop growth model SPASS (Wang 1997; Wang and Engel 2000; Gayler et al. 2002).

The possibility to choose between different sub-routines allows the user to better adapt the model to the specific purpose of an experimental study and to the availability and quality of needed input data. Comparison of different sub-routines for one basic process without changing the others can be helpful to find the sub-routine responsible for differences in simulation results and thus can facilitate the comparison of different approaches to model soil–plant–atmosphere systems. Moreover, Expert-N provides the means to enter user-programmed sub-routines by using a loading procedure for functions from external user-programmed dynamic link libraries (Sperr et al. 1993). Thus, the model system Expert-N is an open modelling framework that allows for the development of completely new soil–plant–atmosphere models.

Expert-N comprises a user friendly data input system to enter soil, crop, management, and weather data (Priesack and Bauer 2003). Additionally, measured time series of soil water and inorganic N contents or crop data can be stored in the database and used to determine the deviation between measurements and simulation.

Soil water processes

In Expert-N several different approaches or options are available for each of the following sub-models needed to simulate water dynamics: potential evapotranspiration, potential evaporation, actual evaporation, soil water flow, surface run-off, snow accumulation or melting, and the lower boundary condition for soil water flow. In the present study soil water flow simulations are based on Richards equation using the numerical solution according to the model HYDRUS 6.0 (Šimunek et al. 1998). Potential evapotranspiration is estimated using a modified Haude approach (DVWK 1996), potential evaporation is determined by taking the crop covered fraction of the soil surface into account, run-off and snow processes are not considered, and at the lower boundary free drainage is assumed.

Transport and turnover processes

Heat transfer and transport of N, i.e. of the N-species urea, ammonium and nitrate, are simulated by numerically solving the corresponding transport equations using the methods of the model LEACHN (Hutson and Wagenet 1992). Similar to the model LEACHN, for the calculation of C- and N-turnover the sub-model is chosen which follows the concept of C and N mineralisation described for SOILN (Johnsson et al. 1987). The degradation of plant residues at the soil surface and their incorporation into the soil during tillage is modelled using the Expert-N approach (Berkenkamp et al. 2002).

Plant processes

For the simulation of crop development and crop growth the corresponding routines of the generic plant models CERES, SPASS, and SUCROS are applied and compared. Since only SUCROS provides a sub-model for sugar beet, this model is used for the sugar beet simulations. In the case of the SUCROS model, N demand of the plant is calculated by the difference between actual and potential N content in the plant, where the potential N content is given by the maximal N contents that the plant organs can reach during the actual development stage. The actual N uptake is limited by the actual root length and modelled considering N uptake by convection and diffusion (McIsaac et al. 1985; Huwe 1992; Vanclooster et al. 1994).

Statistical measures

The normalised root mean square error (NRMSE) is defined according to Wallach and Goffinet (1989) and provides the total difference between predicted and observed values, proportioned against the mean observed value:

$$\text{NRMSE} = \frac{1}{\bar{O}} \sqrt{\frac{\sum_{i=1}^{n}(P_i - O_i)^2}{n}}$$

and the modelling efficiency index (IA) according to Willmott (1982) is given by

$$\text{IA} = 1 - \frac{\sum_{i=1}^{n}(P_i - O_i)^2}{\sum_{i=1}^{n}(|P_i - \bar{O}| + |O_i - \bar{O}|)^2},$$

where P_i and O_i for $1 \leq i \leq n$ are the simulated and measured values that are compared and $\bar{P}, \bar{O}$ denote the corresponding mean values.

Experimental datasets

The experimental data used for the present study are given in detail by Wegehenkel et al. (2004). The data were obtained from field experiments carried out in the experimental station of the Centre for Agricultural Landscape and Land Use Research (ZALF) at Müncheberg, located East of Berlin, Germany. According to the FAO classification the soil type of the studied field site is an Eutric Cambisol. The basic soil properties of the soil profiles at the three field plots that were established for the experimental study is provided by Table 1. Each field plot was differently managed, only dates of seeding and harvest were the same (Wegehenkel et al. 2004). Plot 1 was intensively managed applying inorganic fertilisers and chemical plant protection on a high level, plot 2 was organically managed using only organic fertilisers and non-chemical plant protection and at plot 3 an extensive management was applied using a mixture of organic and inorganic fertilisers and chemical plant protection on a low level (Wegehenkel et al. 2004).

Initial values and parameters

We use the Richards equation for water flow simulations and apply the van Genuchten parameterisation of the soil hydraulic properties (Table 1) as proposed and analysed in detail by Wegehenkel (2005), where also the hydrological measurement techniques at the field plots are

Table 1. Soil properties and soil hydraulic parameters (Wegehenkel 2005).

Horizon	Δz (cm)	sa (%)	si (%)	cl (%)	oc (%)	ρ_b (g cm^{-1})	θ_s (l)	θ_r (l)	n (l)	α (l cm^{-1})	k_s (cm day^{-1})
Plot 1											
Ap	0–30	90	3	7	0.45	1.45	0.385	0.027	2.013	0.021	92
Ael	30–60	90	5	5	0.26	1.50	0.319	0.027	2.179	0.027	162
Bt	60–90	80	8	12	0.10	1.55	0.385	0.065	2.147	0.028	30
C1	90–120	90	6	4	–	–	0.319	0.027	2.379	0.027	162
C2	120–150	90	7	3	–	–	0.319	0.027	2.379	0.027	162
C3	150–225	90	8	2	–	–	0.319	0.027	2.379	0.027	162
Plot 2											
Ap	0–30	85	10	5	0.45	1.45	0.385	0.027	2.013	0.021	92
Ael	30–90	90	5	5	0.26	1.50	0.319	0.027	2.179	0.027	162
Bt1	90–130	80	8	12	0.10	1.55	0.385	0.065	2.147	0.028	30
Bt2	130–170	80	10	10	–	–	0.302	0.027	2.147	0.028	30
C1	170–180	90	5	5	–	–	0.319	0.027	2.379	0.027	162
C2	180–225	90	5	5	–	–	0.319	0.027	2.379	0.027	162
Plot 3											
Ap	0–30	85	9	6	0.45	1.45	0.385	0.027	2.013	0.021	92
Ael	30–100	90	5	5	0.26	1.50	0.319	0.027	2.379	0.027	162
Bt1	100–110	81	6	13	0.10	1.55	0.385	0.065	2.147	0.028	30
Bt2	110–225	80	9	11	–	–	0.385	0.065	2.147	0.028	30

Δz, depth interval; sa, sand fraction; si, silt fraction; cl, clay fraction; oc, organic C content; ρ_b, soil bulk density; θ_s, saturated vol. water content; θ_r, residual vol. water content; n, van Genuchten parameter; α, van Genuchten parameter; k_s, saturated hydraulic conductivity.

described. Where soil bulk densities and organic matter contents were not measured the hydraulic properties were approximated assuming the same value as for the overlying horizon and zero organic C contents.

The parameterisation of the C and N turnover model is taken from a preceding simulation study (Priesack et al. 2001) based on experimental data from the FAM research station near Munich in South Germany (Schröder et al. 2002) using an inverse modelling procedure and measured crop growth data in combination with a simple model for water and N uptake by roots. In this way this parameterisation is carried out independently from the parameterisations of the CERES, SPASS and SUCROS crop growth sub-models.

Also the different crop growth models were calibrated using field data collected at the FAM research station. In order to get optimal correspondence of simulated phenological stages with observed EC stage values defined by Zadoks et al. (1974), the relevant model parameters were varied within the limits proposed by the authors of the original model documentations (Jones and Kiniry 1986; Ritchie 1991; van Laar et al. 1992; Vanclooster et al. 1994; Wang 1997). The model parameterisation is based on observations of germination, emergence, tillering, anthesis and maturity of cereals from 1991 until 1998. After model calibration, phenological stages of cereals can be simulated by the models with a normalised root mean square error (NRMSE) of 0.09 (modelling efficiency IA = 0.99) (CERES), 0.08 (IA = 0.99) (SPASS) and 0.09 (IA = 0.99) (SUCROS).

In a second step, simulation of biomass growth was parameterised using measured data of vegetative aboveground biomass and storage organ biomass. These data were collected at two plots of the FAM research station with optimum growth conditions (in 1995 and 1998). Additionally, in case of cereals the initial conditions of the leaf and root biomass after emergence are slightly modified. The simulations show better correspondence with observed biomass if the initial values of leaves and roots of the seedlings are estimated as 50% of the carbohydrates of the sowing grains rather than using initial values proposed in the original source code of the models.

Using these model adaptions, simulated yields correlate with observations at the calibration plots with a NRMSE of 0.06 (CERES), 0.09 (SPASS), and 0.11 (SUCROS) and simulated aboveground biomass correlate with observations with a NRMSE of 0.13 (CERES), 0.14 (SPASS), and 0.27 (SUCROS). The relatively high NRMSE of 0.27 for the SUCROS model is caused by an overestimation of aboveground biomass growth rates in the early vegetation states. Model testing with yield data collected during six years at five different sites of the FAM research station, showed NRMSE values of 0.18 (IA = 0.97) (CERES), 0.19 (IA = 0.97) (SPASS) and 0.28 (IA = 0.93) (SUCROS).

The adapted parameter values are presented in Table 2, all other model parameters values are taken from the original model documentations. For the simulation study of the Müncheberg field site no further calibration of the growth models was carried out.

Simulation results

For the detailed presentation and discussion of the simulations we will concentrate on the results obtained for the extensively managed plot (plot 3) and only show the main results for the other plots, the intensively managed plot 1 and the organically managed plot 2.

Crop growth and N uptake

All simulations were carried out for the complete crop rotation without resetting of conditions between the different vegetation periods. For all three plots crop growth simulations show results that in most cases compare sufficiently well with observed data for the EC development stages according to Zadoks et al. (1974) and with C storages, but are less accurate for vegetative above ground biomasses and N uptake (Tables 3 and 4, Figures 1–3). Highest crop yields (beet or grains) are simulated by all three crop models for plot 3, i.e. for the extensive agricultural management (Wegehenkel et al. 2004) with the highest N input by mineral and organic fertilisers. Overall, simulations using the CERES sub-model give the best agreement with observed values, which can also be seen from the model efficiencies, i.e. IA values for vegetative aboveground biomass are 0.86, 0.82, 0.88 (CERES), 0.80, 0.79, 0.81 (SPASS) and 0.83, 0.77, 0.85 (SUCROS) for plot 1, 2, and 3 respectively. In particular, vegetative aboveground biomass of winter barley crop is underestimated by SPASS simulations for all three plots (Table 4 and Figure 1). The SUCROS sub-model application results in the highest leaf area indices (LAI) that also stayed high for a longer period. For example,

Table 2. Crop growth model parameter values.

Model	Parameter	Unit	Wheat	Barley	Rye
CERES	P1V	–	3.6	3	4
	P1D	–	2.75	3	2.25
	P5	–	4.5	3	8
	G1	–	6	4	4.2
	G2	–	6	3.8	3.8
	G3	–	2.3	2.3	2.3
SPASS	D_v	day	37	32	33
	D_r	day	27	25	25
	V_{nd}	day	46	30	50
	ω	day^{-1}	0.3	0.3	0.3
	$R_{gfill, max}$	mg grain^{-1} day^{-1}	2.0	2.0	2.0
	ξ_{grain}	grains g^{-1} stem	35	26	25
SUCROS	TbaseV	°C	0	0	0
	DVRV	(°C day)$^{-1}$	0.027	0.030	0.024
	TbaseR	°C	1	2	0
	DVRR^{-1}	(°C day)$^{-1}$	0.030	0.032	0.030

P1V, vernalisation coefficient; P1D, photoperiodism coefficient; P5, grain filling duration coefficient; G1, kernel number coefficient; G2, kernel weight coefficient; G3, spike number coefficient; D_v, D_r, physiological development days before and after anthesis; V_{nd}, vernalisation requirement; ω, daylength sensitivity; $R_{gfill, max}$, maximum grain filling rate; ξ_{grain}, number of grains; DVRV, DVRR, development rate before and after anthesis; TbaseV, TbaseR base temperature of development rate.

Table 3. NRMSE values (l) of crop growth simulations over the period 1.11.92–31.10.98.

Site	Crop model	EC	VB	ST	NU
Plot 1	CERES	0.11	0.44	0.11	0.35
	SPASS	0.18	0.52	0.15	0.34
	SUCROS	0.19	0.55	0.17	0.42
Plot 2	CERES	0.14	0.54	0.31	0.46
	SPASS	0.19	0.53	0.29	0.36
	SUCROS	0.21	0.70	0.20	0.51
Plot 3	CERES	0.14	0.45	0.12	0.23
	SPASS	0.19	0.53	0.21	0.24
	SUCROS	0.21	0.60	0.13	0.33

EC, development stages of all cereal crops; VB, vegetative aboveground biomass; ST, C storage; NU, N uptake.

Table 4. NRMSE values (l) of crop growth simulations at different years for plot 3.

Crop model	Year	EC	VB	ST	NU
CERES	94	0.09	0.20	0.20	0.42
	95	0.07	0.34	0.03	0.34
	96	0.21	0.17	0.25	0.12
	97	–	0.54	0.13	0.11
	98	0.14	0.69	0.06	0.24
SPASS	94	0.10	0.26	0.13	0.25
	95	0.10	0.63	0.05	0.31
	96	0.29	0.26	0.01	0.03
	97	–	0.64	0.40	0.21
	98	0.26	0.21	0.11	0.27
SUCROS	93	–	0.57	0.09	0.21
	94	0.08	0.44	0.20	0.72
	95	0.23	0.74	0.07	0.57
	96	0.19	0.44	0.10	0.64
	97	–	0.64	0.13	0.11
	98	0.36	0.55	0.25	0.06

EC, development stages of cereal crops; VB, vegetative aboveground biomass; ST, C storage; NU, N uptake.

considering the winter wheat crop 93/94 at plot 3 the simulated LAI is above 3 during 72 days for the SUCROS, during 75 days for the CERES and during 46 days for the SPASS sub-model. Application of the SUCROS sub-model also resulted in the highest root biomasses and deepest root depth compared to the other growth sub-models. Related is the simulated high N-uptake of cereal crops by using the SUCROS sub-model that often overestimates observed values (Table 3). Also for plot 3 the N uptake of winter wheat 93/94 and for winter barley 95/96 show this overestimation.

Water balances

Table 5 shows the water balance for the whole simulation period, in Table 6 the annual water balances for plot 3 are given.

For all three plots the simulations of the six-year period with the CERES crop growth sub-model result in the lowest values for evaporation and the highest percolation rates and except for plot 3 in the highest values for actual transpiration. Also the CERES surplus in soil water storage is higher than the surplus calculated with the SPASS and SUCROS sub-models for plot 1 and 3 and is similar to the SPASS result, but higher than that of the SUCROS sub-model version (Table 5). The simulated actual transpiration values obtained using the SPASS crop growth sub-model are considerably lower than the transpiration calculated by the CERES sub-model up to a difference of more than 150 mm over six years (Table 5, plot 1), which is about 10–12% of the simulated transpiration and 4% of the precipitation. The SPASS sub-model simulations show the lowest transpiration, which is related to the shorter periods of high leaf area, resulting also in higher evaporation. The SUCROS sub-model simulations have the lowest percolation values for all plots (Table 5), since transpiration is high due to high LAI values and due to deeper roots that have access to the water of the lower soil compartments. Also from the annual water balances for plot 3 (Table 6) it can be seen, that in cases of winter wheat crop in the years 1994 and 1998 application of the SUCROS crop sub-model leads to high evaporation rates and lowest percolation amounts for all plots. This is especially true for the year 1994 which is characterized by a fairly large precipitation compared to the other years resulting in the highest percolation amounts. In this year the simulated difference in percolation amount between the SPASS and SUCROS versions is 65 mm which corresponds to 9.0% of the precipitation amount. Comparing the SPASS with the CERES version the difference in percolation is 18 mm or 2.5% of the precipitation (Table 6).

Nitrogen balances

The corresponding simulated relative N leaching (cf. Table 7 for the N balance of the whole period) is small and almost negligible for plot 2, the

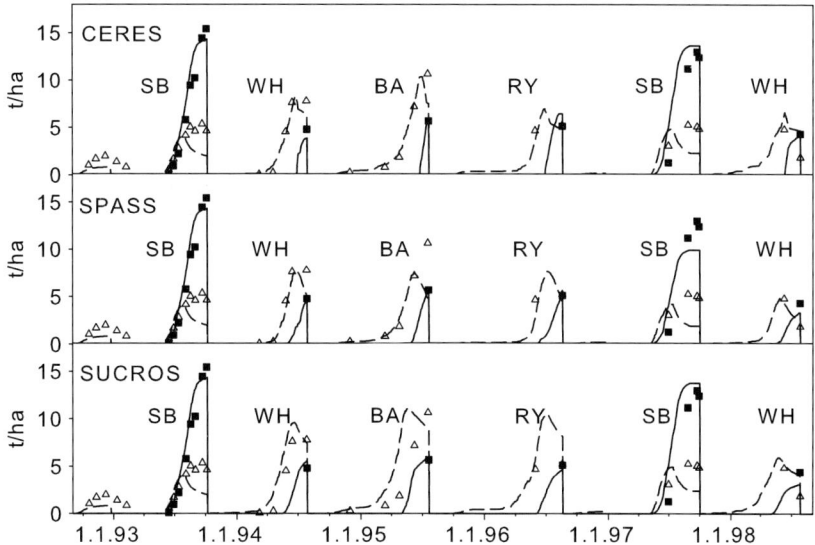

Figure 1. Dynamics of crop biomass for plot 3. Solid line: storage organs [t ha^{-1}]; dashed line: vegetative above ground biomass [t ha^{-1}]; filled squares: measured values (storage organs); open triangles: measured values (vegetative biomass above ground); SB sugar beet, WH winter wheat, BA winter barley, RY winter rye.

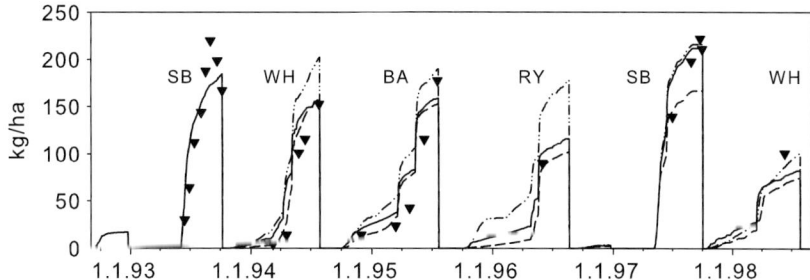

Figure 2. Simulated cumulative N uptake [kg ha^{-1}] versus measured values of total N content in plants [kg ha^{-1}] for plot 3. Solid line: simulated values (CERES); dashed line: simulated values (SPASS); dash-dotted line: simulated values (SUCROS); symbols: measured values; SB sugar beet, WH winter wheat, BA winter barley, RY winter rye.

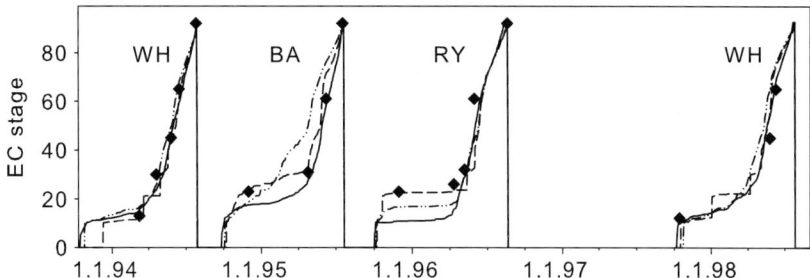

Figure 3. Simulated EC-values [1] versus measured values for plot 3. Solid line: simulated values (CERES); dashed line: simulated values (SPASS); dash-dotted line: simulated values (SUCROS); symbols: measured values; WH winter wheat, BA winter barley, RY winter rye.

Table 5. Water balances in (mm) over the period 1.11.92–31.10.98 with precipitation of 3539 mm.

Site	Crop model	E	T	Q	ΔS
Plot 1	CERES	769	1345	1325	+98
	SPASS	913	1257	1288	+79
	SUCROS	915	1328	1210	+84
Plot 2	CERES	783	1413	1218	+124
	SPASS	964	1235	1214	+125
	SUCROS	932	1381	1107	+117
Plot 3	CERES	704	1334	1399	+100
	SPASS	868	1274	1309	+87
	SUCROS	888	1363	1216	+71

E, Evaporation; T, Transpiration; Q, Percolation; ΔS, Change in soil water storage.

Table 6. Annual water balances in (mm) for Plot 3.

Crop model	Year	P	E	T	Q	ΔS
CERES	93	625	151	177	220	+77
	94	712	96	223	441	−48
	95	507	57	245	265	−60
	96	486	122	214	139	+12
	97	541	183	236	140	−19
	98	655	100	236	273	+45
SPASS	93	625	151	177	220	+77
	94	712	133	204	423	−48
	95	507	107	228	232	−61
	96	486	145	225	104	+12
	97	541	187	232	138	−16
	98	655	148	205	257	+44
SUCROS	93	625	151	177	220	+77
	94	712	126	277	358	−48
	95	507	145	168	246	−55
	96	486	105	282	94	+5
	97	541	184	235	138	−17
	98	655	182	220	207	+45

P, Precipitation; E, Evaporation; T, Transpiration; Q, Percolation; ΔS, Change in soil water storage.

Table 7. Mineral N balances in (kg N ha^{-1}) over the period 1.11.92–31.10.98.

Site	Crop model	J	M	E	U	L	ΔN
Plot 1	CERES	877	310	21	1081	120	−34
	SPASS	875	168	22	982	79	−43
	SUCROS	879	410	28	1196	96	−31
Plot 2	CERES	255	560	18	792	11	−6
	SPASS	253	446	18	691	7	−17
	SUCROS	265	710	22	912	8	+33
Plot 3	CERES	657	423	19	930	142	−12
	SPASS	657	288	20	856	93	−26
	SUCROS	658	542	25	1092	95	−14

J, Input; M, Net mineralisation; E, Gaseous emissions; U, Plant uptake; L, Leaching; ΔN, Change in storage.

average N concentrations in the leachate are estimated to be within 6–11 mg l^{-1} for plot 1 and 3 and within 0.5–1 mg l^{-1} for plot 2. Note the small differences in mineral N input stemming from mineralisation of plant residues at the soil surface, since besides mineral N from atmospheric deposition and from mineral or organic N fertilisers the input also includes mineral N from surface litter decomposition. Note also the differences in net N mineralisation and corresponding mineral N uptake by the crop simulated using SUCROS versus the other sub-model versions (Table 7).

When coupled with the SUCROS growth sub-model simulated N turnover is generally higher indicating that higher amounts of plant residues, mainly from root biomass, but also from incorporated vegetative aboveground biomass, are added to the soil organic matter pools. Gaseous N emissions including N losses due to volatilisation and denitrification are lower than the estimated atmospheric deposition assumed to contribute to the mineral N input at a rate of 15 kg N ha^{-1} per year. The annual balances of mineral N for plot 3 (Table 8) show lower N mineralisation during the drier years 95 and 96 followed by lower N leaching at 2 m depth in the years 96 and 97, but an increased N leaching in the following year 98. The reduced input in the year 98 resulted in lower N turnover and less N uptake. Note the relatively high mineralisation and the high uptake simulated by the SUCROS sub-model version for the year 94. Since all sugar beet crop simulations are carried out by applying the SUCROS crop growth sub-model routines, after the harvest of sugar beet crop in the year 94 the simulations started with the same initial conditions for all three model versions. This is not the case for the sugar beet crop year 97, since then, the system conditions that are simulated by using the different crop growth models deviate already considerably.

Mineralisation-immobilisation turnover

This change in system conditions is mainly due to differences in simulated crop biomass growth that result in different simulated amounts of plant

Table 8. Annual mineral N balances in (kg N ha^{-1}) for plot 3.

Crop model	Year	J	M	E	U	L	ΔN
CERES	93	107	97	5	190	16	−7
	94	105	102	3	178	41	−16
	95	110	52	2	145	24	−9
	96	130	52	2	107	14	+59
	97	132	70	4	229	19	−51
	98	50	42	1	65	36	−10
SPASS	93	107	97	5	190	16	−7
	94	105	79	4	172	40	−31
	95	110	27	2	140	16	−22
	96	130	11	2	101	2	+35
	97	132	42	4	174	5	−9
	98	50	21	2	67	19	−18
SUCROS	93	107	97	5	190	16	−7
	94	105	135	5	230	30	−25
	95	110	99	4	188	14	+2
	96	130	71	3	151	6	+40
	97	132	78	5	227	9	−31
	98	51	56	2	89	24	−9

J, Input; M, Net mineralisation; E, Gaseous emissions; U, Plant uptake; L, Leaching; ΔN, Change in storage.

residues and hence in different contributions to the soil organic matter pools. Because the different crop growth sub-models simulate root biomass growth in different ways, they also differently contribute to the soil organic matter pools. The CERES and SPASS sub-models simulate root decay during the reproductive phase, when the crop invests into the generative plant organs, whereas by the sub-model SUCROS no root decay is described. Moreover, because the root decay rates and also the N allocation patterns differ between CERES and SPASS, root biomass simulated by the SPASS model is lower and has a wider C/N-ratio towards the end of the vegetation period. This leads to less N input to the soil organic matter pools and, furthermore, causes a stronger N immobilisation when the dead root litter is mineralised. Therefore, the simulated net mineralisation is lowest if the SPASS, higher if the CERES and highest if the SUCROS sub-model is applied (Tables 7 and 8). In particular, at plot 3 at the rye harvest in 1996 simulated N input by dead root material is 1 kg N ha^{-1} in case of the SPASS sub-model, 13 kg N ha^{-1} in case of the CERES sub-model, and 40 kg N ha^{-1} in case of the SUCROS sub-model. Until the sugar beet harvest in 1997 calculated net N mineralisation amounts to 26 kg N ha^{-1} (SPASS), 70 kg N ha^{-1} (CERES) or 83 kg N ha^{-1} (SUCROS) resulting in different cumulative N-uptake of the sugar beet crop: 169 kg N ha^{-1} (SPASS), 215 kg N ha^{-1} (CERES) and 220 kg N ha^{-1} (SUCROS). Consequently, with the SPASS simulation the low availability of mineral N limits sugar beet growth and leads to an underestimation of biomass production (Figure 1).

Soil water content

For each crop growth model and each plot trial when evapotranspiration exceeds precipitation, simulations underestimated measured water contents at 60–90 cm soil depth, see for example the SPASS results (Figure 4). During times of high transpiration the simulations using the CERES sub-model showed generally higher water depletion in 0–30 cm soil depth and consequently higher water contents in the deeper soil in 30-60 cm and 60–90 cm depth when compared to the SPASS and SUCROS simulations. Of all three versions the SUCROS sub-model lead to the lowest water contents at 60–90 cm depth. This was mainly caused by root water uptake from deeper roots. Therefore, less water was taken up from the upper soil compartment of 0–30 cm depth contributing to the higher evaporation during periods when the LAI is low. Comparing the simulated with the gravimetrically measured water contents in 0–90 cm depth, the NRMSE values were 0.21 for plot 1 and 0.17 for plot 3 for all crop growth sub-models and ranged from 0.25 to 0.27 for plot 2. Model efficiencies IA obtained for CERES, SPASS and SUCROS were 0.80, 0.81, 0.79 for plot 1, 0.64, 0.63, 0.65 for plot 2, and 0.80, 0.83, 0.83 for plot 3.

Soil N content

Simulated mineral N contents in the rooted soil zone of 0–90 cm depth were similar between the three model versions and were often not in agreement with the measurements in case of plot 1 and plot 3 (Figure 5). Even stronger deviations exist for plot 2. Related NRMSE values ranged from 0.81 to 0.82 for plot 1, from 0.58 to 0.64 for plot 2, and from 0.65 to 0.73 for plot 3. Model efficiencies IA were between 0.68 and 0.69 for plot 1, between 0.56 and 0.63 for plot 2, and between 0.76 and 0.80

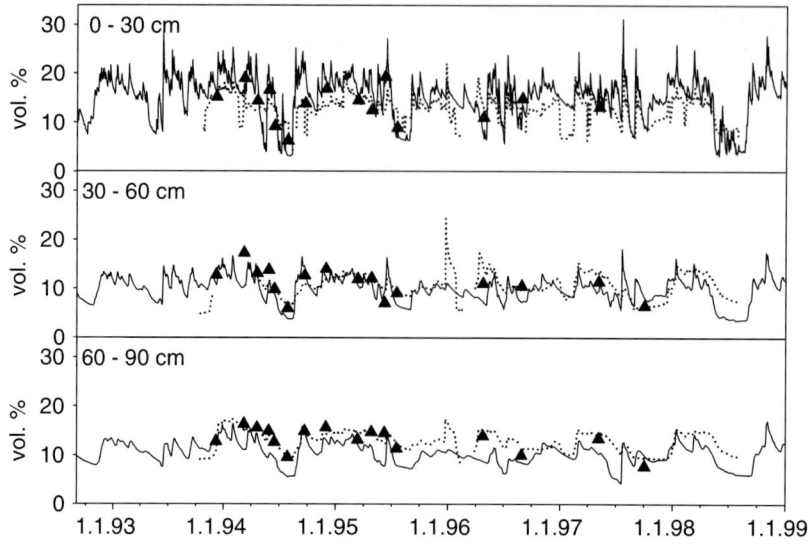

Figure 4. Soil volumetric water contents [%] at plot 3. Solid line: simulated values (SPASS); dotted line: measured values (TDR method); symbols: measured values (gravimetric method).

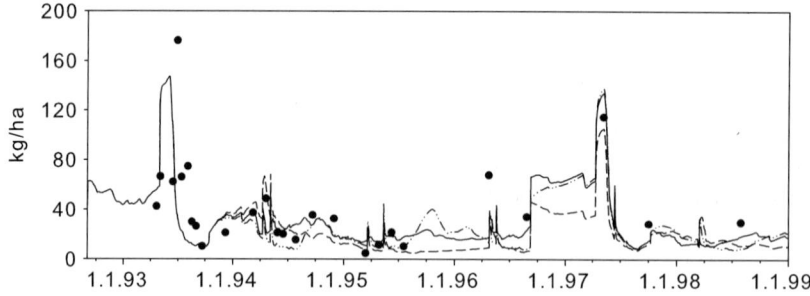

Figure 5. Soil mineral N (nitrate N and ammonium N) contents [kg N ha^{-1}] in 0–90 cm depth at plot 3. Solid line: simulated values (CERES); dashed line: simulated values (SPASS); dash-dotted line: simulated values (SUCROS); symbols: measured values.

for plot 3. Discrepancies between the simulations of the model versions mainly occurred during the periods of net N mineralisation after crop harvest and also after N fertilisation during periods of strong N uptake by the crop (Figure 5). Note, that in contrast to the simulated water contents, where differences between the simulations vanished during the winter half year, the differences in simulated dynamics of soil mineral N contents, once evolved, remained clearly recognisable and only decreased during periods of high N uptake by growing crops. By comparing measured soil nitrate N contents in 0–30 cm, 30–60 cm, and 60–90 cm depth with simulations using the SPASS model version (Figure 6), strong differences can be seen for 0–30 cm and 60–90 cm depth during the first half of 1994 after sugar beet harvest and ploughing in of leaves followed by sowing of winter wheat and application of farmyard manure in October 1993. Model efficiencies IA were 0.93 for 0–30 cm depth, 0.22 for 30–60 cm depth, and 0.51 for 60–90 cm depth. In case of applying the other two model versions, mostly similar differences to measured values occur for 0–30 cm depth (data not shown).

Discussion and conclusions

Due to the differences of the crop growth simulations by the different crop models differences in simulated water balances occur. High leaf area

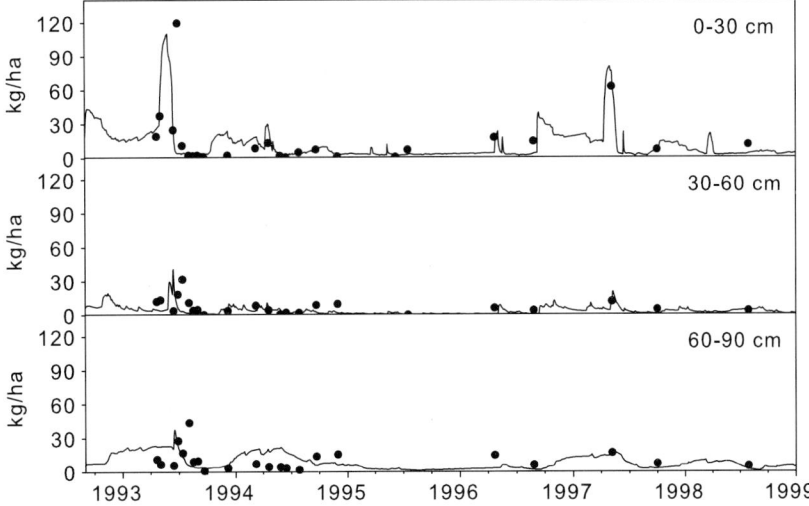

Figure 6. Soil nitrate N contents [kg N ha^{-1}] at plot 3. Solid line: simulated values (SPASS); symbols: measured values.

indices simulated by the CERES crop model lead to the highest simulated actual transpiration rates and correspondingly to the lowest actual evaporation rates. In contrast, in the case of the SUCROS crop model application, the simulation of the higher actual evaporation is mainly caused by the simulation of somewhat shorter periods of high leaf area indices and higher water contents of the top soil. The latter are due to deeper roots that have access to water from the deeper soil compartment and thus take less water from the upper compartment. Compared with the other crop models, this also contributes to the still high transpiration rates and to the estimation of most often lowest percolation rates if the SUCROS model is applied.

Using the proposed parameterisation of the measured soil water retention curves and by applying Richards equation, the soil water flow simulations underestimated soil water contents at 60–90 cm depth of plot 3 irrespective of the crop model chosen. The NRMSE and IA values obtained by comparing measured and simulated volumetric water contents are similar to those that can be obtained from Wegehenkel (2005) and from comparable studies, see Wegehenkel (2005) and citations therein. Based on the analysis of Wegehenkel (2005) we consider the parameterisation of the soil hydraulic properties given by Table 1 as an adequate site-specific adjustment, although subsoil parameters were roughly approximated e.g. by taking identical hydraulic parameters for the C-horizons. Moreover, additional simulations (not shown) suggest that the uncertainty of water balance calculation related to this approximation has to be considered as rather low compared to the uncertainty stemming from using different methods of potential evapotranspiration estimation, see also (Wegehenkel et al. 2004).

Also the NRMSE and IA values for the comparison between simulated and observed nitrate contents are comparable to those obtained from Wegehenkel and Mirschel (2006) that range from 0.94 to 1.49 for NRMSE and from 0.22 to 0.91 for the IA. Similarly the NRMSE values for crop growth (Tables 3 and 4) are within the range of those from Wegehenkel and Mirschel (2006) and similar studies (Asseng et al. 2000; Bellochi et al. 2002; Stoeckle et al. 2003).

The choice of the crop growth model also influences the amounts of N mineralised from soil organic matter (Tables 7 and 8). This can be attributed to different ways the crop models simulate decay and growth of roots, to differences of simulated C:N ratios for the plant organs and to indirect effects of water content on mineralisation rates. The latter are caused by the simulation of different vertical soil water distributions due to different actual evaporation rates and to different vertical distributions of soil water uptake by roots. For example the simulated root decay and following mineralisation of roots during the generative phase of crop growth is one reason of lower simulated net-mineralisation by using the CERES and SPASS crop growth sub-models. Another reason for the simulation of generally higher net-N-mineralisation by the SUCROS model version (Table 8) for plot 3 is the simulation of higher mineralisation rates of the soil organic matter pools in the top soil, because of higher simulated top soil water contents. The overestimation of mineral N contents for plot 3 in 1994 (Figure 5) may be accounted for if the combination effect of crop residue incorporation with farmyard manure application would be a short period of N immobilisation followed by a longer period of N mineralisation, which might occur for a crop residue with a higher C:N ratio.

Generally, the model test has shown the impact of choosing different crop growth models on the simulation of water flow and C–N turnover for a crop rotation. In particular, the simulation study demonstrated the occurrence of only small differences in the calculated water balance that can be attributed to model-specific differences including differences in root growth, vertical root distribution and attained root depth. Stronger differences were detected for the C and N turnover due to the incorporation of plant residues that differ in simulated qualities depending on the chosen crop model. Facing the differences between simulated and measured soil N uptake by the plants, a calibration of C and N mineralisation rates and improvement of root growth model routines seems to be necessary, albeit the actually available limited amount of soil mineral N data at the Müncheberg test sites might hinder the judgement on the best model performance. However, such improved crop growth sub-models with parameterisations that were tested to be useful for different sites and soils in South Germany may also be applicable to predict crop growth for sites in North-East Germany.

Acknowledgements

This research was partially supported by a grant from the German Research Foundation (DFG) within the framework of the Collaborative Research Centre 607 "Growth and Parasite Defense: Competition for Resources in Economic Plants from Agronomy and Forestry". The experimental dataset provided by the Leibniz-Centre for Agricultural Landscape and Land Use Research is gratefully acknowledged. We also thank the anonymous reviewer for his comments that helped to considerably improve the manuscript.

References

Addiscott T.M. and Wagenet R.J. 1985. Concepts of solute leaching in soils: a review of modelling approaches. J. Soil Sci. 36: 411–424.

Asseng S., van Keulen H. and Stol W. 2000. Performance and application of the APSIM NWHEAT model in the Netherlands. Eur. J. Agron. 12: 37–54.

Bellochi G., Silvestri N., Mazzoncini M. and Menini S. 2002. Using the CROPSYST model in continuous rainfed maize (*Zea mays* L.) using alternative management options. Ital. J. Agron. 6: 43–56.

Berkenkamp A., Priesack E. and Munch J.C. 2002. Modelling the mineralisation of plant residues on the soil surface. Agronomie 22: 711–722.

Beven K. 2002. Towards an alternative blueprint for a physically based digitally simulated hydrologic response modelling system. Hydrol. Process. 16: 189–206.

Booltink H.W.G., van Alphen B.J., Batchelor W.D., Paz J.O., Stoorvogel J.J. and Vargas R. 2001. Tools for optimizing management of spatially-variable fields. Agric. Syst. 70: 445–476.

De Willigen P. 1991. Nitrogen turnover in the soil crop system: comparison of fourteen simulation models. Fert. Res. 27: 141–149.

Diekkrüger B., Söndgerath D., Kersebaum K.C. and McVoy C.W. 1995. Validity of agroecosystem models. A comparison of results of different models applied to the same data set. Ecol. Model. 81: 3–29.

DVWK. 1996. Ermittlung der Verdunstung von Land- und Wasserflächen. *Merkblätter zur Wasserwirtschaft* 238, Wirtschafts- und Verlagsgesellschaft Gas und Wasser, Bonn, Germany.

Engel T., Klöcking B., Priesack E. and Schaaf T. 1993. Simulationsmodelle zur Stickstoffdynamik. *Agrarinformatik* 25, Verlag Eugen Ullmer, Stuttgart, Germany.

Engel Th. and Priesack E. 1993. Expert-N, a building block system of nitrogen models as resource for advice, research, water management and policy. In: Eijsackers H.J.P. and Hamers T. (eds), Integrated Soil and Sediment Research: A Basis for Proper Protection, Kluwer Academic Publishers, Dordrecht, The Netherlands, pp. 503–507.

Gayler S., Wang E., Priesack E., Schaaf T. and Maidl F.-X. 2002. Modeling biomass growth, N-uptake and phenological development of potato crop. Geoderma 105: 367–383.

Grant R.F. 1995. Dynamics of energy, water, carbon and nitrogen in agricultural ecosystems: simulation and experimental validation. Ecol. Model. 81: 169–181.

Hansen S., Jensen H.E., Nielsen N.E. and Svendsen H. 1991. Simulation of nitrogen dynamics and biomass production in winter wheat using the Danish simulation model DAISY. Fert. Res. 27: 245–259.

Hutson J.L. and Wagenet R.J. 1992. LEACHM: Leaching Estimation And Chemistry Model: A process-based model of water and solute movement, transformations, plant uptake and chemical reactions in the unsaturated zone. Version 3.0.

Department of Soil, Crop and Atmospheric Sciences, Research Series No. 93-3, Cornell University, Ithaca, NY, USA.

Huwe B. 1992. Deterministische und stochastische Ansätze zur Modellierung des Stickstoffhaushalts landwirtschaftlich genutzter Flächen auf unterschiedlichem Skalenniveau. Mitteilungen 11, Institut für Wasserbau, Stuttgart, Germany.

Johnsson H., Bergström L., Jansson P.E. and Paustian K. 1987. Simulated nitrogen dynamics and losses in a layered agricultural soil. Agric. Ecosyst. Environ. 18: 333–356.

Jones C.A. and Kiniry J.R. (eds), 1986. CERES-Maize. A Simulation Model of Maize Growth and Development. Texas A&M University Press, Temple, TX, USA.

Kersebaum K.C., Lorenz K., Reuter H.I. and Wendroth O. 2002. Modeling crop growth and nitrogen dynamics for advisory purposes regarding spatial variability. In: Ahuja L.R., Ma L. and Howell T.A. (eds), Agricultural System Models in Field Research and Technology Transfer, Lewis Publishers, Boca Raton, FL, USA, pp. 229–252.

McIsaac G., Martin D.L. and Watts D.G. 1985. Users Guide to NITWAT – A Nitrogen and Water Management Model. Agr. Eng. Dpt., University of Nebraska, Lincoln, NA, USA.

Munch J.C., Berkenkamp A. and Sehy U. 2001. The effect of site specific fertilisation on N_2O emissions and N-leaching – measurements and simulations. In: Horst W.J. (ed.), et al., Plant Nutrition – Food Security and Sustainability of Agro-ecosystems, Kluwer Academic Publishers, Dordrecht, The Netherlands, pp. 902–903.

Nicolardot B. and Molina J.A.E. 1994. C and N fluxes between pools of soil organic matter: model calibration with long-term field experimental data. Soil Biol. Biochem. 26: 245–251.

Priesack E., Achatz S. and Stenger R. 2001. Parameterisation of Soil Nitrogen Transport Models by Use of Laboratory and Field Data. In: Shaffer M.J., Ma L. and Hansen S. (eds), Modeling Carbon and Nitrogen Dynamics for Soil Management, CRC Press, Boca Raton, USA, pp. 461–484.

Priesack E. and Bauer C. 2003. Expert-N Datenmanagement. FAM-Bericht 59 Hieronymus, München, Germany.

Ritchie J.T. 1991. Wheat phasic development. In: Hanks J. and Ritchie J.T. (eds), Modeling Plant and Soil Systems, Agronomy Monograph 31, ASA-CSSA-SSSA, Madison, WI, USA, pp. 31–54.

Schaaf T., Priesack E. and Engel T. 1995. Comparing field data from north Germany with simulations of the nitrogen model N-SIM. Ecol. Model. 81: 223–212.

Schröder P., Huber B., Olazábal U., Kämmerer A. and Munch J.C. 2002. Land use and sustainability: FAM research network on agro ecosystems. Geoderma 105: 155–166.

Shaffer M.J., Ma L. and Hansen S. 2001. Modeling Carbon and Nitrogen Dynamics for Soil Management. Lewis Publishers, Boca Raton, FL, USA.

Šimunek J., Huang, K. and van Genuchten, M.T. 1998. The HYDRUS code for simulating the one-dimensional movement of water, heat, and multiple solutes in variably-saturated media. Version 6.0. Research Report No. 144, U.S. Salinity Laboratory, USDA, ARS, Riverside, CA, USA.

Smith P., Smith J.U., Powlson D.S., McGill W.B., Arah J.R.M., Chertov O.G., Coleman K., Franko U., Frolking S., Jenkinson D.S., Jensen L.S., Kelly R.H., Klein-Gunnewiek H., Komarov A.S., Li C., Molina J.A.E., Mueller T., Parton W.J., Thornley J.H.M. and Whitmore A.P. 1997. A comparison of the performance of nine soil organic matter models using datasets from seven long-term experiments. Geoderma 81: 153–225.

Sperr C., Priesack E. and Engel T. 1993. Expert-N: Aufbau, Bedienung und Nutzungsmöglichkeiten des Prototyps. In: Engel T. and Baldioli M. (eds), Expert-N und Wachstumsmodelle. Referate des Anwenderseminars im März 1993 in Weihenstephan. Agrarinformatik 24, Verlag Eugen Ullmer, Stuttgart, Germany, pp. 41–57.

Stenger R., Priesack E., Barkle G. and Sperr C. 1999. Expert-N A tool for simulating nitrogen and carbon dynamics in the soil-plant-atmosphere system. In: Tomer M., Robinson M. and Gielen G. (eds), NZ Land Treatment Collective Proceedings Technical Session 20: Modelling of Land Treatment Systems, New Plymouth, New Zealand, pp. 19–28.

Stoeckle C.O., Donatelli M. and Nelson R. 2003. CROPSYST, a cropping system simulation model. Eur. J. Agron. 18: 289–307.

Svendsen H., Hansen S. and Jensen H.E. 1995. Simulation of crop production, water and nitrogen balances in two German agro-ecosystems using the DAISY model. Ecol. Model. 81: 197–212.

Vanclooster M., Viaene P., Diels J. and Christiaens K. 1994. WAVE a Mathematical Model for Simulating Water and Agrochemicals in the Soil and Vadose Environment. Reference and User's Manual (release 2.0). Institute for Land and Water Management, Katholieke Universiteit Leuven, Leuven, Belgium.

van Ittersum M.K. and Donatelli M. 2003. Modelling cropping systems – highlights of the symposium and preface to the special issues. Eur. J. Agron. 18: 187–197.

van Laar H.H., Goudriaan J. and van Keulen H. (eds) 1992. Simulation of Crop Growth for Potential and Water-limited Production Situations, as Applied to Spring Wheat. Simulation Reports 27, Wageningen, The Netherlands.

Wallach D. and Goffinet B. 1989. Mean squared error of prediction as a criterion for evaluating and comparing system models. Ecol. Model. 44: 299–306.

Wang E. 1997. Development of a Generic Process-oriented Model for Simulation of Crop Growth. Dissertation, Herbert Utz Verlag, München, Germany.

Wang E. and Engel T. 2000. SPASS: a generic process-oriented crop model with versatile windows interfaces. Environ. Model. Software 15: 179–188.

Wegehenkel M. 2005. Validation of a soil water balance model using soil water content and pressure head data. Hydrol. Process. 19: 1139–1164.

Wegehenkel M. and Mirschel W. 2006. Crop growth, soil water and nitrogen balance simulation on three experimental field plots using the Opus model – a case study. Ecol. Model. 190: 116–132.

Wegehenkel M., Mirschel W. and Wenkel K.-O. 2004. Prediction of soil water and crop growth dynamics using the agroecosystem models THESEUS and OPUS. J. Plant Nutr. Soil Sci. 167: 736–744.

Willmott C.J. 1982. Some comments on the evaluation of model performance. Bull. Am. Meteorol. Soc. 64: 1309–1313.

Zadoks J.C., Chang T.T. and Konzak C.F. 1974. A decimal code for the growth stages of cereals. Weed Res. 14: 415–421.

CHAPTER FOURTEEN

Simulating trends in crop yield and soil carbon in a long-term experiment—effects of rising CO_2, N deposition and improved cultivation

Jørgen Berntsen, Bjørn Molt Petersen, and Jørgen E. Olesen

Received: 30 November 2005 / Accepted: 9 June 2006 / Published online: 24 August 2006
© Springer Science+Business Media B.V. 2006

Abstract Measurements of crop yield and soil carbon in the Bad Lauchstädt long-term fertiliser experiment were analysed with the FASSET model. The model satisfactorily predicted yield and soil carbon development in four treatments: no fertiliser, mineral fertiliser, farmyard manure, and farmyard manure plus mineral fertiliser. However, there was a residual between the observed and simulated yield, which was correlated with year. This could be attributed to an increase in observed yields during the last six decades. Scenario analysis showed that the most probable explanation for this yield increase was the use of new crop varieties and/or pesticides, while the increase in atmospheric CO_2 and changes in local N deposition were of lesser importance. The rise in CO_2 thus only explained 9–37% of the yield increase. The observed and simulated developments in soil carbon were quite different in the four treatments. However, the changes within each treatment for different scenarios were much smaller than the substantial difference between treatments. Thus, it was concluded that the type of nutrient applied was more important than development in CO_2 concentration or N deposition in determining soil carbon.

Keywords Atmospheric CO_2 concentration · Bad Lauchstädt · Climate change · FASSET · Long-term experiment · N deposition · Soil carbon · Yield

Introduction

Soil organic matter (SOM) is a major reservoir of carbon (C) within the biosphere. This large reservoir might both be a source and sink of C depending on land use, management, soil types, and climatic conditions. Changes in climatic conditions, especially temperature and moisture, will directly affect the rate of SOM decomposition (Jenkinson et al. 1991). Changes in atmospheric CO_2 concentration, temperature and moisture conditions will also affect the C input as crop growth and production responds to these factors. The main effect of the crops on SOM is through crop residues and their subsequent fate in the soil food web. Part of the crop residues and applied manure will renew the slowly decaying SOM pools. The balance between decay and renewal will determine the long-term trend in SOM content.

J. Berntsen (✉) · B. M. Petersen · J. E. Olesen
Department of Agroecology, Danish Institute of Agricultural Sciences, P.O. Box 50, DK-8830 Tjele, Denmark
e-mail: Jorgen.Berntsen@agrsci.dk

Present Address:
J. Berntsen
Unisense A/S, Brendstrupgårdsvej 21F, 8200 Århus N., Denmark

Long-term experiments are essential for our understanding of how changes in climate, land use, and management influence C sequestration and crop yields (e.g., Jenkinson and Rayner 1977; Christensen and Trentemøller 1995). A fundamental prerequisite in most long-term experiments is that the experimental conditions are kept as constant as possible. However, several anthropogenic factors may hamper this intention as they vary during the experimental period. This, for example, includes the increase in atmospheric CO_2 concentration, changes in local N deposition, new crop varieties, crop protection, and other technical improvements.

Thus, the full potential of such experiments can only be exploited by considering these anthropogenic factors. The most feasible way to conduct such an analysis is probably by the use of well-tested simulation models. Several models have been tested on long-term experiments (e.g. Smith et al. 1997; Jensen et al. 1997) and used for prediction of effects of climate change and management (Falloon et al. 2002).

The whole farm simulation model FASSET (Berntsen et al. 2003) uses a soil organic matter turnover model (Petersen et al. 2005a, b) that has been tested on several long-term experiments.

The purpose of this study was to test the ability of the FASSET model to predict crop yields and changes in soil C in a long-term experiment. In addition, the influence of several anthropogenic factors, such as the rise in atmospheric CO_2, N deposition, new crop protection measures, and varieties were analysed by testing different scenarios.

Materials and methods

Model

The FASSET whole farm model (Berntsen et al. 2003) was used to simulate the long-term experiment. This model contains detailed soil and crop models, described more comprehensive elsewhere (Olesen et al. 2002; Berntsen et al. 2003, 2004, 2005). Briefly, the crop model simulates daily dry matter accumulation and N uptake based on light interception by green leaves, according to Olesen et al. (2002). Daily dry matter production is translocated to the roots and above ground biomass. A fraction of the root biomass is lost by rhizodeposition (Berntsen et al. 2004), which together with top and root residues constitutes organic C and N input for the soil. Daily dry matter production can potentially be limited by nitrogen or water stress (Olesen et al. 2002). The calibration of winter wheat is described in Olesen et al. (2002), while the other crops has been calibrated on data of crop biomass development from Denmark. The crop model's response to rising CO_2 was tested on data from a FACE experiment and compared with other models (Jamieson et al. 2000). The soil model simulates daily changes in water, temperature, turnover of mineral N and organic N coupled to C turnover, and transport of solutes. The soil organic matter model consists of seven discrete C and N pools, two for added organic matter, two for soil microbial biomass, one for soil microbial residues, one for "humus" and an inert pool. The SOM model has been calibrated and tested on a very large dataset, consisting of both long- and short-term experimental data (Petersen et al. 2005a, b). The inclusion of the SOM model in FASSET is described by Berntsen et al. (2004), and it was not re-calibrated or adjusted in the present study.

Experimental data

The model was tested against field data from the German Bad Lauchstädt long-term fertiliser experiment. The experiment is described by Körschens et al. (1994), so only a brief outline will be given here. The experiment was initiated in 1902 and uses a 4-course rotation with sugar beet (*Beta vulgaris* L.), spring barley (*Hordeum vulgare* L.), potato (*Solanum tuberosum* L.), and winter wheat (*Triticum aestivum* L.). Data from four treatments were used in the study:

Plot 1. Organic farmyard manure at a rate of 30 t ha^{-1} every second year plus inorganic NPK fertiliser at varying rates (on average 59 kg N ha^{-1} y^{-1} during the experiment).

Plot 6. Organic farmyard manure at a rate of 30 t ha^{-1} every second year.

Table 1 Soil bulk density, initial carbon content, clay content and volumetric water content at different soil water pressures (pF = log(-cm H$_2$O))

Layer	Bulk density (g cm^{-3})	Carbon (%)	Clay (%)	Volumetric soil water content (%)						
				0	1.8	2.5	3.0	3.4	3.8	4.2
0–20	1.295	1.8	21	50.4	41.2	30.8	21.7	20.7	18.1	17.2
20–45	1.445	1.8	21	50.4	41.2	30.8	21.7	20.7	18.1	17.2
45–125	1.425	1.4	21	46.6	38.7	31.6	26.0	24.1	20.6	19.5

Plot 13. Inorganic NPK fertiliser at varying rates (on average 82 kg N ha^{-1} y^{-1} during the experiment).

Plot 18. No fertilisation.

Initial soil properties (Table 1) were estimated as an average of the measured properties of plot 1 and plot 18. The climatic data contain observations of temperature, precipitation and sunshine hours or global radiation. The sunshine hours were converted to global radiation according to Rietveld (1978). The annual average temperature was 8.6°C and the average annual precipitation 482 mm.

No site-specific calibrations of crop or soil parameters were performed. The model used the observed dates for soil treatment, harvest, mineral fertilisation, and application of farmyard manure. Based on observations of emergence dates, the date of sowing was assumed as 25, 16, 14 and 12 days before the emergence date of beets, spring barley, winter wheat, and potato, respectively. This was estimated from the build-in standard temperature sum from sowing to emergence and the average temperature in the month of sowing.

The composition of farmyard manure was not measured, thus the following composition was assumed: 25% dry matter, 9% C, 0.56% organic N and 0.04% NH$_4^+$.

The observed fresh matter yield was converted to dry matter by assuming a dry matter content of 85%, 85%, 20% and 24% for spring barley, winter wheat, beets, and potato, respectively (Møller et al. 2000).

Scenarios

Several anthropogenic factors are known to have changed during the 20th century. The most important of these are the rise in atmospheric CO$_2$, changes in N deposition, crop protection, and new crop varieties. Five scenarios were constructed to analyse the effect of some of these factors.

(1) Standard. A constant atmospheric CO$_2$ concentration of 321 ppm was assumed (yearly average in scenario 2). An average N deposition of 39 kg N ha^{-1} y^{-1} was assumed (yearly average of scenario 3).

(2) Variable CO$_2$. Atmospheric CO$_2$ concentrations were taken from Amthor (1998) and Keeling and Whorf (2004).

(3) Variable N deposition. Data on N depositions were taken from Asman and Drukker (1988) and Asman et al. (1987). They estimated that the deposition for the investigated area in 1870, 1920, 1950 and 1980 was 26.5, 29.8, 36.5 and 61.0 kg N ha^{-1} y^{-1}, respectively. The yearly N deposition was estimated by linear interpolation between these points. After 1980, the N deposition was assumed constant.

(4) Variable CO$_2$ and N deposition. Combination of scenarios 2 and 3.

(5) Increased radiation use efficiency (RUE). To reflect the increased use of pesticides and the introduction of improved varieties, the radiation use efficiency was increased. For simplicity, this increase was assumed to be linear and was estimated from the wheat yield residual by minimising the squared error between observations and simulations in scenario 1. The atmospheric CO$_2$ concentration and N deposition were as in scenario 1.

Statistics

To compare the simulations and observations, three different statistical analyses were used.

First, a linear regression between observations and simulations was used. Second, the root mean square error (RMSE) was calculated:

$$\text{RMSE} = \left(\frac{1}{N}\sum_{i=1}^{N}(p_i - m_i)^2\right)^{0.5}$$

where m_i is the measured value at the i'th sampling date, p_i is the corresponding simulated value and N in the number of observations. The lower the RMSE, the better does the simulated values agree with the observations. Third, the modelling efficiency (EF) was calculated

$$\text{EF} = \frac{\sum_{i=1}^{N}(m_i - \overline{m})^2 - \sum_{i=1}^{N}(p_i - m_i)^2}{\sum_{i=1}^{N}(m_i - \overline{m})^2}$$

where $\overline{m}$ is the average of all m_i. The maximum value of EF is 1, which indicates that observed and simulated values are identical. If EF is less than zero, the model predicted values are worse than simply using the observed mean of the measured data (Loague and Green 1991). The statistical analysis was performed using the Statistical Analysis System (SAS Institute 1996). The annual trend in yield residuals was analysed by a piecewise linear function:

$$r_x = \begin{cases} s_0 + ax & x < x0 \\ s_1 + bx & x \geq x0 \end{cases} \quad (1)$$

where r_x (g DM m^{-2}) is the yield residual in year x, $x0$ is the year where the curve breaks, a and b are the slopes of the regression line (g DM m^{-2} y^{-1}) and s_0 and s_1 are the intercepts of the regression lines. Note that if a scenario could explain all the trends in yield, the b parameter would equal 0. By comparing the b parameter in the standard scenario with the b parameter in a scenario, it is possible to estimate how much of the yield increase can be explained by the scenario. As the function is continuous in $x0$, only 4 out of the above 5 parameters need to be estimated. The parameters were estimated using the NLIN procedure in the Statistical Analysis System (SAS Institute 1996).

Results

C content

For the treatments that included manure (plots 1 and 6) there was a good agreement between measured and simulated soil C (Fig. 1). The simulation of the unfertilised plot 18 slightly underestimated soil C, while the simulations of the NPK-fertilised plot 13 greatly underestimated soil C.

Yields

The highest correlations between the simulations and observations were observed for spring barley, beet, and winter wheat while the correlation was lower for potatoes (Fig. 2). The modelling efficiency (EF) was greater than zero for spring barley, beet, and winter wheat showing that the model for these crops is better than using a simple average.

The residuals between the observed and simulated crop yield were correlated with year (Fig. 3). The annual trend in observed yield was estimated from Eq. 1. The estimated yield increase in the last period of the century was 3.5, 2.7, 12.0 and 3.3 g DM m^{-2} y^{-1} for spring barley, beets, potato, and winter wheat, respectively (Table 2).

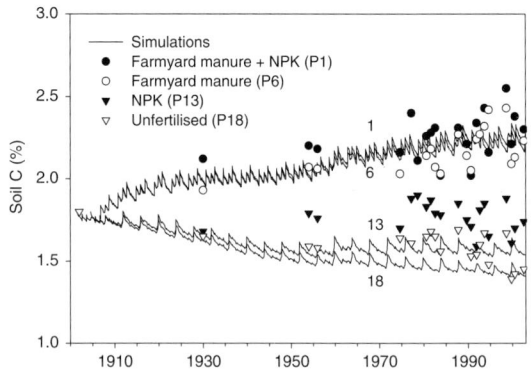

Fig. 1 Development in simulated and observed soil carbon content in 0–20 cm. Different symbols and lines represent different plots. The numbers in the plots refer to simulations of the given plot

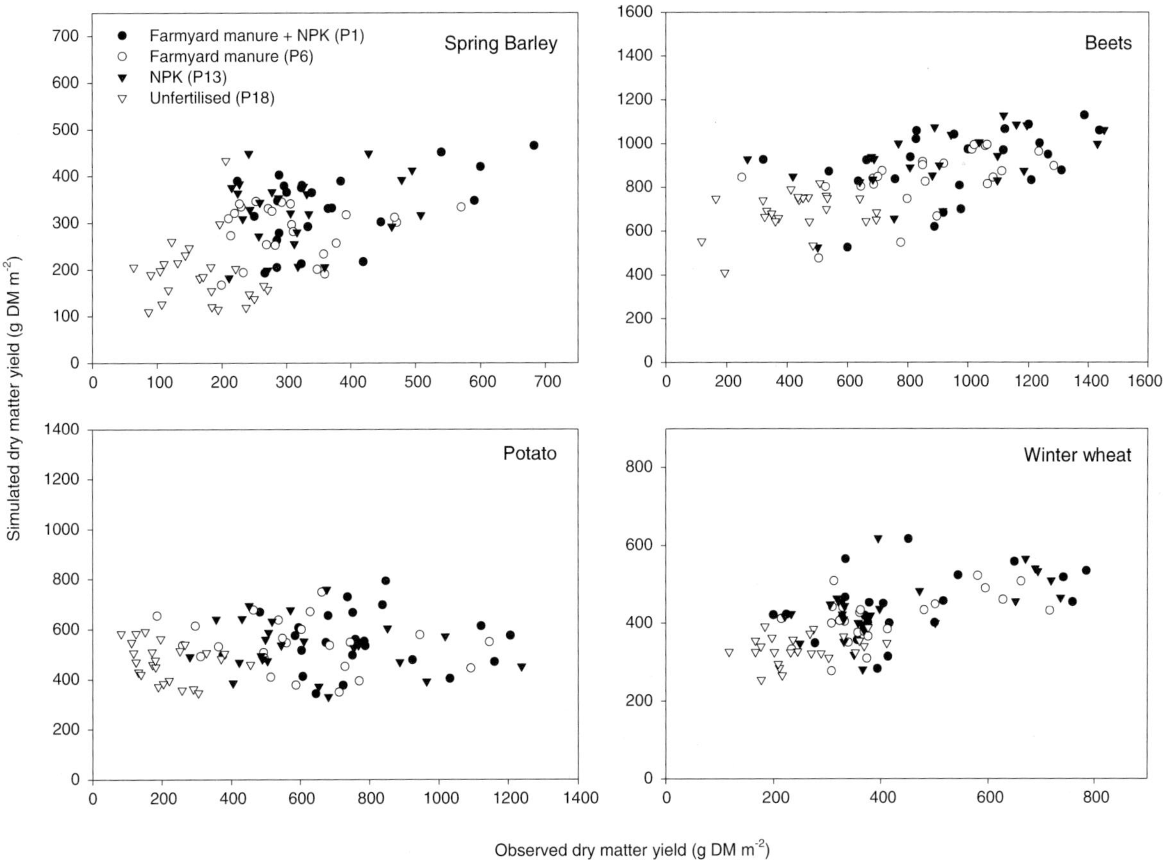

Fig. 2 Simulated dry matter yield (g DM m^{-2}) plotted against observed dry matter yield (g DM m^{-2}) for spring barley, beet, potato and winter wheat, respectively. Symbols represent the four different plots

Scenarios

Based on the above trend in wheat yield residual the crop sub-model was re-calibrated. A linear increase in the maximum RUE parameter of 0.04 g MJ^{-1} y^{-1} was estimated by minimising the wheat yield residual. The increase was assumed to start in 1942 (average of $x0$ in Table 2). These parameters were used in scenario 5 (increased RUE).

In all cases, the simulated values for soil C were higher than for the standard scenario, and most pronounced for scenario 5 with increased radiation use efficiency (Fig 4. and Table 3). However, the effects on the soil C development were in all cases rather small (Table 3).

For nearly all crops, the scenario modifications improved the correlation between the simulated and observed yields (Table 4). This was observed, when the correlation was estimated by linear regression (r^2), RMSE, or modelling efficiency (EF).

Based on Eq. 1, it was calculated that nine percent, 37%, 10% and 21% of the yield increases in spring barley, beets, potato, and winter wheat, respectively, could be explained by an increase in CO_2 (Table 2) (scenario 2). For the N deposition scenario, 11%, 40%, 2% and 15% of the yield increases in spring barley, beets, potato, and winter wheat, respectively, could be explained (Table 2).

Discussion

Soil carbon

The simulations captured the trends of the soil C content rather well, although the soil C in the

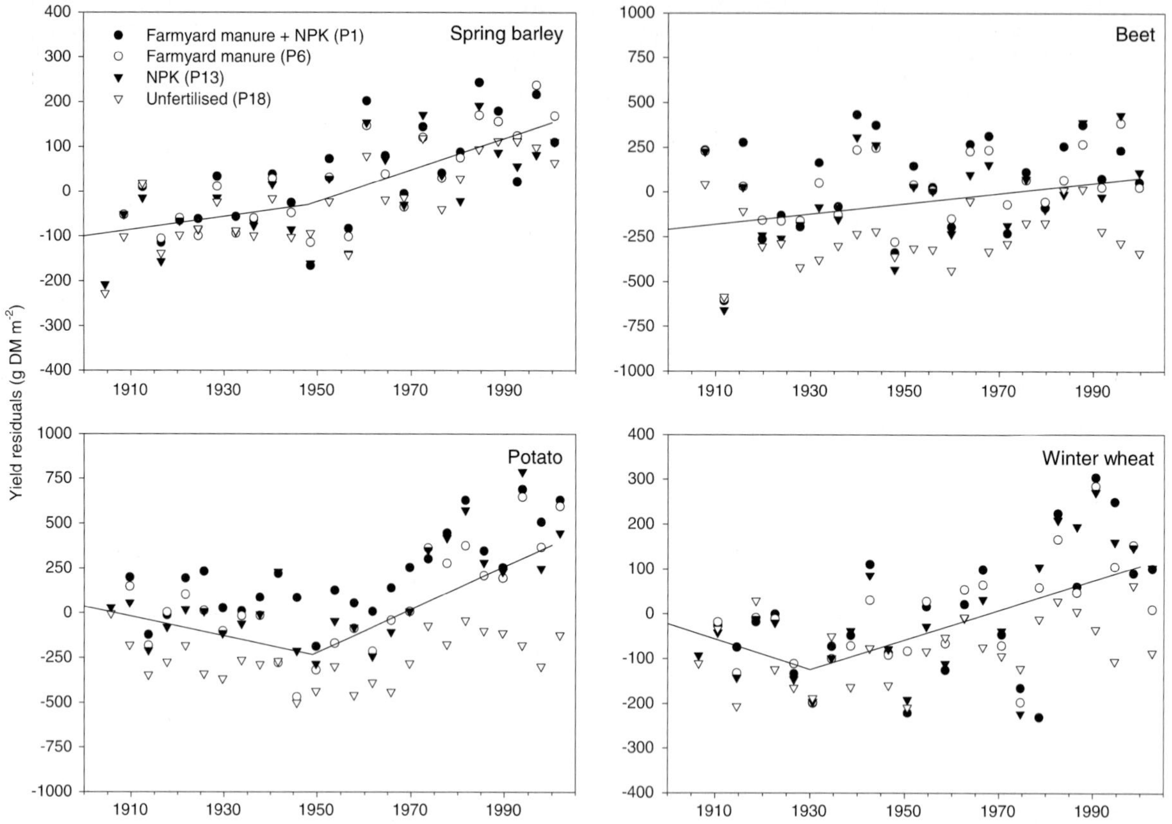

Fig. 3 The residual between observed and simulated yield (g DM m^{-2}) plotted against year for spring barley, beet, potato and winter wheat, respectively. Symbols as in Fig. 2. Regression lines are based on Eq. 1

Table 2 Statistics on the fit of Eq. 1 to the yield residuals. a is the slope (g DM m^{-2} y^{-1}) in the period before $x0$ and b is the slope (g DM m^{-2} y^{-1}) after $x0$

Crop	Parameter	Standard	Variable CO$_2$	Variable N deposition	Var CO$_2$ + N	Changed RUE
Spring Barley						
	a	1.5	0.8	1.1	1.1	1.2
	b	3.5	3.2	3.1	2.8	2.9
	$x0$	1948	1939	1948	1948	1948
Beets						
	a	–	–	2.6	2.2	4.0
	b	2.7	1.7	1.6	0.4	–7.5
	$x0$	1904	1904	1939	1939	1939
Potato						
	a	–4.8	–5.3	–4.9	–5.4	–4.8
	b	12.0	10.8	11.8	10.5	5.9
	$x0$	1949	1949	1950	1949	1956
Winter wheat						
	a	–3.4	–3.2	–3.5	–3.2	–1.6
	b	3.3	2.6	2.8	2.1	-
	$x0$	1930	1930	1930	1930	2003

treatment receiving only mineral fertiliser was underestimated, especially in the standard scenario. The overall best accordance with the measured values was achieved with scenario 5 (increased RUE). So the increased yields, leading to an increased return of crop residues, may have

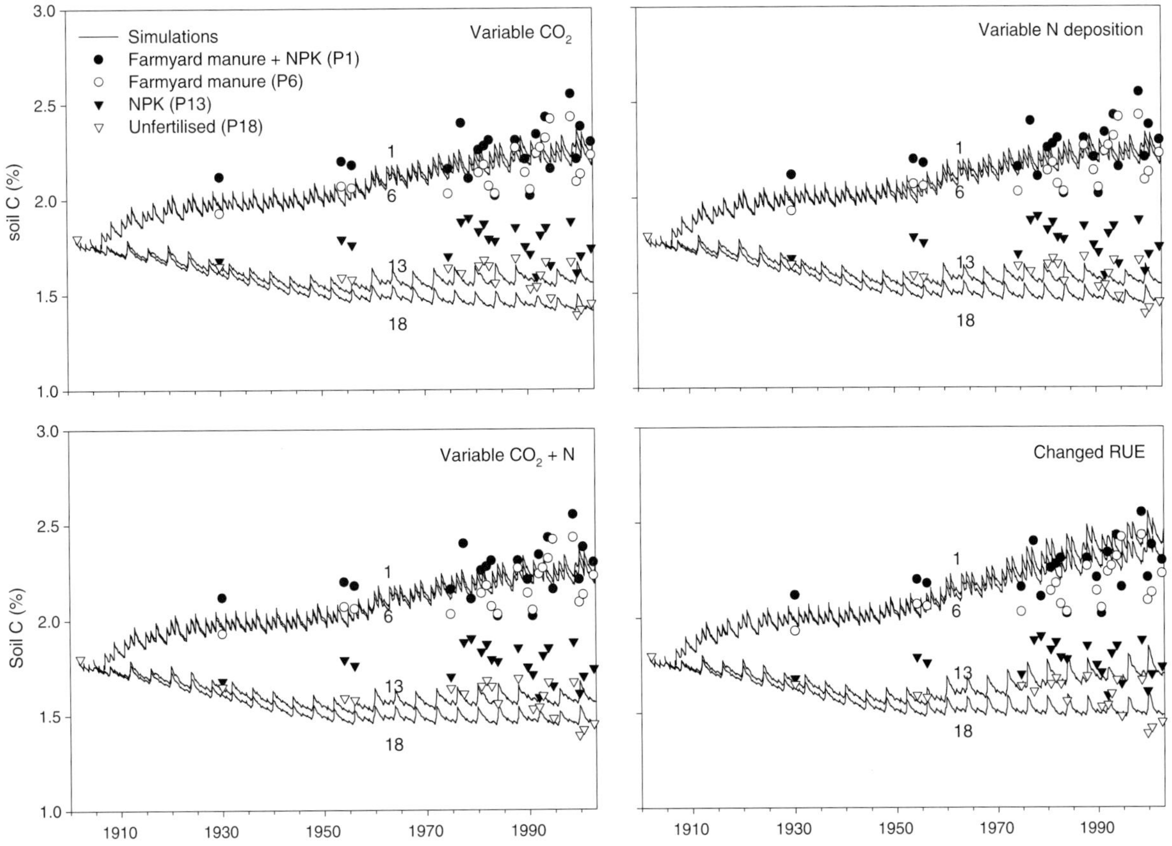

Fig. 4 Simulated soil carbon content in 0–20 cm plotted against years for the four scenarios (see text for explanation). Symbols as in Fig. 1

Table 3 Average soil carbon (C %) during the period 1984–2002. For observations the standard error are shown in parentheses

Plot	Observations	Standard	Variable CO₂	Variable N deposition	Var CO₂ + N	Changed RUE
Farmyard manure + NPK (P1)	2.32 (0.15)	2.25	2.25	2.25	2.26	2.39
Farmyard manure (P6)	2.26 (0.16)	2.21	2.21	2.21	2.22	2.31
NPK (P13)	1.71 (0.11)	1.57	1.58	1.57	1.59	1.70
Unfertilized (P18)	1.48 (0.11)	1.45	1.45	1.47	1.48	1.51

had a noticeable effect on the soil C level in the second half of the 20th century. However, the variation between scenarios within a single treatment was always smaller than the variation between experimental treatments. Thus, the effect of fertiliser and manure was much greater than the effect of changes in environment and cultivation practices.

The use of increased RUE to reflect both yield improvements mediated by pesticides and improved varieties of course represents a simplification. In a study of older wheat varieties, Shearman et al. (2005) estimated that the preanthesis RUE increased 0.012 g DM MJ^{-1} y^{-1} during the period 1970–1995. This is smaller than the increase of 0.04 g DM MJ^{-1} y^{-1} in maximum RUE estimated in this study. However, the values cannot be directly compared because our maximum RUE was reduced by growth limiting factors such as water and nitrogen. In addition, the

Table 4 Statistical evaluation of the model simulations for observed yields and soil C for each scenario. a and b are interception and slope, respectively

Crop	Parameter	Standard	Variable CO_2	Variable N deposition	Var CO_2 + N	Changed RUE
Spring Barley						
	a	170***	164***	165***	156***	143***
	b	0.39***	0.41***	0.42***	0.46***	0.53***
	r^2	0.27	0.31	0.35	0.40	0.37
	RMSE	105	101	97	92	99
	EF	0.22	0.27	0.33	0.39	0.29
Beets						
	a	589***	556***	615***	583***	480***
	b	0.32***	0.34***	0.29***	0.32***	0.69***
	r^2	0.41	0.40	0.37	0.38	0.44
	RMSE	349	251	262	257	253
	EF	0.21	0.38	0.32	0.34	0.36
Potato						
	a	501***	476***	505***	481***	480***
	b	0.04	0.08*	0.04	0.08*	0.31***
	r^2	0.01	0.06	0.01	0.05	0.27
	RMSE	291	279	291	279	264
	EF	−0.05	0.03	−0.05	0.03	0.13
Winter wheat						
	a	286***	254***	286***	254***	148***
	b	0.31***	0.40***	0.33***	0.42***	0.96***
	r^2	0.40	0.50	0.42	0.50	0.60
	RMSE	152	114	124	125	115
	EF	0.30	0.45	0.36	0.35	0.45
Soil C						
	a	−0.16	−0.16	−0.12	−0.13	−0.20
	b	1.04***	1.04***	1.02***	1.03***	1.10***
	r^2	0.85	0.85	0.85	0.84	0.85
	RMSE	0.16	0.16	0.15	0.15	0.14
	EF	0.72	0.72	0.73	0.73	0.78

*$P<0.05$, ***$P<0.001$

estimate includes root production which was not included in the study by Shearman et al. (2005). The improvement of crop management and genetics has also influenced the size and duration of the leaf area and the harvest index. Such changes in C allocation were not covered by this study, although they may affect the C sequestration in the soil.

The modelled trends in soil C for the different treatments depends on the balance between input of crop residues and manure, the fraction of these inputs that are incorporated in the "humus" pool (NOM pool, see Petersen et al. 2005a), and the decay of this pool, determined by its size, temperature and water conditions. So the fact that the trends are reflected rather well by the model, does not in itself imply that the C and N fluxes are represented with high accuracy, but only that the above balance is well described.

Yields

The low correlation between simulated and observed yield for potato (Fig. 2 and Table 4) is mainly due to the high, simulated yields in the unfertilised treatment. This indicates that the FASSET N-response for potato (taken from Duchenne et al. 1997) is underestimated. However, this can only be solved by making new N fertilisation experiments with the variety used in the experiment and testing the model on these data. In addition, potatoes are known to be especially susceptible to potato late blight (*Phytophthora infestans*). The high year-to-year variations of this disease are very difficult to capture with field scale models.

The composition of manure was assumed constant during the entire period, as there were no measurements of the composition available. If the composition did change significantly this might affect the simulations and conclusions. However, there appeared to be no significant deviations between observed and simulated crop yield in plots with and without manure.

Several other authors have simulated yields for the Bad Lauchstädt experiment (e.g. Jensen et al. 1997; Li et al. 1997). However, the yield trends were not analysed. The trends in yield residuals (Fig. 3) could be attributed to the well-known yield increase especially after 1940 (Amthor 1998). This general increase is caused by a number of different direct factors such as increased fertilisation, pesticides, new varieties and other technical improvements. In addition, several indirect factors, such as CO_2 increase and N deposition, are known to affect yields. Amthor (1998) estimated an increase of about 7.5 g DM m^{-2} y^{-1} for UK wheat yields. Chloupek et al. (2004) estimated an increase of 6.0, 8.9 and 9.7 g DM m^{-2} y^{-1} during the last 40 years for the Czech Republic, EU and Germany, respectively. Olesen et al. (2000) found increases between 8.5 and 12.8 g DM m^{-2} y^{-1} in Denmark after 1970. However, these estimates include effects of increasing N fertilisation, which is not included in the above residuals, as the simulation uses the exact same N fertilisation levels as the experiment. According to Chloupek et al. (2004), the effect of fertilisers constitutes about 64%, 70%, 64% and 55% of the annual yield increase for winter wheat, barley, potato, and beet, respectively. Thus, subtracting the N fertiliser effect brings the above estimates close to the 3.3 g DM m^{-2} y^{-1} estimated in this study.

Scenarios

Weigel et al. (2000) estimated an N deposition for Bad Lauchstädt of 50–58 kg N ha^{-1} y^{-1} using an indirect estimate and 65 kg N ha^{-1} y^{-1} with a direct method. This is in good agreement with the 61 kg N ha^{-1} y^{-1} used in the current study.

Vleeshouwers and Verhagen (2002) predicted an increase in C sequestration in European soils of 1 g C m^{-2} y^{-1} due to an expected CO_2 rise in the period 2008–2012, while incorporation of straw and application of manure was estimated at 15 and 150 g C m^{-2} y^{-1}, respectively. A similar conclusion was made by Smith (2005), who considered the impact of CO_2 fertilisation of crops very small compared to deliberate efforts to increase soil carbon by appropriate agricultural management.

For nearly all crops, the scenarios improved the correlation between the simulated and observed yields (Table 4). The general tendency was that the RUE scenario had the highest correlation, followed by the variable CO_2 + N, the variable CO_2 and the variable N scenario. This indicates that the development in crop protection and new varieties has been more important for crop yield improvements than the response to increasing CO_2 levels and N deposition. A similar conclusion regarding CO_2 concentrations was made by Amthor (1998), who concluded that "future CO_2 increases will play only a minor role in future yield increases if technology and management continue to increase crop yields at nearly the present rates". However, as noted by Olesen and Bindi (2002), this only holds in situations where yields are not limited by environmental conditions.

Conclusion

The modelling efficiency (EF) showed that the FASSET model described the yield of spring barley, beets and winter wheat significantly better than a simple average. In addition, the simulated soil C agreed well with the observations. The variation in soil C between plots was greater than the estimated variation in the scenarios. Thus, the effect of different fertiliser types was much greater than the analysed anthropogenic factors.

The yield residuals showed significant trends. Scenario analyses showed that this could mostly be attributed to improved technology and only to a lesser degree to changes in CO_2 concentrations and N deposition.

Thus, future studies of climate change need to consider the improvements and changes in technology, varieties and management. Model forecasts assuming current yields and practice may miss major factors impacting the C sequestration in agricultural soils.

Acknowledgements Willem Asman is thanked for data and help with the N deposition scenarios. The work was supported by the Danish Ministry of Food, Agriculture and Fisheries under the research programme "Agriculture from a holistic resource perspective". We also wish to thank Margit Schacht for greatly improving the readability of the paper.

References

Amthor JS (1998) Perspective on the relative insignificance of increasing atmospheric CO_2 concentration to crop yield. Field Crops Res 58:109–127

Asman WAH, Drukker B (1988) Modelled historical concentrations and depositions of ammonia and ammonium in Europe. Atm Env 22:725–735

Asman WAH, Drukker B, Janssen AJ (1987) Estimated historical concentrations and depositions of ammonia and ammonium in Europe and their origin (1870–1980). Report R–87–2. Instituut voor Meteorologie en oceanografie, Utrecht. The Netherlands

Berntsen J, Hauggaard-Nielsen H, Olesen JE, Petersen BM, Jensen ES, Thomsen A (2004) Modelling dry matter production and resource use in intercrops of pea and barley. Field Crops Res 88:69–83

Berntsen J, Petersen BM, Jacobsen BH, Olesen JE, Hutchings NJ (2003) Evaluating nitrogen taxation scenarios using the dynamic whole farm simulation model FASSET. Agr Syst 76:817–839

Berntsen J, Petersen BM, Olesen JE, Eriksen J, Søegaard K (2005) Simulation of residual effects and nitrate leaching after incorporation of different ley types. Eur J Agron 23:290–304

Chloupek O, Hrstkova P, Schweigert P (2004) Yield and its stability, crop diversity, adaptability and response to climate change, weather and fertilisation over 75 years in the Czech Republic in comparison to some European countries. Field Crops Res 85:167–190

Christensen B, Trentemøller U (1995) The Askov Long-Term Experiments on Animal Manure and Mineral Fertilizers. 100th Anniversary Workshop. SP report no 29. Danish Institute of Plant and Soil Science

Duchenne T, Machet JM, Martin M (1997) Potatoes. In: G Lemaire (ed) Diagnosis of the nitrogen status in crops. Springer, Berlin Heidelberg, Germany, pp. 119–130

Falloon P, Smith P, Szabó J, Pásztor L (2002) Comparison of approaches for estimating carbon sequestration at the regional scale. Soil Use Manage 18:164–174

Jamieson PD, Berntsen J, Ewert F, Kimball BA, Olesen JE, Pinter PJ, Porter JR, Semenov MA (2000) Modelling CO_2 effects on wheat with varying nitrogen supplies. Agr Eco Env 82:27–37

Jenkinson DS, Adams DE, Wild A (1991) Models estimates of CO_2 emissions from soil in response to global warming. Nature 351:304–306

Jenkinson DS, Rayner JH (1977) The turnover of soil organic matter in some of the Rothamsted classical experiments. Soil Sci 123:298–305

Jensen LS, Mueller T, Nielsen NE, Hansen S, Crocker GJ, Grace PR, Klír J, Körschens M, Poulton PR (1997) Simulating trends in soil organic carbon in long-term experiments using the soil-plant-atmosphere model DAISY. Geoderma 81:29–44

Keeling CD, Whorf TP (2004) Atmospheric CO_2 records from sites in the SIO air sampling network. In Trends: A Compendium of Data on Global Change. Carbon Dioxide Information Analysis Center, Oak Ridge National Laboratory, U.S.A

Körschens M, Stegemann K, Pfefferkorn A, Weise V, Müller A (1994) Der Statische Düngerungsversuch Bad Lauchstädt nach 90 Jahren. Teubner, Stuttgart, Germany

Li C, Frolking S, Crocker GJ, Grace PR, Klír J, Körchens M, Poulton PR (1997) Simulating trends in soil organic carbon in long-term experiments using the DNDC model. Geoderma 81:45–60

Loague K, Green E (1991) Statistical and graphical methods for evaluating solute transport models: Overview and application. J Contam Hydrol 7:51–73

Møller J, Thøgersen R, Kjeldsen AM, Weisbjerg MR, Søegaard K, Hvelplund T, Børsting CF (2000) Feedstuff table. Composition and feeding value of feedstuffs for cattle. (in Danish). Report no. 91. Danish Agricultural Advisory Service. Denmark

Olesen JE, Bindi M (2002) Consequences of climate change for European agricultural productivity, land use and policy. Eur J Agron 16:239–262

Olesen JE, Bøcher PK, Jensen T (2000) Comparison of scales of climate and soil data for aggregating simulated yields of winter wheat in Denmark. Agr Eco Env 82:213–228

Olesen JE, Petersen BM, Berntsen J, Hansen S, Jamieson PD, Thomsen AG (2002) Comparison of methods for simulation effects of nitrogen on green area index and dry matter growth in winter wheat. Field Crop Res 74:131–149

Petersen BM, Berntsen J, Jensen LS (2005a) CN-SIM - a model for the turnover of soil organic matter. I: Long term carbon development. Soil Biol Biochem 37:359–374

Petersen BM, Jensen LS, Hansen S, Pedersen A, Henriksen TM, Sørensen P, Trinsoutrot-Gattin I, Berntsen J (2005b) CN-SIM - a model for the turnover of soil organic matter. II: Short term carbon and nitrogen development. Soil Biol Biochem 37:375–393

Rietveld MR (1978) A new method for estimating the regression coefficients in the formula relating solar radiation to sunshine. Agric Meteorol 19:243–252

SAS Institute (1996) SAS/STAT Software: Changes and enhancements through Release 6.11. SAS Institute Inc., Cary, NC, USA

Shearman VJ, Sylvester-Bradley R, Scott RK, Foulkes MJ (2005) Physiological processes associated with wheat yield progress in the UK. Crop Sci 45:175–185

Smith P (2005) Limited increase of agricultural soil carbon and nitrogen stocks due to increased atmospheric CO_2 concentrations. J Crop Imp 13:393–399

Smith P, Smith JU, Powlson DS, McGill WB, Arah JRM, Chertov OG, Coleman K, Franko U, Frolking S, Jenkinson DS, Jensen LS, Kelly RH, Klein-Gunnewiek H, Komarov AS, Li C, Molina JAE, Mueller T, Parton WJ, Thornley JHM, Whitmore AP (1997) A comparison of the performance of nine soil organic matter models using datasets from seven long-term experiments. Geoderma 81:153–225

Vleeshouwers LM, Verhagen A (2002) Carbon emission and sequestration by agricultural land use: a model study for Europe. Global Change Biol 8:519–530

Weigel A, Russow R, Körschens M (2000) Quantification of airborne N-input in long-term field experiments and its validation through measurements using ^{15}N isotope dilution. J Plant Nutr Soil Sci 163:261–265

CHAPTER FIFTEEN

Comparison of methods for the estimation of inert carbon suitable for initialisation of the CANDY model

Martina Puhlmann[1,*], Katrin Kuka[2] and Uwe Franko[2]

[1]*Leibniz-Centre for Agricultural Landscape Research (ZALF) e. V., Institute of Landscape Systems Analysis, Eberswalder Straße 84, 15374, Müncheberg, Germany;* [2]*Department of Soil Science, UFZ – Centre for Environmental Research Leipzig-Halle, Theodor-Lieser-Str. 4, 06120, Halle, Germany; *Author for correspondence (e-mail: martina.puhlmann@zalf.de; fax: +49-334-3282334)*

Received 25 August 2005; accepted in revised form 23 January 2006

Key words: CANDY, Carbon and nitrogen dynamics, Inert carbon, Model initialisation, Modelling

Abstract

Almost all soil organic carbon turnover models rely on a partitioning of total organic carbon into an inert and a decomposable pool. The quantification of these pools has a large impact on modelling results. In this study several methods to estimate inert carbon in soils, based either on total soil organic matter or physical protection, were assessed with the objectives of (1) minimising errors in carbon and nitrogen dynamics and (2) ensuring usability for sites with marked differences in site conditions. CANDY simulations were carried out by varying solely the method for calculating the size of the inert carbon pool used to initialise the model. Experimental data from Bad Lauchstadt and Müncheberg were used for the simulation. The data were made available for modellers at a workshop held at Müncheberg (Germany) in 2004. The results concerning not only carbon but also nitrogen dynamics were analysed by applying selected statistical methods. It was shown that even in short-term simulations model initialisation procedure may influence the simulation results considerably. Three methods of estimating inert carbon were identified as being the most appropriate. These methods are either based on soil texture or pore-space classes and therefore account for the physical protection of soil organic matter. Thus, physical protection seems to be of major importance. By extending the scope of the investigation into nitrogen dynamics, additional support for the applicability of a selected method was obtained.

Introduction

The simulation model CANDY (Carbon and Nitrogen Dynamics, Franko et al. 1995, 1997; Franko 1996, 1997) has been developed in order to provide information about carbon stocks in soils, organic matter turnover, N uptake by crops, leaching and water quality. A comparison of soil organic carbon models in 1997 showed CANDY to be in the group with the best performance (Smith et al. 1997).

In most long-term organic matter models (Hansen et al. 1991; Franko et al. 1995; Coleman and Jenkinson 1996; Parton 1996, etc.), soil organic matter is partitioned in several pools with different turnover rates. In CANDY the decomposable soil organic matter can be subdivided into biologically active and stabilised soil organic matter. The decomposable and the inert organic matter contribute to the total stock of carbon in the soil. The inert organic matter is considered to be stable and, therefore, does not participate in turnover processes.

As in other models (Falloon et al. 2000; Bruun and Jensen 2002) the manner in which CANDY is initialised influences the simulation results considerably. Model initialisation may be responsible for the behaviour of a soil as a source or sink of atmospheric CO_2 (Falloon and Smith 2000). In order to initialise the CANDY model, a value for decomposable carbon (C_{DEC}) has to be specified by the user. Because it is not possible to measure the decomposable pool directly, in CANDY C_{DEC} can be calculated from the history of the plot and the site-specific turnover conditions, or from the difference between an organic carbon (C_{ORG}) measurement and the estimated inert carbon (C_I). Thus, the amount of C_{DEC} is affected by uncertainties in estimating C_I. For example, if decomposable carbon is too high, the mineralisation of carbon and nitrogen may be overestimated, leading to a misinterpretation of organic matter dynamics and a surplus of mineral nitrogen in soil.

In the literature, several methods for calculating the amount of C_I are described. They use either the relation between C_I and soil texture (Körschens 1980; Körschens et al. 1998; Rühlmann 1999) or estimate C_I as a part of the whole amount of soil organic carbon (Falloon et al. 1998, 2000). A new approach by Kuka et al. (2006), based on pore-space classes, shows that soil organic matter (SOM) localised in micro-pores is stabilised over a long time. Thus, this part of organic carbon should be very similar to the inert pool of the CANDY model.

In the investigation reported here, we tested all of these methods on their applicability for initialising the inert carbon pool in the CANDY model. Our objective was to find a method which fulfils the following criteria: (1) minimises errors in carbon and nitrogen dynamics and (2) provides usability for sites with marked differences in site conditions. In order to achieve our objective CANDY simulations were carried out by varying solely the method of estimating the initial value for C_I. Simulation results for soil organic carbon and soil mineral nitrogen of the sites 'Bad Lauchstädt' and 'Müncheberg' were compared with observations. From among the huge number of statistical methods that can be found in literature (Willmott and Wicks 1980; Fox 1981; Addiscott and Whitmore 1987; Loague and Green 1991; Smith et al. 1996, etc.) we used the mean, the root mean square error (RMSE; Fox 1981), the mean bias error (MBE; Addiscott and Whitmore 1987) and the index of agreement (IA; Willmott and Wicks 1980) for comparing predicted and observed values and assessing the C_I estimations.

Materials and methods

CANDY model

CANDY consists of a modular system of submodels and a data base system for model parameters, initial values, weather data, soil management data, and measurement values. The submodels of CANDY are described briefly below and in detail by Franko et al. (1995).

In the soil temperature model the heat flow equation is solved based on a statistical approach for the calculation of the soil surface temperature.

The hydrological model is based on a capacity approach, and takes into consideration the draining of water by gravitation forces, interception of water by crops, potential and actual evapotranspiration, surface runoff and snow cover dynamics.

The crop model in its standard version consists of parameters describing the temporal development of crop height, soil cover and rooting depth as piecewise linear functions. Nitrate uptake by plants is calculated by a sigmoidal function for the distribution of the N demand over vegetation time. After harvest, a yield-dependent amount of organic matter from roots and plant residues is recycled to the soil as fresh organic matter with crop-specific quality parameters.

The organic matter turnover model includes the soil nitrogen model. Soil organic matter is subdivided in several compartments: up to six different pools of added organic matter, two pools of decomposable soil organic matter (one active and one stabilised) and an inert soil organic matter pool, which is independent from climate and agriculture and stays constant over time. Turnover dynamics of all degradable carbon pools follow first-order kinetics but are influenced by soil temperature, soil moisture and aeration conditions indicated by soil texture and depth of the soil layer. Nitrogen dynamics are connected to the carbon fluxes via the C/N-ratio of the pools concerned. Mineral nitrogen is divided into a nitrate and an ammonium pool. The nitrogen-related processes are: plant uptake, mineral nitrogen input including atmospheric deposition, input of organic manure or plant residues, nitrate leaching, nitrification, denitrification and mineralisation of soil organic matter leading to nitrogen mineralisation or immobilisation.

The model inputs can be classified as parameters and scenario data. The parameters are:
- plant development characteristics (standard) – crop height, vegetation time, maximum root depth, etc.;
- soil data – soil density, particle density, field capacity, permanent wilting point, saturated hydraulic conductivity, amount of clay and silt;
- organic matter characteristics – C/N-ratio, dry matter content, turnover time, SOM reproduction coefficient.

The scenario data include:
- initial conditions for carbon, nitrogen and soil moisture;
- agricultural management data – emergence, harvesting, fertilisation, tillage, yield, N-uptake;
- weather data – precipitation, air temperature and global radiation preferably on a daily basis.

Datasets from Bad Lauchstädt and Müncheberg and simulation with CANDY

The datasets from Bad Lauchstädt (crop rotation) and Müncheberg (plot 1) used for the model simulation are described in detail in Franko et al. (2007) and Mirschel et al. (2007). In addition to these data we used the organic carbon measurements presented in Table 1. For Müncheberg no value at the beginning of the simulation was available. Therefore, the first value in Table 1 was estimated with a trend line using the four measured values. The simulation for both sites has been fitted to meet the starting C_{ORG} values. For the mineral nitrogen measurements the reader is referred to Franko et al. (2007) and Mirschel et al. (2007).

The simulation period for the Bad Lauchstädt crop rotation (BL) lasted from 1 September 1996 to 31 December 2003 and at Müncheberg plot 1 (MÜ) from 1 September 1992 to 31 December

Table 1. Measured soil organic carbon values for Bad Lauchstädt and Müncheberg.

Bad Lauchstädt		Müncheberg	
Date	C_{ORG}	Date	C_{ORG}
13 August 1997	2.13	01 September 1992	0.63[a]
23 September 1999	2.14	03 May 1993	0.58
27 October 2000	2.19	21 September 1993	0.67
23 October 2001	2.03	17 July 1995	0.51
09 October 2002	2.07	01 October 1997	0.56
22 September 2003	2.06	–	–

[a]Estimated with a trend line using the four measured values

1998. Input data such as soil parameters, daily meteorological data and management information were taken from the datasets. In some cases data for the nitrogen uptake by crops, the vegetation time, the carbon input via straw, and the C/N-ratio of straw were not specified but deducible from the datasets. If no data were available or deducible – for example, for C and N input with roots and crop residues – CANDY standard parameters (C/N-ratio, dry matter, rate and synthesis coefficients, etc.) were used. The soil parameters for wilting point and field capacity were adapted to measurements of soil moisture.

In order to initialise the CANDY model, a value for C_{DEC} in soil has to be specified. According to the equation

$$C_{DEC} = C_{ORG} - C_I, \quad (1)$$

with C_{ORG} = total organic carbon in soil and C_I = inert carbon. C_{DEC} were calculated using the starting C_{ORG} values from Table 1 and C_I values were calculated with the methods mentioned below. This procedure resulted in five initial values for C_{DEC}. With each initial value a CANDY simulation was started. No other parameters or scenario data were changed. Therefore, differences in simulated carbon and nitrogen dynamics came from the different initial values used.

Calculation of inert carbon (C_I)

Körschens (1980) and Körschens et al. (1998) found a relationship between C_I and the content of clay and fine silt which can be described with the equation:

$$C_{\text{I-KÖ}} = a \cdot b \quad (2)$$

where b is the percentage of soil particles $< 6\ \mu m$ and a is usually taken as 0.04. According to Schulz (1997) the regression coefficient a has a range from 0.04 to 0.05. In our investigation we used both 0.04 ($C_{\text{I-KÖI}}$) and 0.05 ($C_{\text{I-KÖII}}$) to define the possible range of C_I.

Rühlmann (1999) suggested an equation to describe the influence of soil texture on the carbon content of long-term bare fallow soils:

$$C_{\text{I-RÜ}} = 0.017 \cdot c - 0.001 \cdot \exp(0.075 \cdot c) \quad (3)$$

where c is the percentage of soil particles $<20\ \mu m$ and $C_{\text{I-RÜ}}$ (%) is the soil organic carbon content of long-term bare fallow soils which were assumed to be similar to the size of the inert organic carbon pool.

Based on Falloon et al. (1998, 2000) C_I (tonne C per hectare) can be estimated from the total stock of carbon using the following equation (Eq. 4) and its upper (Eq. 5) and lower (Eq. 6) confidence levels:

$$C_{\text{I-FAL}} = 0.049 \cdot C_{\text{ORG}}^{1.139} \quad (4)$$

$$C_{\text{I-FAL}(+95)} = 0.1733 \cdot C_{\text{ORG}}^{1.4624} \quad (5)$$

$$C_{\text{I-FAL}(-95)} = 0.01384 \cdot C_{\text{ORG}}^{0.8156} \quad (6)$$

where C_{ORG} (in tonnes C per hectare) is the total amount of organic C in the soil.

An approach to estimate organic carbon in micro-pores based on the hypothesis that the stabilisation of soil organic matter is a result of its location at places with low biological activity, which in turn is caused by a limitation of oxygen, was suggested by Kuka et al. (2006). In our investigation with the CANDY model the organic carbon in micro-pores is considered to be inert.

The total organic carbon is distributed to the different pore-space classes – micro-, meso- and macro-pores – according to their surface area. The pore-space classes used (micro, meso and macro) are related to wilting point (WP), field capacity (FC) and pore volume (PV), respectively. The inner surface area of each pore class is calculated from the volume of the considered pore-space class V_m [WP, FC–WP, PV–FC (in cubic metres)] and the equivalent pore radius R_m (in metres). R_m for the micro-, meso- and macro-pores was set to 5×10^{-8}, 10×10^{-8} and 500×10^{-8} m, respectively.

$$A_m = 2 \cdot \frac{V_m}{R_m}, \quad m \in \{\text{micro; meso; macro}\}. \quad (7)$$

Considering organic carbon in micro-pores ($C_{\text{I-MIP}}$) as 'inert', its amount can be calculated according to the equation:

$$C_{\text{I-MIP}} = C_{\text{ORG}} \cdot \frac{A_{\text{micro}}}{A_{\text{micro}} + A_{\text{meso}} + A_{\text{macro}}}, \quad (8)$$

with $C_{\text{I-MIP}}$ (M.%) mass of carbon in micro-pores, C_{ORG} (M.%) mass of total organic carbon in soil and $A_{\text{micro, meso, macro}}$ (in square metres) inner surface of pore class.

Statistical methods

For this investigation we selected the following statistical methods to obtain information about differences deriving from varying C_I initial values.

The root mean square error (RMSE; Fox 1981) and the mean bias error (MBE; Addiscott and Whitmore 1987) provide information about the average difference between predicted (P_i) and observed (O_i) values.

$$\text{RMSE} = \sqrt{\frac{\sum_{I=1}^{n}(P_I - O_I)^2}{n}} \quad (9)$$

$$\text{MBE} = \sum_{I=1}^{n}\frac{P_I - O_I}{n}, n = \text{number of samples}. \quad (10)$$

A lower RMSE indicates a more accurate simulation. The lower limit of RMSE is 0. MBE can take positive or negative values. Its calculation does not include a square term, thus, predicted values below and above the observed values cancel out, and the result gives an indication of the bias error.

The 'index of agreement' (IA) suggested by Willmott and Wicks (1980) is intended to be descriptive and is both a relative and a bounded measure.

$$\mathrm{IA} = 1 - \frac{\sum_{i=1}^{n}(P_I - O_I)^2}{\sum_{i=1}^{n}(|P_i - \overline{O}| + |O_I - \overline{O}|)^2},$$

$$0 \leq \mathrm{IA} \leq 1, \qquad (11)$$

with n = number of samples and $\overline{O}$ = mean of the observed data.

Results and discussion

Calculated inert carbon

The C_I values used to calculate initial C_{DEC} values for the simulation of the Bad Lauchstädt and Müncheberg sites and the total organic carbon content are shown in Figure 1. Apart from the remarkable differences in total organic carbon between the two sites, Figure 1 shows that with the C_{I-FAL} method, which is simply based on total C_{ORG}, the lowest C_I values are obtained. However, the lower and upper confidence levels of C_{I-FAL} cover nearly the whole range of possible inert carbon. This is in complete agreement with the statement of Falloon et al. (1998) 'that the confidence limits of the model are wide'. For the methods $C_{I-KÖI}$, $C_{I-KÖII}$ and $C_{I-RÜ}$, which are all based on soil texture and suggest a kind of physical protection of soil organic matter, the following order for both sites was found: $C_{I-RÜ} < C_{I-KÖI} < C_{I-KÖII}$. Whereas, with the new C_{I-MIP} approach the highest inert carbon value of all methods was calculated in the case of the Bad Lauchstädt site. With respect to the Müncheberg site C_{I-MIP} is located between $C_{I-RÜ}$ and $C_{I-KÖI}$. Thus, a clear ranking of the C_{I-MIP} method with respect to the amount of calculated inert carbon is not possible. Similar to the $C_{I-KÖI}$, $C_{I-KÖII}$ and $C_{I-RÜ}$ approaches, the C_{I-MIP} approach is suggesting physical protection, but the latter uses pore-space classes instead of soil texture and is also dependent on the total amount of C_{ORG}. Therefore, it shows a more dynamic behaviour. For example, changes in pore-space class distribution because of a modified tillage regime result in a different C_{I-MIP} value.

Simulation results and statistics

To give the reader an impression of the model outputs produced with different initial C_I values,

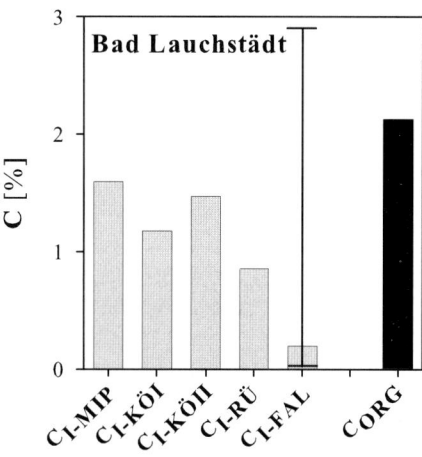

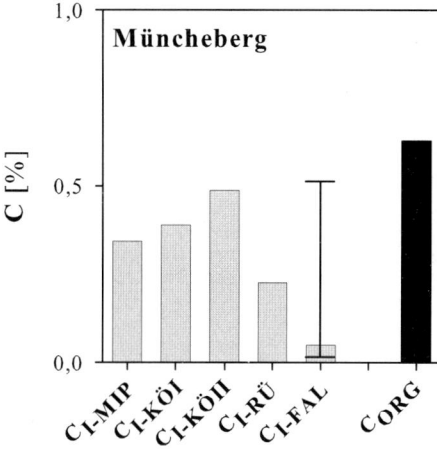

Figure 1. Calculated inert carbon (C_I) and total soil organic carbon (C_{ORG}) for Bad Lauchstädt and Müncheberg. C_{I-MIP} = Kuka et al. 2006; $C_{I-KÖI}$ and $C_{I-KÖII}$ = Körschens 1980 and Körschens et al. 1998 with $a = 0.04$ and 0.05, respectively; $C_{I-RÜ}$ = Rühlmann 1999; C_{I-FAL} = Falloon et al. 1998, 2000. I = Confidence interval for C_{I-FAL}.

the simulation results for C_{ORG} and N_{MIN} are presented in Figures 2 and 3, respectively. In the case of C_{ORG} there is a clear differentiation – increasing with time – between the model runs. The distinct influence of different initial C_I values on N_{MIN} in soil, especially for the Bad Lauchstädt site, emphasises the importance of a careful model initialisation even for short simulation periods.

For Bad Lauchstädt the carbon and nitrogen dynamics and the differences between the simulations are higher than for Müncheberg. In addition to mineralisation, this is may be due to other processes such as leaching or denitrification. In fact, nitrogen leaching is very low at the Bad Lauchstädt

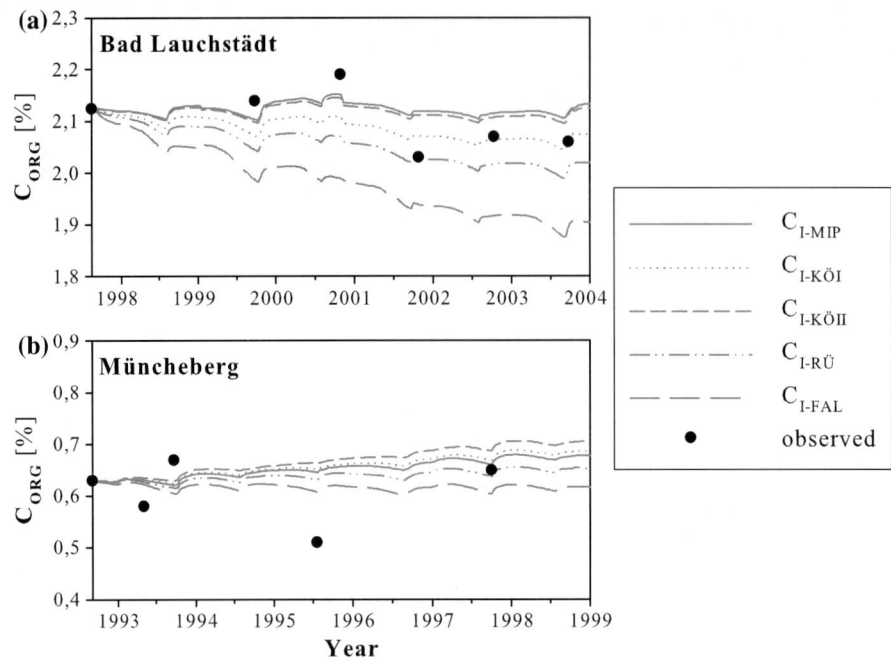

Figure 2. Simulation results for organic carbon (C_{ORG}) dynamics using different initial C_I values for Bad Lauchstädt (**a**) and Müncheberg (**b**). $C_{I\text{-}MIP}$ = Kuka et al. 2006; $C_{I\text{-}KÖI}$ and $C_{I\text{-}KÖII}$ = Körschens 1980 and Körschens et al. 1998 with $a = 0.04$ and 0.05, respectively; $C_{I\text{-}RÜ}$ = Rühlmann 1999; $C_{I\text{-}FAL}$ = Falloon et al. 1998, 2000.

site (see Kersebaum et al. 2007) due to a high water storage capacity and low precipitation (see Franko et al. 2007). Thus, the other nitrogen-related processes have to be adequately taken into account for an assessment of model initialisation.

The examination of the mean (m_C, m_N) and the mean bias error (MBE_C, MBE_N) for carbon and nitrogen, respectively (Tables 2, 3), indicates that in most cases the observed values are exceeded by the corresponding predicted values. Only for organic carbon and $C_{I\text{-}KÖI}$, $C_{I\text{-}RÜ}$ and $C_{I\text{-}FAL}$ of the Bad Lauchstädt site an underestimation can be found. For the Müncheberg site, the MBE_C and MBE_N are the lowest for $C_{I\text{-}FAL}$ and $C_{I\text{-}KÖII}$, respectively. For Bad Lauchstädt, the lowest mean bias errors are obtained for $C_{I\text{-}KÖI}$ and $C_{I\text{-}MIP}$. Consequently, no method seems to be clearly better than the others in predicting initial values for C_I.

Looking at the carbon root mean square error (RMSE, Figure 4) for Bad Lauchstädt, the method for estimating C_I proposed by Falloon et al. (1998, 2000) differs clearly from the other C_I methods, which account for physical protection of soil organic matter. Obviously, the $C_{I\text{-}FAL}$ method

underestimates the inert part of soil organic matter for Bad Lauchstädt (see Figure 1), leading to an overestimation of the decomposable part, followed by a fast decline in soil organic matter and a surplus of mineral nitrogen (see Figure 2). For Müncheberg there is no substantial difference in $RMSE_C$ between any of the methods used in this investigation. In the case of mineral nitrogen, greater differences in the root mean square error can be found. $C_{I\text{-}MIP}$ and $C_{I\text{-}KÖII}$ are in the leading group for the Bad Lauchstädt site and, accompanied by $C_{I\text{-}KÖI}$, in the leading group for the Müncheberg site.

The index of agreement (IA, Figure 5) more precisely suggests that for both sites the $C_{I\text{-}FAL}$ method decreases in accuracy compared to the others with respect to predict C_I in the initialisation of the CANDY model. Falloon et al. (2000) pointed out that their C_I method was not expected to be valid for soils with a large C_I content. The 'Haplic Chernozem' (FAO 1994) of Bad Lauchstädt may belong to that category. However, Schmidt et al. (1999) detected charred organic carbon (up to 45% of the bulk organic carbon) in German chernozemic soils.

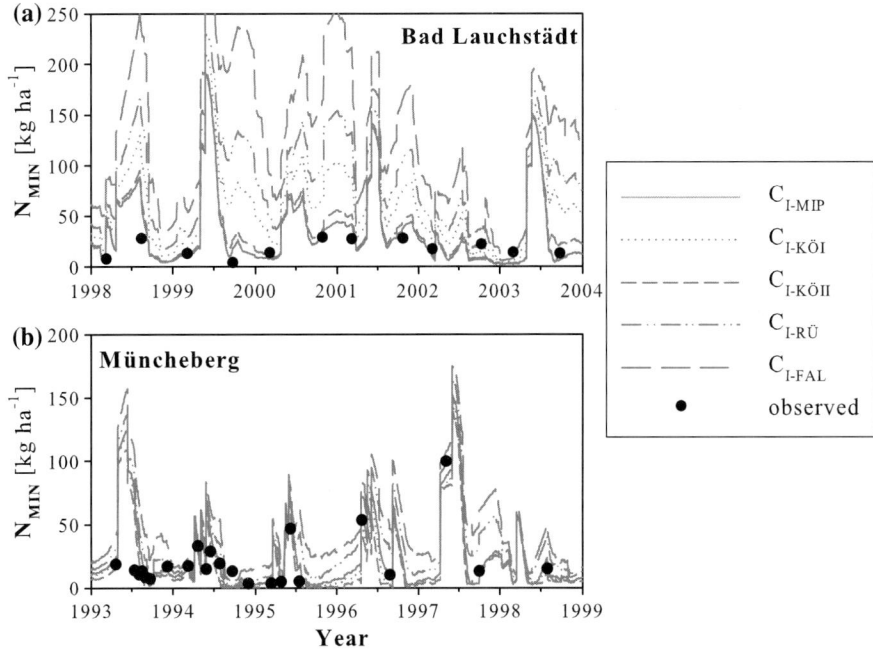

Figure 3. Simulation results for mineral nitrogen (N_{MIN}) dynamics using different initial C_I values for Bad Lauchstädt (**a**) and Müncheberg (**b**). C_{I-MIP} = Kuka et al. 2006; $C_{I-KÖI}$ and $C_{I-KÖII}$ = Körschens 1980 and Körschens et al. 1998 with $a = 0.04$ and 0.05, respectively; $C_{I-RÜ}$ = Rühlmann 1999; C_{I-FAL} = Falloon et al. 1998, 2000.

Table 2. Quantitative statistical measures for simulations with different initial inert carbon (C_I) values for the Bad Lauchstädt site. C_{I-MIP} = Kuka et al. 2006; $C_{I-KÖI}$ and $C_{I-KÖII}$ = Körschens 1980 and Körschens et al. 1998 with $a = 0.04$ and 0.05, respectively; $C_{I-RÜ}$ = Rühlmann 1999; C_{I-FAL} = Falloon et al. 1998, 2000

Bad Lauchstädt[a]	m_C[b]	MBE_C[b]	m_N[b]	MBE_N[b]
C_{I-MIP}	2.12	0.027	23.6	5.49
$C_{I-KÖI}$	2.08	−0.018	51.6	33.5
$C_{I-KÖII}$	2.11	0.021	28.8	10.7
$C_{I-RÜ}$	2.03	−0.060	75.8	57.7
C_{I-FAL}	1.94	−0.149	124	106
O[c]	2.09	–	18.1	–

[a] $n_C = 5$; $n_N = 12$.
[b] m, Mean; MBE, mean bias error; subscript C, carbon (%); subscript N, mineral nitrogen (kg ha^{-1}).
[c] O, Observed.

Table 3. Quantitative statistical measures of simulations with different initial inert carbon (C_I) values for the Müncheberg site. C_{I-MIP} = Kuka et al. 2006; $C_{I-KÖI}$ and $C_{I-KÖII}$ = Körschens 1980 and Körschens et al. 1998 with $a = 0.04$ and 0.05, respectively; $C_{I-RÜ}$ = Rühlmann 1999; C_{I-FAL} = Falloon et al. 1998, 2000

Müncheberg[a]	m_C[b]	MBE_C[b]	m_N[b]	MBE_N[b]
C_{I-MIP}	0.64	0.036	28.5	8.07
$C_{I-KÖI}$	0.64	0.040	25.3	4.86
$C_{I-KÖII}$	0.65	0.050	20.9	0.53
$C_{I-RÜ}$	0.63	0.024	38.8	18.4
C_{I-FAL}	0.61	0.006	50.8	30.4
O[c]	0.60	–	20.4	–

[a] $n_C = 4$; $n_N = 23$.
[b] m, Mean; MBE, mean bias error; subscript C, carbon (%); subscript N, mineral nitrogen (kg ha^{-1}).
[c] O, Observed.

The new C_{I-MIP} approach proves to be as good as $C_{I-KÖI}$ and $C_{I-KÖII}$. $C_{I-RÜ}$ is located between the leading group and C_{I-FAL}. The usefulness of the approach by Körschens (1980) and Körschens et al. (1998) was also pointed out by Ludwig et al. (2003), however there are some indications that this method is not generally applicable. Problems arise particularly for soils with a very high content of fine particles (FP = clay + fine silt) and low C_{ORG} content, a situation that can be found in the long-term experiment from Prague-Ruzyně with FP = 41.2% (J. Klir, personal communication) and

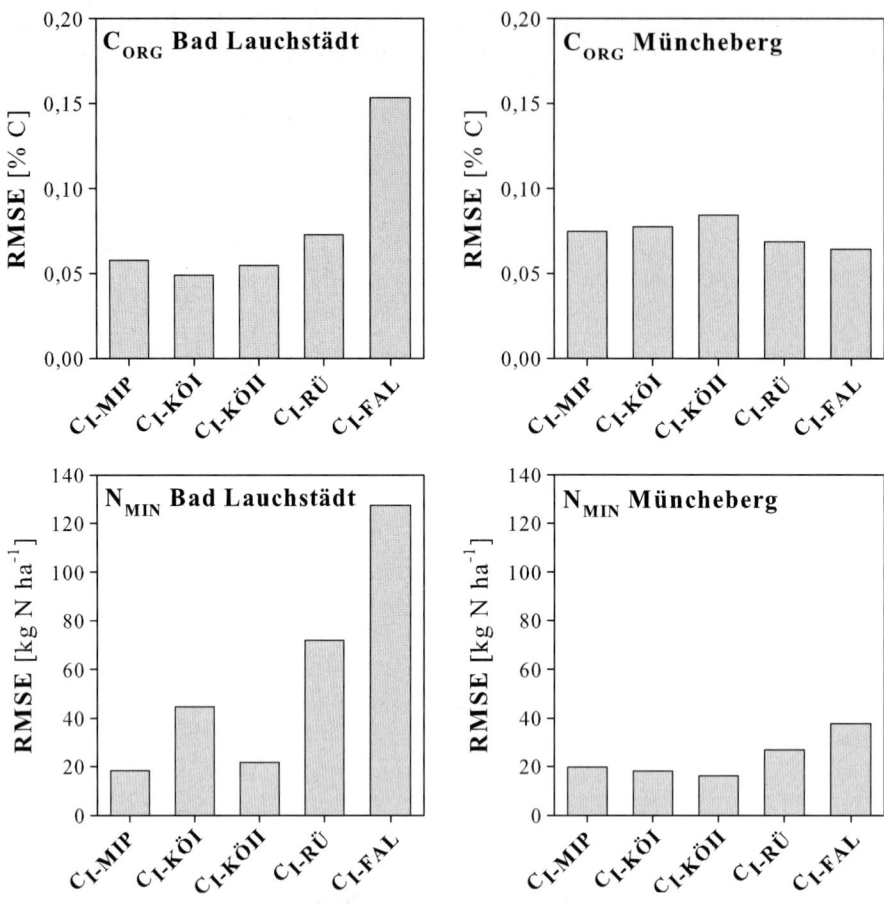

Figure 4. Root mean square error (RMSE) for organic carbon (%) and mineral nitrogen (kg ha^{-1}) of simulations with different initial C_I values for the sites Bad Lauchstädt and Müncheberg. $C_{I\text{-MIP}}$ = Kuka et al. 2006; $C_{I\text{-KÖI}}$ and $C_{I\text{-KÖII}}$ = Körschens 1980 and Körschens et al. 1998 with $a = 0.04$ and 0.05, respectively; $C_{I\text{-RÜ}}$ = Rühlmann 1999; $C_{I\text{-FAL}}$ = Falloon et al. 1998, 2000).

C_{ORG} values from 1.1 to 1.4% (Kubat et al. 2003). Other examples are the long-term experiments of Giessen with $C_{ORG} = 0.99\%$ and FP = 25% (Boguslawski and Debruck 1976) and Gembloux with $C_{ORG} = 0.92\%$ and FAT = 24% (Droeven et al. 1982). In all these cases the approach by Körschens (1980) and Körschens et al. (1998) leads to an overestimation of C_I.

Conclusion

Our results show that an important factor in the CANDY model initialisation procedure is the method chosen for the calculation of C_I as this will have a strong influence on modelling results, even in short-term simulations. The coupled investigation of carbon and nitrogen dynamics leads to a more secure decision on the applicability of a C_I method than simply looking at carbon dynamics because an underestimation of inert carbon causes an overestimation of decomposable carbon and a surplus of mineral nitrogen in the soil (or vice versa).

Of the C_I methods considered, the $C_{I\text{-FAL}}$ method, based simply on total soil organic carbon, is shown to be the least appropriate for a soil with a high amount of inert carbon, such as the Bad Lauchstädt 'Haplic Chernozem', whereas the $C_{I\text{-KÖI}}$ and $C_{I\text{-KÖII}}$ methods as well as the new $C_{I\text{-MIP}}$ approach, all of which allow for the effects of physical protection of soil organic matter, proved to be successful in estimating the inert organic carbon pool in soils of both of the sites investigated. Hence, physical protection seemed to be of major importance for simulating carbon and nitrogen dynamics.

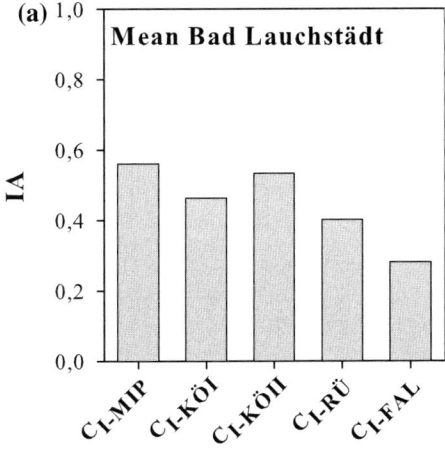
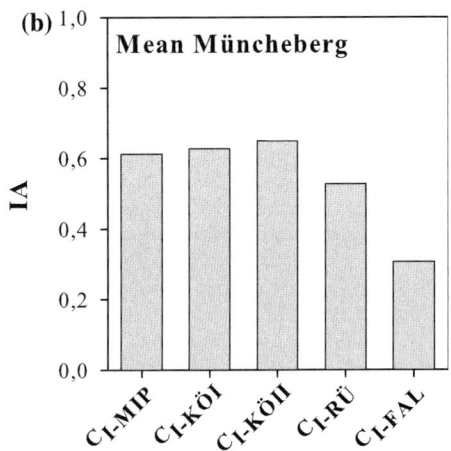
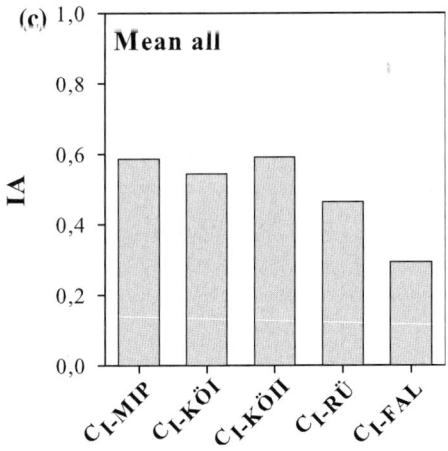

Figure 5. Mean index of agreement (IA) of simulations with different initial C_I values for the sites Bad Lauchstädt (a) and Müncheberg (b) (in each case: mean of organic carbon and mineral nitrogen) and for both sites (c) (mean of organic carbon and mineral nitrogen and sites). $C_{I\text{-MIP}}$ = Kuka et al. 2006; $C_{I\text{-KÖI}}$ and $C_{I\text{-KÖII}}$ = Körschens 1980 and Körschens et al. 1998 with $a = 0.04$ and 0.05, respectively; $C_{I\text{-RÜ}}$ = Rühlmann 1999; $C_{I\text{-FAL}}$ = Falloon et al. 1998, 2000.

In contrast to the methods which primarily consider soil texture ($C_{I\text{-KÖI}}$, $C_{I\text{-KÖII}}$ and $C_{I\text{-RÜ}}$) the $C_{I\text{-MIP}}$ approach uses pore-space classes and also includes the total amount of C_{ORG}. Changes in pore-space class distribution of one soil, because of a modified tillage or a changed organic manure regime, would result in a different $C_{I\text{-MIP}}$ value for the soil. Conseuently, this $C_{I\text{-MIP}}$ method is able to express differences in soil structure and organic matter supply at one site. Further studies, with a wider range of soil types and soils with different tillage regimes and soil organic matter levels, are required to see whether the new $C_{I\text{-MIP}}$ approach is generally applicable.

Acknowledgements

We thank Drs. W. Mirschel and K.C. Kersebaum for providing additional C_{ORG} data of the Müncheberg site, Mr. A.D. Liston for improving the English and two anonymous reviewers for helpful comments on an earlier draft of this manuscript. We are also grateful to the organisers of the workshop.

References

Addiscott T.M. and Whitmore A.P. 1987. Computer simulation of changes in soil mineral nitrogen and crop nitrogen during autumn, winter and spring. J. Agric. Sci. Cambridge 109: 141–157.

von Boguslawski E. and Debruck J. 1976. Ergebnisse aus einem langjährigen Stallmist-Schafpferchversuch in Rauisch-Holzhausen. Z. Acker-u. Pflanzenbau 143: 223–242.

Bruun S. and Jensen L.S. 2002. Initialisation of soil organic matter pools of the Daisy model. Ecol. Model. 153: 291–295.

Coleman K. and Jenkinson D.S. 1996. Roth C-26.3-A model for the turnover od carbon in soil. In: Powlson D.S., Smith P. and Smith J.U. (eds), Evaluation of Soil Organic Matter Models Using Existing Long-term Datasets, Vol. 38. NATO ASI series, Series I, Springer-Verlag, Heidelberg, pp. 237–246.

Droeven G., Rixhon L., Crohain A. and Raimond Y. (1982). Long term effects of different systems of organic matter supply on the humus content and on the structural stability of soils regard to the crop yields in loamy soils. Soil degradation. In:Balkema A.A. (ed.), Proc Land Use Sem Soil Degradation, PUDOC, Wageningen, pp. 203–222.

Falloon P. and Smith P. 2000. Modelling refractory soil organic matter. Biol. Fertil. Soils 30: 388–398.

Falloon P., Smith P., Coleman K. and Marshall S. 1998. Estimating the size of the inert organic matter pool from total soil organic carbon content for use in the Rothamsted carbon model. Soil Biol. Biochem. 30: 1207–1211.

Falloon P., Smith P., Coleman K. and Marshall S. 2000. How important is inert organic matter for predictive soil carbon modelling using the Rothamsted carbon model?. Soil Biol. Biochem. 32: 433–436.

FAO 1994. FAO-UNESCO soil map of the world. Revised legend. FAO, Rome.

Fox D.G. 1981. Judging air quality model performance: a summary of the AMS Workshop on dispersion model performance. Bull. Am. Meteorol. Soc. 62: 599–609.

Franko U. 1996. Modelling approaches of soil organic matter within the CANDY system. In: Powlson D.S., Smith P. and Smith J.U. (eds), Evaluation of Soil Organic Matter Models Using Existing Long-term Datasets, Vol. 38, NATO ASI Series, Series I, Springer, Berlin Heidelberg New York, pp. 247–254.

Franko U. 1997. Modellierung des Umsatzes der organischen Bodensubstanz. Arch. Acker-Pfl. Boden 41: 527–547.

Franko U., Oelschlägel B. and Schenk S. 1995. Simulation of temperature-, water- and nitrogen dynamics using the model CANDY. Ecol. Model. 81: 213–222.

Franko U., Crocker G.J., Grace P.R., Klír J., Körschens M., Poulton P.R. and Richter D.D. 1997. Simulating trends in soil organic carbon in long-term experiments using the CANDY model. Geoderma 81: 109–120.

Franko U., Puhlmann M., Kuka K., Böhme F. and Merbach I. 2007. Dynamics of water, carbon and nitrogen in an agricultural used Chernozem soil in Central Germany. In: Kersebaum KC, Hecker J-M, Mirschel W, Wegehenkel M (eds) Modelling water and nutrient dynamics in soil-crop systems. Springer, Dordrecht, pp 245–258.

Hansen S., Jensen H.E., Nielsen N.E. and Svendsen H. 1991. Simulation of nitrogen dynamics and biomass production in winter wheat using the Danish simulation model DAISY. Fert. Res. 27: 245–259.

Kersebaum K.C. (2007). Modelling nitrogen dynamics in soil-crop systems with HERMES. In: Kersebaum KC, Hecker J-M, Mirschel W, Wegehenkel M (eds) Modelling water and nutrient dynamics in soil-crop systems. Springer, Dordrecht, pp 147–160.

Klír J. 2005. Personal communication.

Körschens M. 1980. Beziehungen zwischen Feinanteil, Ct- und Nt-Gehalt des Bodens. Arch. Acker- Pfl. Boden 24: 585–592.

Körschens M., Weigel A. and Schulz E. 1998. Turnover of soil organic matter (SOM) and long-term balances – tools for evaluating sustainable productivity of soils. Z. Pflanzenernähr. Bodenk. 161: 409–424.

Kubat J., Klír J. and Pova D. 2003. The dry matter yields, nitrogen uptake and the efficacy of nitrogen fertilisation in long-term field experiments in Prague. Plant Soil Environ. 49: 337–345.

Kuka K., Franko U., Rühlmann J., Martens R., Vogt M. and Kalbitz K. 2006. Modelling the impact of pore space distribution on carbon turnover. Ecol. Model. (submitted).

Loague K. and Green R.E. 1991. Statistical and graphical methods for evaluating solute transport models: overview and application. J. Contam. Hydrol. 7: 51–73.

Ludwig B., John B., Ellerbrock R., Kaiser M. and Flessa H. 2003. Stabilization of carbon from maize in sandy soil in a long-term experiment. Eur. J. Soil Sci. 54: 117–126.

Mirschel W., Wenkel K.-O., Wegehenkel M., Kersebaum K.C., Schindler U. and Hecker J.-M. (2007). Müncheberg field trial data set for agro-ecosystem model validation. In: Kersebaum KC, Hecker J-M, Mirschel W, Wegehenkel M (eds) Modelling water and nutrient dynamics in soil-crop systems. Springer, Dordrecht, pp 219-243.

Parton W.J. (1996). The CENTURY model. In: Powlson D.S., Smith P. and Smith J.U. (eds), Evaluation of Soil Organic Matter Models Using Existing Long-term Datasets, Vol. 38, NATO ASI Series, Series I, Springer, Berlin Heidelberg New York, pp. 283–293.

Rühlmann J. 1999. A new approach to estimating the pool of stable organic matter in soil using data from long-term field experiments. Plant Soil 213: 149–160.

Schulz E. 1997. Charakterisierung der organischen Bodensubstanz (OBS) nach dem Grad ihrer Umsetzbarkeit und ihre Bedeutung für Transformationsprozesse für Nähr- und Schadstoffe. Arch. Acker-Pfl. Boden 41: 465–483.

Schmidt M.W.I., Skjemstad J.O., Gehrt E. and Kögel-Knabner I. 1999. Charred organic carbon in German chernozemic soils. Eur. J. Soil Sci. 50: 351–365.

Smith J.U., Smith P. and Addiscott T. 1996. Quantitative methods to evaluate and compare soil organic matter (SOM) models. In: Powlson D.S., Smith P. and Smith J.U. (eds), Evaluation of Soil Organic Matter Models Using Existing Long-term Datasets, Vol. 38, NATO ASI Series, Series I, Springer, Berlin Heidelberg New York, pp. 181–199.

Smith P., Smith J.U., Powlson D.S., McGill W.B., Arah J.R.M., Chertov O.G., Coleman K., Franko U., Frolking S., Jenkinson D.S., Jensen L.S., Kelly R.H., Klein-Gunnewiek H., Komarov A.S., Li C., Molina J.A.E., Mueller T., Parton W.J., Thonley J.H.M. and Whitmore A.P. (1997). A comparison of the performance of nine soil organic matter models using datasets from seven long-term experiments. Geoderma 81: 153–225.

Willmott C.J. and Wicks D.E. 1980. An empirical method for the spatial interpolation of monthly precipitation within California. Phys. Geogr. 1: 59–73.

CHAPTER SIXTEEN

Müncheberg field trial data set for agro-ecosystem model validation

Wilfried Mirschel[1,*], Karl-Otto Wenkel[1], Martin Wegehenkel[1], Kurt Christian Kersebaum[1], Uwe Schindler[2] and Jens-Martin Hecker[1]

Abstract A 6-year experimental data set for three field plots located at the Müncheberg Experimental Station with mainly sandy soils and the soil type Eutric Cambisol is documented in detail. These plots were managed at different levels of intensity, such as intensive management on high level, organic management without chemicals in fertilization and pest management, and without ploughing and extensive management. The data set contains coherent data for soil water and soil nitrogen content, crop growth and weather at a different time solution. The soil and weather processes have been intensively monitored. Field management, as well as measurement methods are described in detail. The crop growth conditions within the whole time period are analysed. This data set was one basis for the international workshop "Modelling water and nutrient dynamics in soil-crop systems" in Müncheberg, Germany, in 2004, organized by the Institute of Landscape Systems Analysis of the ZALF Müncheberg. The usage of different independent measurements of soil water content at all three field plots enables to assess the quality of soil water data and the different measurement methods.

Keywords Agro-ecosystem model testing, Crop growth, Field experiment, Management practices, Nitrogen, Soil water, Time series

Introduction

In the year 1991, the *Climate Impact Research* focus granted by the German Federal Ministry of Consumer Protection, Food and Agriculture was established. Within this research focus, the project "Basic knowledge and models for assessment of climate change effects on soil, biomass accumulation and yield for an economic crop rotation" was carried out in the Leibniz-Centre for Agricultural Landscape Research (ZALF) e. V. Müncheberg, Germany. Within the project, two different experimental set-ups for sampling of coherent data sets, especially for agro-ecosystem model verification and validation, were established. In the first set-up, experiments in closed climate chambers under controlled conditions were carried out. In the second set-up, field experiments at the experimental station of ZALF with different management intensities were established. These field experiments were mid-term experiments established as a four-field crop rotation starting in 1992. These experiments were

---

Wilfried Mirschel, Karl-Otto Wenkel, Martin Wegehenkel, Kurt Christian Kersebaum, Uwe Schindler and Jens-Martin Hecker
*Corresponding author
[1] Institute of Landscape Systems Analysis, Leibniz-Centre for Agricultural Landscape Research (ZALF) e. V. Müncheberg, Eberswalder Strasse 84, D-15374 Müncheberg, Germany
[2] Institute of Soil Landscape Research, Leibniz-Centre for Agricultural Landscape Research (ZALF) e. V. Müncheberg, Eberswalder Strasse 84, D-15374 Müncheberg, Germany

continued over a time period of one and a half crop rotation cycles.

Data sets obtained from three plots of these experiments at the location Müncheberg were one of three data sets used within the International Workshop "Modelling water and nutrient dynamics in soil-crop systems" in Müncheberg, Germany, in June 2004, organized by the Institute of Landscape Systems Analysis of the ZALF Müncheberg. The field plots were located on the non-irrigated part of the experiments. Each of them represented one of the three different management intensities within the experiment. Continuously recording data-logger stations for meteorological data and soil water contents and pressure heads were established near and between the experimental trials. Other soil and crop variables were measured at weekly to monthly intervals. During the first half of the experiment measurements were more frequent than in the second half.

Before the start of the workshop, participants were provided with the data sets described in this paper and were requested to use their own model or own chosen model to simulate the time courses for soil water and evaporation, for crop growth and for nitrogen dynamics in soil and crops. At the workshop, participants presented their simulation results and comparisons with the measurements from soil and crop in the papers of this special issue. Similarities and differences between approaches presented at the workshop are summarized in the model comparison paper by Kersebaum et al. (2007).

This paper focuses on a 6-year (1993–1998) data set of field measurements used for the workshop. Not all data are presented here in detail. The full data set, especially the daily weather values, the daily values for water content measured by TDR and the daily pressure head values, may be obtained from the corresponding author.

Experimental site

The three plots, from which data were used for the model comparison were located at the non-irrigated part of the four-field crop rotation experiment. This experiment was established at the Experimental Station for Agriculture (about 40 ha) of ZALF at Müncheberg. Müncheberg is located about 40 km east of Berlin, between Berlin and the river Oder. The experiment with an area of 32.193 m² (four field trials) has the coordinates 52°515'N and 14°07E, has an altitude of 62 m above sea level and has no slope. The soil at the field trials is a Eutric Cambisol according to the FAO classification with a soil quality index[1] of 26. The groundwater table is deeper than 12 m and there is no artificial drainage.

The climate is characterized by high temperatures in early summer/summer with an increasing frequency of early summer drought periods and cool winters, often without snow. For the time period 1951–2003, the mean annual air temperature was 8.46°C and the mean annual precipitation sum was 530.8 mm. The last decade showed an increase of the mean annual temperature (8.97°C), an increase of the mean negative annual climatic water balance (−156.6 mm year^{-1}) and an increase in the frequency of annual precipitation sums. The dynamic of weather elements within the whole year is shown in Fig. 1 as a mean for a 35 year time period (1967–2001).

Experimental design

The field experiment, at which the three plots were located, had a split-block design with the treatment factors (1) management practise and (2) irrigation. The factor (1) was subdivided into an (a) intensive management on high level, (b) organic management without chemicals in fertilization and pest management and without ploughing and (c) extensive management. The factor (2) was subdivided into a variant (a) with irrigation and (b) without irrigation.

The field experiment consisted of four field trials (27m × 294.5m each) for the crop rotation sugar beet–winter wheat–winter barley–winter rye–catch crop. Each field trial was subdivided into six plots (21m × 45m each) for the six variants and had boarder strips between the trials and between the plots. Each plot consisted of four 3 m wide strip-areas for realizing all necessary measurements including the hand harvests and also two 3 m wide strip-area for harvest using a harvester or combine. At the last ones, the crops can grow undisturbed without any measurements during

---

[1] The soil quality index, which ranges from 1 to 100 is assumed on the basis of the parent material of the soil, its pedogenetic development stage and the hydrological boundary conditions. Lowest values are attributed to the poor diluvial sandy soils and highest values with the chernozoms from loess. The soil quality index was developed for the land evaluation of agricultural used land in Germany starting in the 1930s.

Fig. 1 Climatic diagram for the meteorological station Müncheberg as a mean of 35 years (1967–2001) time period (summer day - maximum air temperature > 25°C; frozen day – minimum air temperature < 0°C; ice day – maximum air temperature < 0°C)

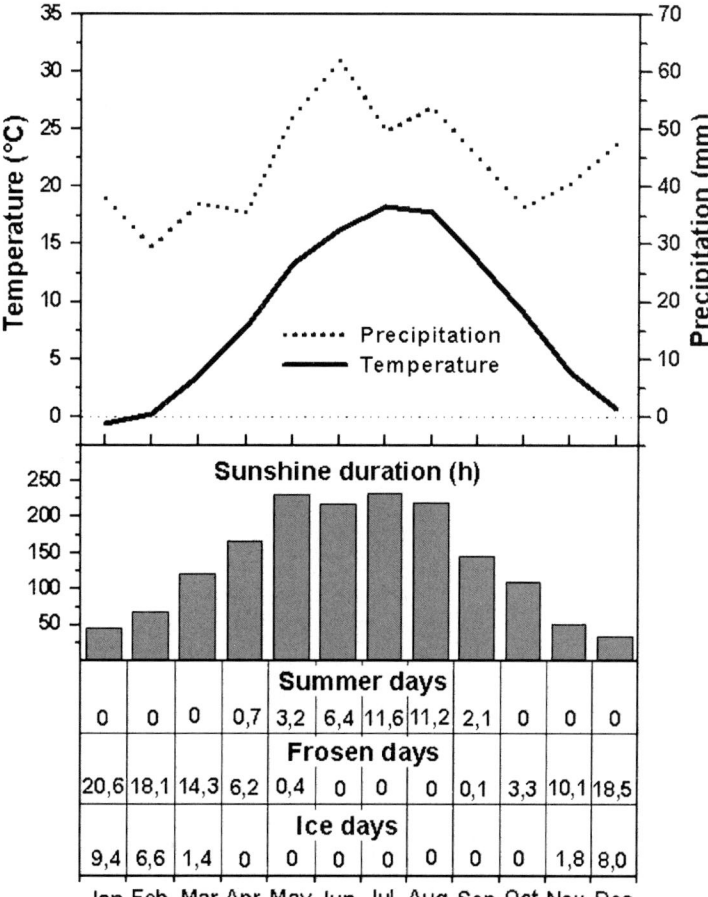

the vegetation period. Figure 2 shows the experimental design of the field experiment including the locations of plot 1, plot 2 and plot 3. These plots were not irrigated, but treated with the three different management options. Approximate locations of the periodic crop, soil and meteorological sampling points are shown in Fig. 2. The automatic recording micro-meteorological station was located at the southern part of the field experiment. The measurement frequency ranged from 10-minutely and hourly for the meteorological and certain soil measurements made by automatic data logger, to approximately monthly or some times annually for soil and crop measurements (Table 1). For a data use within the model comparison workshop, most of measurements were integrated to daily values. During the first crop rotation period, the sampling intervals for soil and crop data were more frequent.

Management practices

The three plots were treated with different intensities of management practices, especially in tillage, fertilization and pest management. The integrated management at plot 1 with a traditional intensive tillage was realized on a high level using only mineral fertilizer and chemicals for pest management. Green manure from catch crops, and straw and leaves as residuals from cereals and sugar beet were the only organic materials for the soil fertility. Within the organic management at plot 2 with a reduced ploughing regime only non-chemical organic fertilizers (farmyard manure and liquid manure) and biological, as well as tillage methods against pest and weeds were used. The extensive management at plot 3 with a moderate ploughing regime used both inorganic and organic fertilizers and chemicals for pest management. The level of intensity was not so high compared with the integrated management variant.

The cropped catch crops, the selected cultivars and the seed densities differed between all three management variants and were in accordance with each management variant. For instance, the seed density in the integrated management variant was higher compared with the organic variant because of

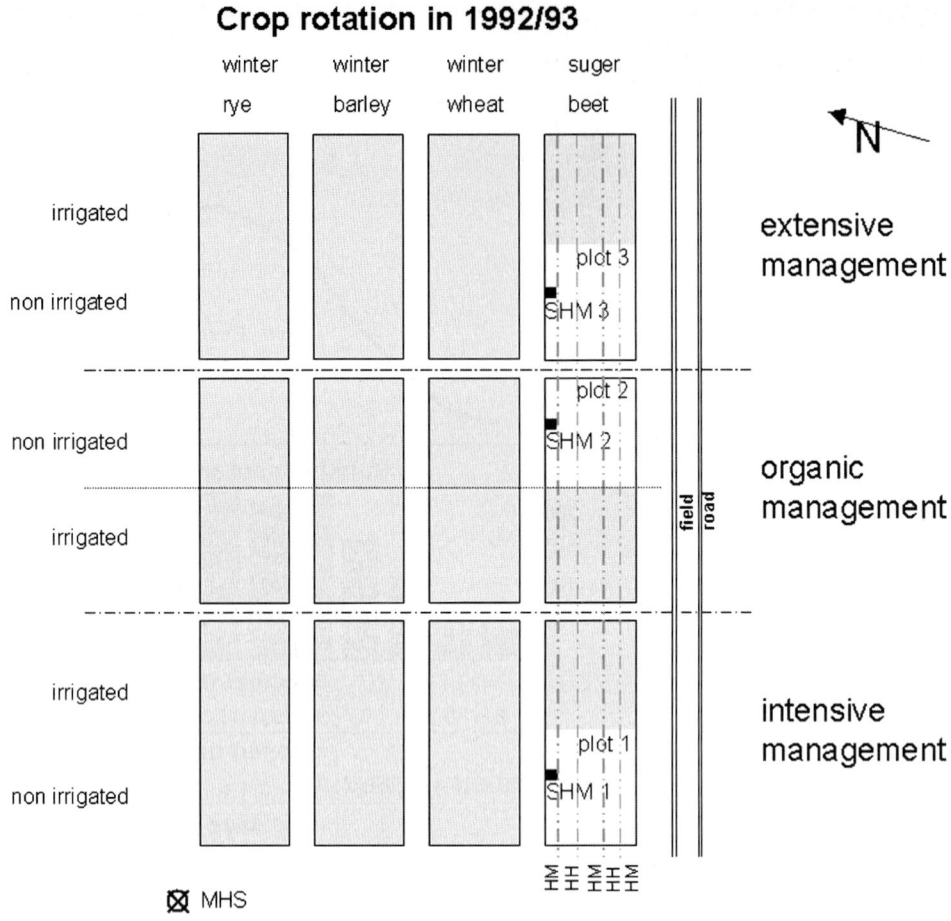

Fig. 2 Scheme of the experimental design of the experiment "Versuch 004" (SHM1, SHM2 and SHM3 – soil hydrological measurement points at plot 1, plot 2 and plot 3, respectively; MMS – micro-meteorological station; HM – areas for periodical manual harvests; HH – areas for harvest by harvester/combine)

a better nitrogen supply and a higher pest management level. In all three management variants the sowing dates were the same. Sowing, fertilization, and tillage practices realized at all three plots, which are relevant for an effective model comparison, are shown in Table 2.

Additional management information especially on irrigation and pest management practices are given in Wenkel and Mirschel (1995).

Measurements

Sampling of soil, crop and meteorological values was realized using both manual and automatic methods. The automatic methods were used for sampling soil and meteorological values and were based on different sensors. The sensors were connected with data loggers at the plots, or they were connected online with a PC for data condensing. The crop data were sampled using manual methods only.

Soil data

For sampling soil data, manual and automatic methods were used. For analysing water and nitrogen contents in the laboratory at every plot randomized soil probes were taken in three different soil layers (0–30 cm, 30–60 cm and 60–90 cm) in replicates using an auger as basis for a mixed probe. For continuous measurements of soil water, automatically measured soil-hydrologic measurement points were installed at each plot.

Table 1 Sampling interval, depth/high and years of measurement for variables of plot 1, plot 2 and plot 3 included into the workshop data set

Variable	Sampling Interval	High/Depth (cm)	Years of measurement
Weather			
Precipitation, sum	10 min	100	1992–1998
Air temperature, minimum	10 min	200	1992–1998
Air temperature, average	10 min	200	1992–1998
Air temperature, maximum	10 min	200	1992–1998
Air temperature, at 14:00 h	10 min	200	1992–1998
Air temperature, minimum	10 min	20	1992–1998
Air temperature, average	10 min	20	1992–1998
Air temperature, maximum	10 min	20	1992–1998
Rel. air humidity, average	10 min	200	1992–1998
Rel. air humidity, at 14:00 h	10 min	200	1992–1998
Global radiation, average	10 min	200	1992–1998
Windspeed, average	10 min	250	1992–1998
Soil			
Temperature (under grass)	10 min	5, 20, 50,	1992–1998
Temperature (under crop rotation)	1 h	90, 120, 150, 200, 300	1992–1998
Nitrate content	14–60 days	0–30, 30–60, 60–90	1993–1995
	1–2 per year	0–30, 30–60, 60–90	1996–1998
Ammonium content	14–60 days	0–30, 30–60, 60–90	1993–1995
	1–2 per year	0–30, 30–60, 60–90	1996–1998
Water content, gravimetry	14–60 days	0–30, 30–60, 60–90	1993–1995
	1–2 per year	0–30, 30–60, 60–90	1996–1998
Water content, TDR	10 min	30, 60, 90	1993–1998
	1 h	120, 150, 200	1996–1998
Water pressure	10 min	30, 60, 90	1993–1998
	1 h	150, 200, 300	1993–1998
	1 h	120	1995–1998
Crop			
Ontogensis	7 30 days	–	1994–1998
Plants per m^2	1–3 per year	–	1994–1998
Tillers per m^2	1–3 per year	–	1994–1998
Ears per m^2	1–3 per year	–	1994–1998
Biomass, steam and leaves	1–6 per year	–	1992–1998
Biomass, root	1–4 per year	–	1992, 1994–1995
Biomass, storage organ	1–6 per year	–	1993–1998
Sugar in beet	1–5 per year	–	1993, 1997
Carbon content, steam and leaves	1–6 per year	–	1993–1997
Carbon content, root	4 per year	–	1994–1995
Carbon content, storage organ	1–6 per year	–	1993–1997
Nitrogen content, steam and leaves	1–6 per year	–	1993–1997
Nitrogen content, root	4 per year	–	1994–1995
Nitrogen content, storage organ	1–6 per year	–	1993–1997

Profile description The physico-chemical properties for the soil profiles for the three plots are given in Table 3. For the deeper horizons some values are not available. Plot 1, plot 2 and plot 3 differ in the order and in the thickness of horizons. In the Ap- and Ael-horizons the texture proportions are approximately equal in all plots. In the deeper horizons the differences between the plots are more distinct. In organic carbon, total nitrogen, bulk density and pH there are not significant differences between the three plots. Additional information on soil characteristics and soil profiles for the field trials, at which the three plots are located can be found in Schindler (1980).

Table 2 Management practices[a] at the three field plots

Crop	Cultivar	Row distance (cm)	Sowing date	Seed grains (m^{-2})	Emergence date	Harvest date	Residue management	Fertilizer type[b]	Amount (kg ha^{-1})	Date of application	Tillage (depth in cm)	Date tillage
Plot 1												
Oil radish	Gilda	45	1992-09-03		1992-09-09	ploughed up	Green manure				Ploughing (25)	1992-09-02
Sugar beet	Gilda	45	1993-04-26		1993-06-01	1993-10-06	Green manure	AUS	N:80	1993-05-01	Ploughing (18)	1993-05-19
			1993-05-21				Leaves ploughed in	IOF	P:30 K:160 Mg:50	1993-10-12		
Winter wheat	Bussard	10	1993-10-15	327	1993-10-25	1994-07-29	Straw ploughed in	AUS	N:50	1994-04-06	Cultivator drill (10)	1993-10-15
								AUS	N:40	1994-05-04		
								ANL	N:50	1994-05-27		
								IOF	K:120	1994-09-12		
Winter barley	Grete	10	1994-09-26	264	1994-10-03	1995-07-21	Straw ploughed in	AUS	N:35	1995-03-20	Cultivator (15)	1994-08-10
								AUS	N:40	1995-05-11	Rotary tiller (10)	1994-08-22
								AUS	N:30	1995-05-17	Cultivator (10)	1994-09-19
								AUS	N:40	1995-05-30		
								IOF	P:30 K:160 Mg:50	1995-09-11		
Winter rye	Clou	10	1995-10-02	306	1995-10-10	1996-08-21	Straw ploughed in	AUS	N:40	1996-04-19	Cultivator (15)	1995-07-24
								AUS	N:30	1996-05-17	Cultivator (15)	1995-09-09
								AUS	N:30	1996-06-05	Cultivator (8)	1995-09-14
								ANL	N:60	1996-09-09		
Oil radish			1996-09-05		1996-09-13		Green manure				Ploughing (23)	1996-09-04
											Cultivator drill (10)	1996-09-05
Sugar beet	Sonja	42.8	1997-04-03	11	1997-05-05	1997-09-23	Leaves ploughed in	IOF	P:45 K:171 Mg:118	1997-03-06	Rotary tiller (10)	1997-03-26
								AUS	N:70	1997-04-10		
								ANL	N:60	1997-06-02		
Winter wheat	Bussard	10	1997-10-08	384	1997-10-24	1998-07-27	Straw ploughed in	IOF	P:30 K:114 Mg:12	1997-10-07	Cultivator drill (10)	1997-10-08
								AUS	N:50	1998-03-16	Cultivator (8)	1998-08-17

Crop	Variety								
Plot 2			1992-09-03			Green manure			1992-09-02
Yellow mustard	Gilda	45	1993-04-26 1993-05-21		Ploughing up 1993-10-06	Green manure Leaves ploughed in	FM FM	N: 198 N: 66	Ploughing (25) Ploughing (18) 1993-05-19
Sugar beet	Ramiro	12	1993-10-15	1993-06-01	1994-07-29	Straw exported	LM	N:30	Cultivator drill (10) 1993-10-15
Winter wheat	Berit	10	1994-09-26	1993-10-25	1995-07-21	Straw exported	LM LM	N:30 N:37	Cultivator (15) 1994-08-10 Rotary tiller (10) 1994-08-22
Winter barley	Rapid	12	1995-10-02	1994-10-03	1996-08-21	Straw	LM FM	N:64 N: 198	Ploughing (22) 1994-09-19 Cultivator (15) 1995-07-24 Cultivator (15) 1995-09-09 Ploughing (25) 1995-09-14
Winter rye			1996-09-05	1995-10-05		Green manure			Cultivator drill (23) 1996-09-04 Ploughing (10) 1996-09-05
Yellow mustard	Sonja	42.8	1997-04-03	1997-05-05	1997-10-01	Leaves ploughed in	IOF	P:45 K:149 Mg:30	Rotary tiller (10) 1997-03-26
Sugar beet	Ramiro	10	1997-10-08	1997-10-24	1998-07-27	Straw exported	FM LM	N: 66 N:42	Ploughing (22) 1997-10-08 Cultivator (8) 1998-08-17
Winter wheat									
Plot 3			1992-09-03			Green manure			1992-09-02
Phacelia Sugar beet	Gilda	45	1993-04-26 1993-05-21	1993-06-01	ploughing up 1993-10-06	Green manure Leaves ploughed in	FM AUS	N: 198 N:80	Ploughing (25) Ploughing (18) 1993-05-19
Winter wheat	Greif	10	1993-10-15	1993-10-25	1994-07-29	Straw exported	FM	N: 66	Cultivator drill (10) 1993-10-15
							AUS AUS IOF	N:40 N:50 K:120	1994-04-06 1994-05-04 1994-09-12
Winter barley	Noveta	10	1994-09-26	1994-10-03	1995-07-21	Straw exported	AUS AUS IOF	N:35 N:60 P:30 K:160 Mg:50	Cultivator (15) 1994-08-10 Rotary tiller (10) 1994-08-22 Ploughing (22) 1994-09-19
Winter rye	Rapid	10	1995-10-02	1995-10-05	1996-08-21	Straw exported	AUS AUS	N:30 N:45	Cultivator (15) 1995-07-24 Cultivator (15) 1995-09-09 Cultivator (8) 1995-09-14

(Continued)

Table 2 Management practices[a] at the three field plots—Cont'd

Crop	Cultivar	Row distance (cm)	Sowing date	Seed grains (m^{-2})	Emergence date	Harvest date	Residue management	Fertilizer type[b]	Amount (kg ha^{-1})	Date of application	Tillage (depth in cm)	Date tillage
Phacelia			1996-09-05				Green manure				Ploughing (23) Cultivator drill (10)	1996-09-04 1996-09-05
Sugar beet	Sonja	42.8	1997-04-03	11	1997-05-05	1997-10-01	Leaves ploughed in	FM	N: 198	1996-09-04	Rotary tiller (10)	1997-03-26
								ANL	N:40 kg N/ha	1996-09-09		
								IOF	P:45 K:171 Mg:118	1997-03-06		
								AUS	N:70	1997-04-10		
								ANL	N:40	1997-06-02		
Winter wheat	Greif	10	1997-10-08	383	1997-10-24	1998-07-27	Straw exported	FM	N: 66	1997-10-08	Cultivator drill (10)	1997-10-08
								AUS	N:35	1998-03-16	Cultivator (8)	1998-08-17

[a] In 1993, at all three plots were realized three irrigations (1993-5-13, 1993-5-18 and 1993-5-21) with 6 mm each as help for emergence of sugar beet.
[b] AUS -ammonium urea solution; ANL -ammonium nitrate lime; IOF -Inorganic fertilizer; FM -Farmyard manure with a nitrogen content of 0.6 % in fresh mass; LM – liquid manure

Table 3 Soil properties measured at plot 1, plot 2 and plot 3

Horizon	Depth (cm)	Sand (%)	Silt (%)	Clay (%)	Organic Carbon (%)	Total nitrogen (%)	C:N	pH (KCl)	Bulk density (g cm^{-3})
Plot 1									
Ap	0–30	83	9	8	0.66	0.054	12.1	6.1	1.45
Ael	30–60	86	8	6	0.16	0.015	11	6.1	1.5
Bt	60–90	72	14	14	0.08	0.007	11.1	6.3	1.55
C1	90–110	83	10	7	n.a	n.a	n.a	n.a	n.a
C2	110–160	92	7	1	n.a	n.a	n.a	n.a	n.a
C3	160–210	98	1	1	n.a	n.a	n.a	n.a	n.a
Plot 2									
Ap	0–30	83	9	8	0.58	0.05	11.6	6.5	1.45
Ael	30–90	93	6	1	0.13	0.013	10	6.5	1.5
Bt1	90–130	78	12	10	n.a	n.a	n.a	n.a	1.55
Bt2	130–170	80	11	9	n.a	n.a	n.a	n.a	n.a
C1	170–180	97	2	1	n.a	n.a	n.a	n.a	n.a
C2	180–225	98	1	1	n.a	n.a	n.a	n.a	n.a
Plot 3									
Ap	0–30	83	9	8	0.62	0.054	11.4	6.3	1.45
Ael	30–100	93	6	1	0.13	0.013	10	6.3	1.5
Bt1	100–110	65	18	17	n.a	n.a	n.a	n.a	1.55
Bt2	110–225	84	7	9	n.a	n.a	n.a	n.a	n.a

n.a. – not available

Water content and water pressure head At all plots manual and automatic sampling methods for measuring the volumetric soil water content or water pressure heads were realized. Volumetric soil water and water pressure heads were measured continuously using automatic recording *Time Domain Reflectometry with Intelligent Micro Elements* (TRIME-TDR) probes and tensiometers installed in soil-hydrologic measurement points at each plot.

In the experiment TRIME-EZ probes from IMKO (IMKO Micromodultechnik GmbH, Ettlingen, Germany) were installed vertically in depths of 30 cm, 60 cm, 90 cm, 120 cm, 150 cm and 200 cm (Fig. 3). This installation scheme enables also the estimation of the total soil water down to 150 cm depth including the amount of stored soil water available for soil evaporation and crop transpiration. This installation scheme can lead to effects, such as some underestimations in the soil water dynamics due to the fact, that the heads of the probes with a diameter of 63 mm can act as small barriers, which can delay percolation. The test and the field calibration of the TRIME-probes at the three field plots with a resulting accuracy of ± 2.5 vol% is described in detail in Wegehenkel (1998). The application and validation of a soil water balance model using the results of these TRIME-probes and tensiometers at the three field plots including a detailed evaluation of the suitability of these soil hydrological measurement techniques for the purpose of model validation, as well as an assessment of the data quality of the field plots is described in Wegehenkel (2004).

For a continuous measurement of soil water tension, tensiometers were installed at soil depths of 30 cm, 60 cm, 90 cm, 120 cm, 150 cm, 200 cm and 300 cm in a distance of 2.5 m from the access path to reduce boundary effects. The tensiometers each consisted of a porous ceramic cup (20 mm in diameter, 50 mm in length), glued to a plexiglas connector (~60 mm in the length), which then was threaded with a temperature-compensated pressure transducer. In the experiment, tensiometers from UGT (Umwelt-Geräte-Technik GmbH, Müncheberg, Germany) were used. The tensiometers in 30 cm and 60 cm depths were installed vertically and removed during frost periods or during soil tillage. For the 90–300 cm depths, the tensiometers were installed via a mechanical support pipe to allow insertion from the surface through sheath tubes installed at specific angels. To reduce artificial infiltration along the sheath tubes, they were

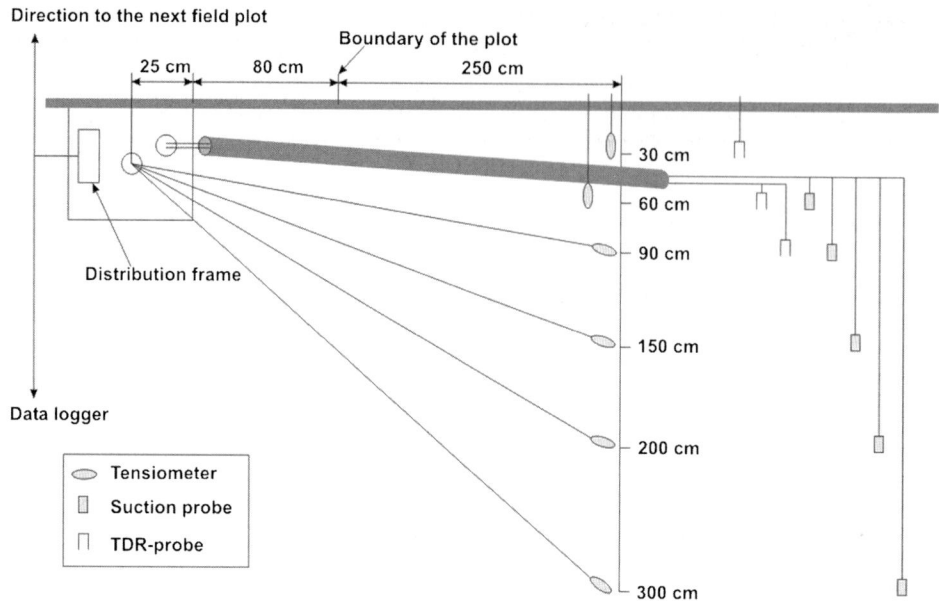

Fig. 3 Sensor installation scheme for the soil hydrologic measurement points installed at each field plot (Modified from Wenkel and Mirschel 1995)

made shorter by about 30 cm than the tensiometers so that the end of the tensiometer tube would seal well to the soil above the porous cup. In addition, at the surface the sheath tubes were covered using a closed waterproof chamber. The tensiometer installation design is shown in Fig. 3.

The automatic measurement methods for soil water content and soil water pressure gave continuous data sets, except instrument failures or frost periods. At the experimental plots with sandy soils and low plant available water capacities, the measurements of water pressure heads with tensiometers showed with difficulties in the soil layers shallower than 60 cm depth because of soil drying effects during spring and summer. The results were gaps in the water pressure head data sets. Figure 4 shows an example for water pressure head courses in different soil depths at plot 1 in 1993. The tensiometer in 30 cm depth was broken down at the end of August because of not enough water in the upper 30 cm soil layer and was reactivated after a rewetting. For the tensiometer in 60 cm depth, the breakdown point was nearly reached some days later, but the rewetting process improved the measurement conditions just in time. In Fig. 4, in addition, there are shown gaps in water pressure head courses at all measurement depths caused by instrument failure for tensiometers in 60 cm depth at the end of September, in 90 cm depth in May/June, August and December, in 150 cm depth in June and December and in 200 cm depth in December, or by frost periods for tensiometers in 30 cm and 60 cm soil depths from middle of November up to the end of the year. During the data quality assessment, such data were removed from the data set (Wegehenkel 2004).

Separate sets of gravimetric water content measurements were also made on the soil nitrogen samples described in the next section. The frequency of sampling differed year by year. The measurements for all three plots and soil layers are shown in Table 4.

Nitrogen Manual samples for determining the soil nitrate and soil ammonium contents were taken at different sample time intervals for the 0–30 cm, 30–60 cm and 60–90 cm soil depths using the "N_{min} Method" according to Wehrmann and Scharpf (1979). Mixed samples at each plot, formed in the field, consisted randomized 12–14 subsamples (replicates). The subsamples were taken using a "Pürkhauer" half-cylindrical soil auger that was pounded into the soil up to 90 cm depth. For sample analysis in the laboratory, probes with a weight of 150 g taken from the mixed samples, were crashed and riddled, and analysed on the same day.

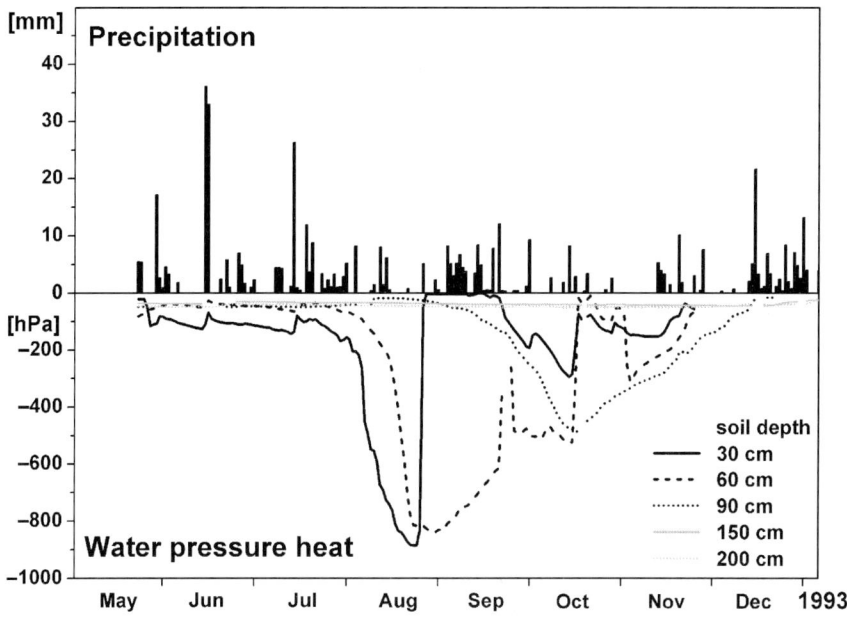

Fig. 4 Water pressure head courses in different soil depths at plot 1 from second sugar beet sowing date (21 May 1993) up to the end of the year 1993 in the result of water uptake by plant and precipitation

Table 4 Soil nitrate, soil ammonium and soil water content in 0–30 cm, 30–60 cm and 60–90 cm soil depths from manual sampling for all three plots

Date	Water content (vol%)			Nitrate content (kg ha^{-1})			Ammonium content (kg ha^{-1})		
	0–30 cm	30–60 cm	60–90 cm	0–30 cm	30–60 cm	60–90 cm	0–30 cm	30–60 cm	60–90 cm
Plot 1									
1993-04-21	10.6	11.6	13.7	16.3	6.7	4.8	2.2	0.5	0.0
1993-05-03	9.1	11.7	13.5	77.8	8.4	6.2	123.4	10.3	2.9
1993-07-13	15.2	13.2	14.0	9.6	29.8	49.9	4.3	1.0	1.0
1993-08-03	9.9	9.3	10.8	2.9	9.6	20.2	7.7	5.3	3.8
1993-08-17	7.1	7.2	11.0	3.4	27.8	34.6	9.6	6.2	4.3
1993-08-31	12.0	5.8	8.2	3.4	13.9	20.2	5.3	2.4	1.9
1993-09-21	13.7	10.9	11.6	3.2	3.5	10.7	3.7	1.9	0.8
1993-12-07	17.6	14.1	13.6	9.1	9.6	10.1	7.7	3.4	3.4
1994-03-08	20.2	19.0	16.8	8.0	10.2	8.6	9.4	5.4	5.3
1994-04-19	14.4	14.0	17.3	12.5	8.2	8.4	20.6	3.7	2.4
1994-05-27	17.7	12.5	15.2	4.1	1.8	5.0	10.7	4.7	3.7
1994-06-15	8.9	9.8	14.2	9.4	1.2	3.6	19.2	4.7	3.6
1994-07-27	3.0	3.5	8.2	9.4	2.0	1.9	9.8	5.5	4.7
1994-09-21	14.9	13.7	14.4	7.1	11.9	16.4	5.9	1.3	0.4
1994-11-30	18.1	13.8	16.6	2.4	7.7	14.4	1.0	0.1	0.0
1995-03-13	14.3	12.5	14.6	0.5	0.4	0.8	3.1	1.8	0.7
1995-04-26	12.9	13.7	16.2	0.0	0.0	0.1	4.9	2.2	2.3
1995-06-07	17.8	17.2	12.3	7.2	1.4	0.5	39.4	8.6	3.4
1995-07-17	7.9	8.1	11.2	2.5	0.2	0.5	2.6	1.0	1.0
1996-04-22	10.9	10.7	13.6	26.4	12.0	20.5	27.0	6.1	6.2
1996-08-26	13.3	7.6	7.7	6.6	2.4	3.2	3.5	1.4	0.6
1997-05-05	13.0	11.6	15.6	68.8	14.2	12.6	31.2	2.8	2.9
1997-10-01	7.4	6.7	7.1	6.6	3.0	4.3	6.4	3.5	3.1
1998-07-29	10.1	5.1	9.1	8.2	1.8	4.2	6.8	2.0	1.7

(*Continued*)

Table 4 Soil nitrate, soil ammonium and soil water content in 0–30 cm, 30–60 cm and 60–90 cm soil depths from manual sampling for all three plots—Cont'd

Date	Water content (vol%)			Nitrate content (kg ha^{-1})			Ammonium content (kg ha^{-1})		
	0–30 cm	30–60 cm	60–90 cm	0–30 cm	30–60 cm	60–90 cm	0–30 cm	30–60 cm	60–90 cm
Plot 2									
1993-04-21	9.8	11.1	13.5	32.2	16.3	10.1	1.0	0.0	0.0
1993-05-03	9.5	11.7	13.1	24.0	13.9	12.0	5.3	1.4	1.4
1993-06-14	21.2	16.4	20.0	5.3	19.2	20.6	5.8	3.8	3.8
1993-06-28	14.0	12.4	15.9	3.8	4.3	22.6	7.2	3.4	3.8
1993-07-13	13.7	12.0	11.4	5.3	18.2	22.6	2.9	0.5	0.0
1993-08-03	9.9	8.0	8.9	1.4	9.1	9.1	5.8	4.3	4.3
1993-08-17	5.6	6.4	10.4	1.7	4.8	9.8	6.2	4.7	2.8
1993-08-31	7.5	3.8	8.3	6.2	2.9	5.3	3.4	2.9	2.9
1993-09-21	11.3	10.2	8.3	1.0	1.0	3.0	3.2	2.9	1.3
1993-12-07	15.9	13.5	14.7	2.4	4.3	3.8	5.3	2.9	2.9
1994-03-08	18.3	17.5	18.1	5.0	7.4	6.7	5.5	3.7	1.7
1994-04-19	14.5	13.5	17.1	2.2	2.6	4.7	7.1	4.7	2.6
1994-05-27	16.8	13.1	15.7	0.0	0.0	2.3	7.0	4.0	3.0
1994-06-15	9.4	9.9	15.4	0.4	0.4	2.6	7.1	52.0	3.1
1994-07-27	6.3	8.0	10.6	8.2	0.6	1.3	5.8	1.7	1.4
1994-09-21	14.8	12.8	16.3	9.1	13.1	14.6	6.6	1.4	0.4
1994-11-30	16.6	13.8	16.0	2.5	5.3	12.2	4.4	2.5	1.3
1995-03-13	14.4	12.9	16.1	0.0	0.0	0.0	48.4	1.1	0.2
1995-04-26	12.2	12.7	17.3	0.0	0.0	0.0	3.7	1.8	1.3
1995-06-07	19.3	11.5	14.3	1.0	0.0	0.0	8.6	5.8	2.4
1995-07-17	11.1	9.6	13.7	4.8	0.0	0.0	1.7	0.1	0.0
1996-04-22	10.0	10.7	17.1	5.0	5.9	19.8	1.7	1.6	1.4
1996-08-26	15.4	11.8	13.7	19.0	7.7	5.6	7.3	1.9	1.3
1997-05-05	12.5	12.0	15.0	25.0	14.4	12.1	6.4	1.3	1.4
1997-10-01	6.7	6.7	8.4	5.3	2.9	4.4	4.9	1.9	1.7
1998-07-29	11.9	7.4	11.2	11.8	4.6	8.8	6.7	2.8	3.2
Plot 3									
1993-04-21	11.2	11.6	14.8	19.20	12.00	11.04	0.48	0.00	0.00
1993-05-03	8.5	12.1	14.8	37.44	13.44	6.72	9.12	0.00	0.00
1993-06-14	18.6	15.7	16.3	24.96	3.84	5.76	22.08	2.88	2.88
1993-06-28	14.6	13.8	14.9	119.81	18.43	27.65	5.28	3.36	1.92
1993-07-13	13.3	10.2	11.7	11.04	31.68	16.80	2.88	2.88	0.96
1993-08-03	11.1	11.1	13.0	2.40	11.04	43.68	7.68	6.24	3.84
1993-08-17	4.8	4.6	7.9	2.04	4.20	9.12	5.88	4.80	3.84
1993-08-31	11.0	5.2	8.7	2.40	4.80	9.60	4.80	3.36	1.44
1993-09-21	12.5	10.1	8.0	0.60	0.24	0.96	3.60	3.12	1.80
1993-12-07	15.2	12.8	12.8	2.40	3.84	3.36	4.80	3.36	3.36
1994-03-08	19.0	17.2	16.3	8.64	8.40	6.96	8.88	2.88	1.80
1994-04-19	14.4	13.1	15.6	13.32	4.44	4.44	21.84	3.00	1.92
1994-05-27	16.6	13.8	14.9	2.04	0.72	4.32	9.24	2.88	1.92
1994-06-15	9.2	9.8	12.7	0.72	2.16	3.36	6.60	3.96	3.24
1994-07-27	6.3	5.9	9.6	5.16	2.04	1.80	2.88	2.16	1.44
1994-09-21	13.8	12.6	14.9	7.56	8.88	13.68	4.20	1.08	0.24
1994-11-30	16.8	14.0	15.7	1.44	9.72	15.36	3.96	1.68	0.60
1995-03-13	14.5	11.9	13.2	0.00	0.00	0.00	3.72	0.96	0.12
1995-04-26	12.5	12.1	14.7	0.00	0.00	0.00	6.24	2.88	2.40
1995-06-07	19.3	7.0	14.5	0.96	0.00	0.00	12.00	5.76	2.88
1995-07-17	8.9	9.0	11.4	7.20	0.12	0.00	2.16	0.72	0.12
1996-04-22	11.1	11.2	13.9	18.12	6.24	14.64	20.16	5.16	3.72
1996-08-26	14.8	10.5	10.1	14.76	4.08	6.00	6.72	1.56	1.32
1997-05-05	13.2	11.4	13.4	63.00	12.00	16.80	19.32	1.44	2.28
1997-10-01	7.1	6.5	7.7	7.44	5.04	7.08	6.12	1.68	1.44
1998-07-29	10.3	6.0	7.9	11.88	3.84	5.04	6.60	2.04	0.84

The soil extraction was carried out with KCl- solution and the suspension was shaken for about 1 h. The filtered extract was spectro-photometric analysed at 540 nm for NO_3-N content and at 660 nm for NH_4-N content according to ISO14256 (Anonymous 1997).

The soil nitrogen content (kg N ha^{-1} layer^{-1}) was calculated using the measured content (mg NO_3-N or NH_4-N (100 g)$^{-1}$ field-wet soil), the soil bulk density (g cm^{-3}), the volumetric water content of the field-wet soil (cm^3 cm^{-3}) and the depth of the soil layer (30 cm). The bulk densities used for the nitrogen und water content calculations for all three plots were taken from Table 3. The NO_3-N and the NH_4-N measurements for all three plots and soil layers are shown in Table 4.

Solute (nitrogen) leaching was quantified based on soil hydrological measurements below the zero flux plane down to 5 m soil depth. Once a day tension and water content values were measured. The relationship between tension and water content was fitted and the relative hydraulic conductivity in dependence on water content was calculated. Afterwards, relative deep drainage rates were quantified based on the water content dynamics and the hydraulic conductivity using Darcy's law. In loamy soils, the hydraulic gradient has to be taken into consideration. In sandy soils, using the unit gradient is sufficient. The assumption of steady state conditions is valid, because water content differences within daily intervals are negligible below the zero flux plane. For reliable deep drainage rates, the relative deep drainage was calibrated to the water balance in a winter period (drainage = precipitation − evapotranspiration ± water content difference). Using soil water samplers at the same depths, nitrate concentrations were analysed. Taking into consideration deep drainage dynamics, nitrogen losses were estimated.

As an example, in Fig. 5 the nitrogen content in soil water is shown in 150 cm depth at plot 1, in 200 cm depth at plot 2 and for 150 cm depth at plot 3 as time courses.

Temperature Soil temperatures were measured at 10 min resolution using "PT100" epoxy-embedded thermistors with an accuracy of ± 0.2 K. For depths at 5 cm, 20 cm and 50 cm, the thermistors were located about 2.5 m far from the micro-meteorological station under grass cover, not in the actual cropped plots. For depths at 90 cm, 120 cm, 150 cm, 200 cm and 300 cm, the termistors, together with the tensiometers, were installed at each soil-hydrological measurement point of the three plots under crops used within the crop rotation.

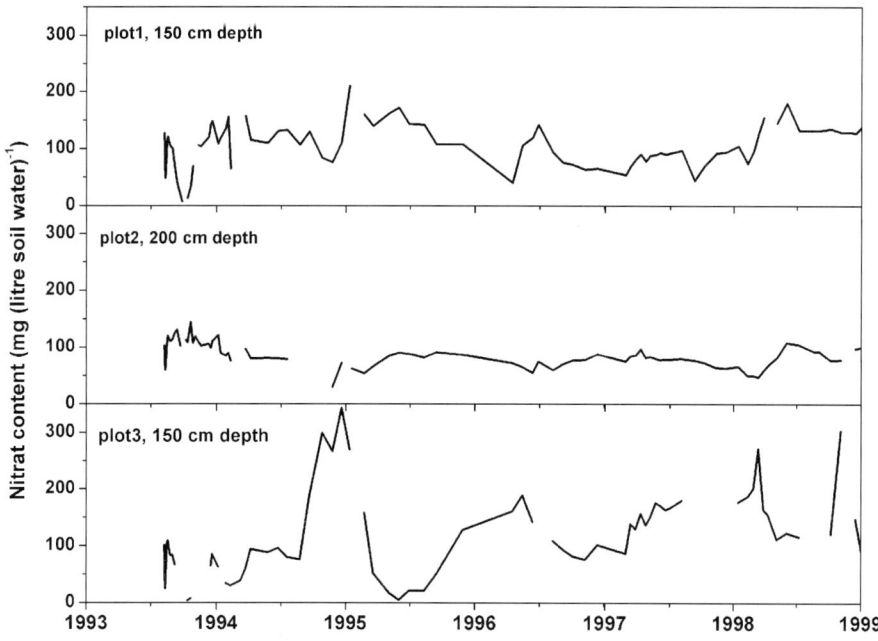

Fig. 5 Time courses for nitrogen content in soil water in 150 cm depth at plot 1, in 200 cm depth at plot 2 and in 150 cm depth at plot 3 over a 6-years time period

Crop data

The crop growth stages were estimated weekly using the BBCH-scale (Hack et al. 1972), a decimal code system prevalent in Germany, almost identical to the decimal code system according to Zadoks (1974).

For cereals, the number of plants, tillers and ears were counted many times per year on the basis of 1 m² sections in nine randomized replicates within every plot. For sugar beet, the number of plants was counted based on a 9.6 m long row section (4 m²) in nine randomized replicates. After counting, a plot average was calculated.

The ontogenesis measurements, as well as the measurements for plant, tiller and ear numbers for all three plots are given in Table 5.

The areas for manual harvests (see Fig. 2) were the basis for the periodical determination of aboveground and root biomass. At each plot, three randomized replicates for cereals and sugar beets and six randomized replicates for catch crops were taken for mixed probes per sampling date. Sampling sections of 0.25 m² for cereals, of 1 m² (a 2.4 m long row section) for sugar beets, and of 1 m² for catch crops were the sampling basis. After combining replicate subsamples, the harvested material was separated into stem/leaves and ears (including grain, glume and rest of the ear) for cereals and into beet and leaves (including petioles) for sugar beet, and the fresh biomass was measured. In addition, for cereals only the root biomass was measured. For this, the soil of the upper 30 cm layer was taken and the roots were washed out. For determination

Table 5 Crop decimal code for ontogenesis, numbers of plant, tiller and ear, sugar in beet, stem and leaf, root and storage organ dry biomasses for periodical hand harvests as well as for the final harvest by harvester/combine for all three plots

					Biomass (kgha⁻¹)							
					Harvested by hand				Harvested by harvester/combine			
									Grain/beet		Straw/leaf	
Date	Ontogenesis (DC)	Plants (m⁻²)	Tillers S (m⁻²)	Ears (m⁻²)	Stem & Leaf Dry	Root Dry	Storage organ (beet/ear) Dry	Sugar in beet (%)	Dry	Fresh	Dry	Fresh
Plot 1												
1992-10-19					1225	163						
1992-11-10					1453	271						
1992-12-08					1636	433						
1993-01-11					1110	352						
1993-02-08					650							
1993-07-13					459		119	5.2				
1993-08-03					2048		1425					
1993-08-17					3265		2932	9.55				
1993-08-31					4156		5150	11.2				
1993-09-21					5858		7898	12.35				
1993-10-06					4324		8455	13.55	84.55	368.27	43.24	296.48
1994-03-09	13				50	16						
1994-04-19	30	195	436		235	52						
1994-05-26	45		440		4283	801						
1994-06-14	65			307	7949	879	1353					
1994-07-26	92			342	5393		7690					
1994-07-29									45.16	451.62		
1994-11-30	23	279	670		233	56						
1995-03-16		303	1169		711	175						
1995-04-24	31	235	967		1535	269						
1995-06-07	61			463	5633	539	2494					
1995-07-17	92			577	5678		7321					
1995-07-21									56.52	565.23		
1995-11-29	23	241	959									
1996-04-10	26	198	1277									
1996-05-06	32		1147									
1996-05-30	61			412	4125		598					
1996-08-21	92								68.12	681.19	40.16	401.61
1997-06-30					3155		1217					

Table 5 Crop decimal code for ontogenesis, numbers of plant, tiller and ear, sugar in beet, stem and leaf, root and storage organ dry biomasses for periodical hand harvests as well as for the final harvest by harvester/combine for all three plots—Cont'd

					Biomass (kg ha^{-1})							
					Harvested by hand				Harvested by harvester/combine			
									Grain/beet		Straw/leaf	
Date	Onto-genesis (DC)	Plants (m^{-2})	Tillers S (m^{-2})	Ears (m^{-2})	Stem & Leaf Dry	Root Dry	Storage organ (beet/ear) Dry	Sugar in beet (%)	Dry	Fresh	Dry	Fresh
1997-08-27					6416		12108					
1997-09-23					4652		12869		116.39	557.36	42.08	294.87
1997-10-01					4208		11639	15.76				
1997-10-27	12	270										
1998-05-25	45											
1998-06-09	65			333	5657		1554					
1998-07-27									34.9	348.96	23.99	
Plot 2												
1992-10-19					1063	146						
1992-11-10					1283	184						
1992-12-08					1165	184						
1993-01-11					957	228						
1993-02-08					676							
1993-06-14					371		84					
1993-06-28					1013		575	9.5				
1993-07-13					1911		1744	9.25				
1993-08-03					3070		4574					
1993-08-17					4309		7924	12.625				
1993-08-31					4846		10524	14.4				
1993-09-21					5337		13279	13.9				
1993-10-06					4521		9467	14.38	94.67	372.02	45.21	262.82
1994-03-09	13.00				56	15						
1994-04-19	30.00	238	658		286	64						
1994-05-26	47.00		358		2479	808	380					
1994-06-14	65.00			209	4275	812	873					
1994-07-26	92.00			282	2973		4544					
1994-07-29									32.45	324.52		
1994-11-30	23	290	1142		360	71						
1995-03-13		351	1553		892	224						
1995-04-24	31	268	1119		1728	302						
1995-06-07	61			371	5443	576	2175					
1995-07-17	92			386	5343		4734					
1995-07-21									30.2	302.04		
1995-11-29	23.00	178	883									
1996-04-10	26.00	162	1018									
1996-05-06	32.00		708									
1996-05-30	61.00			279	2627		489					
1996-08-21	92.00								24.5	244.99	38.99	389.93
1997-06-30					1786		856					
1997-08-27					3529		9029					
1997-09-23					2648		10502		117.62	510.37	29.66	178.27
1997-10-01					2966		11762	16.64				
1997-10-14	12	246										
1998-05-25	45											
1998-06-09	65			231	2452		1126					
1998-07-27									14.3	142.96	19.4	
Plot 3												
1992-10-19					1039	63						
1992-11-10					1697	85						
1992-12-08					1997	103						
1993-01-11					1385	82						

(*Continued*)

Table 5 Crop decimal code for ontogenesis, numbers of plant, tiller and ear, sugar in beet, stem and leaf, root and storage organ dry biomasses for periodical hand harvests as well as for the final harvest by harvester/combine for all three plots—Cont'd

					Biomass (kgha^{-1})							
					Harvested by hand				Harvested by harvester/combine			
									Grain/beet		Straw/leaf	
Date	Onto-genesis (DC)	Plants (m^{-2})	Tillers S (m^{-2})	Ears (m^{-2})	Stem & Leaf Dry	Root Dry	Storage organ (beet/ear) Dry	Sugar in beet (%)	Dry	Fresh	Dry	Fresh
1993-02-08					824							
1993-06-14					552		120					
1993-06-28					1688		960	8.25				
1993-07-13					2816		2212	9.25				
1993-08-03					4178		5790					
1993-08-17					5035		9420	12.445				
1993-08-31					4610		10225	15.2				
1993-09-21					5369		14446	13.29				
1993-10-06					4627		15427	13.85	154.27	600.81	46.27	263.46
1994-03-09	13				67	26						
1994-04-19	30	211	475		255	53						
1994-05-26	45		616		4532	1054						
1994-06-14	65			332	6488	920	1135					
1994-07-26	92			463	5013		7568					
1994-07-29									47.97	479.72		
1994-11-30	23	276	812		264	40						
1995-03-13		285	1330		817	166						
1995-04-24	31	266	1225		1863	276						
1995-06-07	61			391	4829	516	2374					
1995-07-17	92			614	7369		8953					
1995-07-21									56.81	568.07		
1995-11-29	23.00	343	1500									
1996-04-10	26.00	281	2067									
1996-05-06	32.00		1175									
1996-05-30	61.00			385	4028		658					
1996-08-21	92.00								51.35	513.49	52.56	525.63
1997-06-30					3074		1287					
1997-08-27					5284		11209					
1997-09-23					5076		12983		124.19	587.74	48.55	272.99
1997-10-01					4855		12419	16.54				
1997-10-14	12	311										
1998-05-25	45											
1998-06-09	65			331	4121		1707					
1998-07-27									43.68	436.75	18.65	

of dry biomass, 1000 g fresh biomass, probes were taken from each biomass fraction subsample. Sugar beets were coarsely chopped prior to drying. The fresh biomass probes were dried up to dry biomass at about 60°C during 2–3 days using a special drying oven. The dry biomass for stem and leaf, for root and for the storage organ for periodical hand harvests, as well as for the final harvest by harvester/combine for all three plots are given in Table 5.

After drying of samples for determination of total plant nitrogen and total plant carbon in the laboratory, the probes were milled. The extraction of the total plant nitrogen was realized using the Kjeldahl method. The determination of nitrogen compounds, transferred into ammonium, were based on the indophenol blue reaction, a modified Berthelot reaction, using the spectro-photometric analysis at 578 nm (photometer EPOS 5060 produced by Eppendorf). The total plant carbon was determined using the element analysis method. About 100 mg of the probe were incinerated within a oxygen supply at a temperature of 1250°C, i.e. the carbon compounds were oxidized. The measurement of the CO_2 gas was realized in an infrared cell using the element analyser CNS 2000 produced by Leco.

The sugar from beet was measured using the automatic light–electric polarimeter "POLAMAT S". This method of sugar determination is based on the principle that the 100°S point of the linear sugar division is fixed for the strain value of 40.764°, which is characteristic for a normal sugar solution at a temperature of 20°C and a measurement at 546.1 nm in the case, if the sugar solution is polarimetered within a 200 mm long pipe (Riegler and Schiek 1970). The automatic light-electric polarimeter "POLAMAT S" has a measurement accuracy of less than 0.05°S (Strube and Scholze 1970).

The total plant nitrogen and plant carbon measurements for stem and leaf, for root and for storage organ for all three plots are shown in Table 6 and the sugar in beet for the cropping years 1993 and 1997 is given in Table 5.

Weather data

Weather data at field experiment were collected every 10 min (see Table 1) by the automated micro-meteorological station FMA 86 (type "Weihenstephan", produced by Lambrecht GmbH Göttingen, Germany),

Table 6 Total plant nitrogen and plant carbon contents for stem and leaf, for root and for storage organ for all three plots

Date	Carbon content (kg C ha^{-1})			Nitrogen content (kg C ha^{-1})		
	Stem and leaf	Root	Storage organ (beet/ear)	Stem and leaf	Root	Storage organ (beet/ear)
Plot 1						
1992-10-19						
1992-11-10						
1992-12-08						
1993-01-11						
1993-02-08						
1993-07-13	165		45	20		2
1993-08-03	809		621	58		16
1993-08-17	1185		1280	91		31
1993-08-31	1564		2361	89		27
1993-09-21	2238		3611	159		62
1993-10-06	1774		3770	114		59
1994-03-09	22	7		3	0	
1994-04-19	107	19		12	1	
1994-05-26	1928	157		95	5	
1994-06-14	3502	261	627	134	8	27
1994-07-26	2502	0	3548	27		150
1994-11-30	102	28		11	1	
1995-03-16	323	73		17	2	
1995-04-24	668	113		39	4	
1995-06-07	2555	342	1173	57	7	46
1995-07-17	2253		3121	29		124
1996-05-30	1839		283	83		15
1997-06-30	1162		497	125		22
1997-08-27	3060		5065	148		127
1997-09-23	1774		5384	119		115
1997-10-01	1568		4778	105		102
1998-06-09				69		30
Plot 2						
1992-10-19						
1992-11-10						
1992-12-08						
1993-01-11						
1993-02-08						
1993-06-14	145	0	33	19		1
1993-06-28	456	0	256	24		6
1993-07-13	662	0	686	54		18
1993-08-03	1176	0	1995	65		38

(*Continued*)

Table 6 Total plant nitrogen and plant carbon contents for stem and leaf, for root and for storage organ for all three plots—Cont'd

Date	Carbon content (kg C ha^{-1})			Nitrogen content (kg C ha^{-1})		
	Stem and leaf	Root	Storage organ (beet/ear)	Stem and leaf	Root	Storage organ (beet/ear)
1993-08-17	1605	0	3320	83		56
1993-08-31	1898	0	4832	118		70
1993-09-21	2075	0	6051	112		81
1993-10-06	1771	0	3685	100		45
1994-03-09	24	6	0	3	0	0
1994-04-19	129	22	0	10	1	0
1994-05-26	1071	164	170	34	2	8
1994-06-14	1899	150	402	30	2	15
1994-07-26	1353	0	2054	10	0	67
1994-11-30	162	0	0	16	1	0
1995-03-13	390	93	0	25	3	0
1995-04-24	755	121	0	35	4	0
1995-06-07	2528	247	1009	41	4	28
1995-07-17	2074	0	2047	23	0	60
1996-05-30	1175	0	232	48	0	13
1997-06-30	640	0	349	55	0	12
1997-08-27	1982	0	3803	75	0	64
1997-09-23	955	0	4379	58	0	67
1997-10-01	1065	0	4863	65	0	74
1998-06-09				21		18
Plot 3						
1992-10-19						
1992-11-10						
1992-12-08						
1993-01-11						
1993-02-08						
1993-06-14	191	0	46	28	0	2
1993-06-28	744	0	417	53	0	11
1993-07-13	1027	0	848	88	0	24
1993-08-03	1657	0	2502	110	0	34
1993-08-17	1987	0	4037	121	0	66
1993-08-31	1812	0	4685	132	0	88
1993-09-21	2116	0	6742	110	0	89
1993-10-06	1634	0	6020	89	0	78
1994-03-09	29	10	0	4	1	0
1994-04-19	115	20	0	13	1	0
1994-05-26	1992	284	0	95	6	0
1994-06-14	2898	242	528	89	5	21
1994-07-26	2301	0	3448	22	0	130
1994-11-30	116	0	0	13	1	0
1995-03-13	363	68	0	21	2	0
1995-04-24	808	134	0	39	3	0
1995-06-07	2245	246	1115	66	4	46
1995-07-17	2773	0	3888	36	0	141
1996-05-30	1790	0	311	73	0	17
1997-06-30	1146	0	532	118		22
1997-08-27	2495	0	4736	106		92
1997-09-23	1866	0	5453	115		107
1997-10-01	1784	0	5228	109		102
1998-06-09				68	0	32

which is located directly near the field experiment (see Fig. 2). The measured weather data and used sensors are listed in Table 7.

The thermistors in 20 cm and 200 cm hights were installed radiation protected within a ventilated Assmann pipe. All thermistors for soil temperature were epoxy-embedded and had a direct soil contact. The basis of relative humidity measurement was the change of the length of hairs within a radiation and precipitation protected hair hygrometer with potentiometer. The wind speed was measured using a three-cup anemometer on the top of the meteorological station (2.5 m high). The precipitation was sampled using a heated and wind-shielded gauge (type HP 3) with a sampling area or 200 cm^{-2} and a 0.1 mm resolution. Here, the see-saw technique according Joss-Tognini with a read contact is used.

The collected data were aggregated to hourly and/or daily data sets automatically. The data can be stored on a memory chip up to 1 month or can be transferred online to a PC. After a monthly revision on plausibility and correction the weather data were stored in a weather data base.

Table 7 Measured weather data and sensors used at the micrometeorological station FMA 86

Value	Unit	Instrument	Accuracy	Range
Air temperature	°C	Thermistor PT100	± 0.2 K	−30–60
Relative air humidity	%	Pernix-hygrometer	± 3%	5–100
Wind speed	ms^{-1}	Anemometer	± 0.2 ms^{-1}	0–60
Global radiation	W m^{-2}	Pyranometer CM 6B	± 10 Wm^{-2}	0–1400
Precipitation	mm	See-saw technique HP 3	± 0.1 mm	–
Soil temperature	°C	Thermistor PT100	± 0.2 K	−30–60

Table 8 shows the annual average of air temperature, the annual precipitation sum and the annual sunshine duration sum for the period 1992–1998 in comparison with the 35-year average (1997–2001).

For analysing the crop growth conditions from driving forces, the dynamic of weather values during the growing period is much more important than annual averages and annual sums. For air temperature, precipitation and sunshine duration Fig. 6 shows the deviation of monthly weather values from the 35-year (1967–2001) averages as dynamic over the whole year for all years 1992–1998. In addition, the ± 20% zone of the monthly mean long-term weather values is given in Fig. 6.

Analysis of growth conditions

In the following section the weather (driving forces), soil and management conditions and specifics during the experimental period 1992–1998 considered in the workshop and most important for biomass accumulation and yield formation will be described and analysed in a short manner.

Weather analysis

A comparison of all years included in the data set relevant for the workshop shows that in all years except, 1993, 1995 and 1996, the annual average air temperature was significantly higher in comparison with the 35-year average. This indicates an increasing temperature trend in the last decade. The year 1996 with a annual average air temperature of only 6.58°C was significantly below the long-term average air temperature, because of two very cold and long winter

Table 8 Annual average of air temperature, annual precipitation sum and annual sunshine duration sum for 1992–1998 and as 35 year average (1997–2001) for the experimental station Müncheberg

	1992	1993	1994	1995	1996	1997	1998	Average
Annual average air temperature (°C)	9.23	8.18	9.52	8.42	6.58	8.83	9.00	8.65
Annual precipitation sum (mm)	445.8	625.2	712.4	506.8	481.9	541.1	654.6	527.8
Sunshine duration (h)[a]	1796	1620.7	1766	1770	1539	1861	1616	1656.4

[a] The sunshine duration was taken from the weather station Müncheberg of the German Weather Service (DWD)

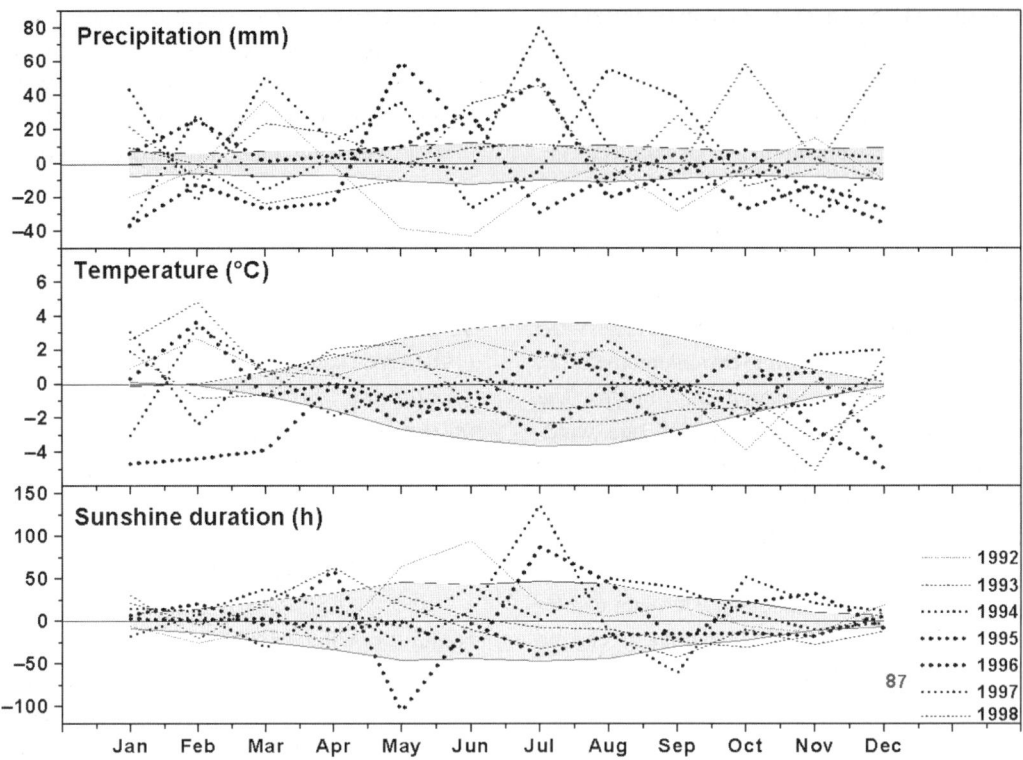

Fig. 6 Deviations of monthly weather values for precipitation, average air temperature and sunshine duration from the 35-year (1967–2001) averages as time course over the whole year for 1992–1998 (grey zone - ±20% zone of the monthly mean long-term weather values)

periods in order, i.e. from 1995 to 1996 and from 1996 to 1997, and a cool summer period in 1996 (Fig. 6). In addition, 1996 also was a dry year with the lowest sum of sunshine duration during the last decade of the last century. At all three plots; the continuous frost period from the middle of November 1995 up to 20 March 1996 (see Fig. 6) led to an ontogenesis delay especially for the late sown winter cereals on one hand, and on the other hand to a reduction in the plant density because of frost-lifting of plants (Table 5).

The comparison of the monthly precipitation courses (Fig. 6) shows that there were more years with precipitation deficits in the last quarter of the year than with enough water supply by rain at that time. A dry period between September and November leads to dry soil conditions and this can be the reason for a germination delay of winter cereals.

The positive deviation tops of monthly precipitations from the 35-year average are a hint for extreme precipitation events with surface run-off, like in July 1997. There were three precipitation events with high intensities (27 mm, 19 mm and 42 mm) within three sequent days. During 1992–1998, the years 1992 and 1996 were dry years. In 1992, the dry period was between April and October, the growing season, that was the reason for very low yields. In 1996, the whole winter and spring period, beginning in October 1995, was characterized by a precipitation deficit. This led to a decrease in soil water and to bad water supply conditions for the field crops. The years 1992 and 1993 have an early summer drought period characterized by precipitation deficit and high temperatures. Such situations negatively influence the biomass accumulation and grain filling processes, especially for winter wheat and winter rye, and delay the emergence of sowing sugar beets (Table 2).

Except 1996, all years showed higher winter temperatures in comparison to the long-term average (Fig. 6) that is typical for the last decade and seems to be a hint for a possible climate change. This warmer winter periods accelerated the ontogenesis and the biomass accumulation of winter cereals and catch crops.

Analysis of soil conditions

Water and nitrogen in soil are the most important soil variables, which are taken into account in agro-ecosystem models for describing crop growth and plant development processes. A sufficient water and nitrogen availability during the whole growing period is very important for a stress-free biomass accumulation, especially for sandy soils like at the Experimental Station of Müncheberg. Here the risk of nutrition leaching during winter is very high, especially in situations without any crop canopy, i.e. under bare soils. This is the reason for including catch crops into the crop rotation to fix nitrogen, after winter rye harvested in August and before sugar beets sown in April/May of the following year. The catch crops are used as green manure and support the mineralization process. At the Müncheberg experimental site mostly low soil nitrate and ammonium levels in the root zone can be observed in early spring. At all three plots the nitrogen content in March was less than 50 kg ha^{-1} (4.8–49.7 kg ha^{-1}, see Table 4). The lowest level was measured in March 1995. In that year, a nitrate content of < 1.7 kg ha^{-1} was measured (even 0.0 kg ha^{-1} at plot 2 and plot 3) and low ammonium contents also. Usually, high leaching rates during winter periods with high precipitation sums are the reason for such low levels. For example, after a very wet year 1994 with an annual precipitation sum of 696 mm (132 % of the long-term average) and a wet 5 months winter period (from November 1994 to March 1995) with a precipitation sum of 260 mm (135 % of the long-term average), the soil water level was at field capacity in all layers and the percolation rates were high.

At the Müncheberg site, more or less strong water stress situations for the crops can be observed during the growing period in every year. For the year 1993, the interactions between the soil water dynamic in the rooting soil layer, the water uptake by plants and the precipitation is shown in Fig. 4 using water pressure head courses measured by tensiometers. The time courses of the pressure heads show the soil depth dependent water dynamic in 30 cm, 60 cm, 90 cm, 150 cm and 200 cm depths. The dynamic of water content is less, the deeper the soil depth. In 30 cm, the influence of infiltrated precipitation water and the water uptake by plants on water content is very high and in cases of high evapotranspiration rates the water content reaches the wilting point and the tensiometer breaks down. This can be seen in Fig. 4 at the end of August. It is also shown that with a decrease of water content in the upper soil depth, the water uptake by plants in the deeper layer increased. This process continued via the deeper soil layers. In a precipitation-free case, a soil drying driven by transpiration-induced water uptake by plants up to the rooting depth can be obtained starting at the upper soil layer. The deeper the soil layer, the time-delayed is the decrease of soil water content. In Fig. 4, this is shown comparing the water pressure head courses in 30 cm, 60 cm, and 90 cm depths. This is the basis for obtaining the movement of the water uptake frontline induced by plants over the whole growing period and over the soil depths. The unchanged water pressure head courses in 150 cm and 200 cm depths from May to December 1993 show that there was no percolation and no water uptake by plants. For the latter, it means that the roots did not reach 150 cm depth. After harvest of sugar beets in October 1993, the water uptake by plants was finished and the soil layers filled up continuously up to the field capacity by infiltrated water. This process started at upper soil layer, so that the deeper soil layers were filled up later (see Fig. 4).

If precipitation during winter is at the long-term average level or higher than at Müncheberg experimental fields, usually the soil profile is at about field capacity in spring (at the end of March).

Management analysis

The agro-management realized during the experiment seems not to be optimal in every case beginning with the selected crops within the crop rotation. For example, 75% cereals in a crop rotation under organic management is not typically because of a not fully balanced nitrogen cycle and an increased weed stress. Under non-irrigated conditions, the Müncheberg site with sandy soils and water stress periods is not suitable for cropping sugar beets in normal case because of low yields. Under irrigation the sugar beets cropping is possible with moderate yields only.

In 1993, at all plots an unusual management in sugar beet cropping was necessary, a twice sowing. Because of a very high water deficit during 1992 and a dry spring in 1993 (only 50% of the average precipitation in March and April) the first sown sugar beet, sown on 26 April 1993, did not emerge. The two irrigations of 6 mm each as help for emergence were without any success. After ploughing, the sugar beet were sown again on 21 May, also with a 6 mm irrigation as help for emergence. The very late second sowing and emergence dates leading to a significant shortage of the sugar beet growing period were the reasons for fresh matter yields of only 36.8 t ha^{-1} and 37.2 t ha^{-1} at plot 1 and plot 2, respectively.

Because winter wheat followed after sugar beets in the crop rotation, it was necessary to find a compromise between a late sowing date for wheat and an early harvest date for sugar beets, i.e. for both crops the management was not optimal for high yields.

The seed density and the cultivars for mean crops and catch crops used at each plot were different to adapt the typically cultivar characteristics to the management intensities. Among winter cereal cultivars, there were more straw dominant and yield dominant ones. For the intensive management on high level (plot 1), the seed density usually was higher because of a better nutrient supply and a more effective pest management. For winter wheat the seed density at plot 1 is higher by 43 and 118 grains per m^2 in 1993 and 1997, respectively, (Table 2).

At plot 2 (organic management), there were problems with weed populations, especially *Apura spicaventil* and *Papaver rhoeas*. The weed density increased year by year as a result of non-optimal chosen crop rotation within the organic management and non-optimal pest management. The existing high competition between cereal plants and weed plants resulted in a low yield level (Table 5).

Because at plot 2 and at plot 3 the straw from cereals was exported, for a better organic matter supply in soil, it was necessary to add organic fertilizer. Both farmyard and liquid manure were used. For farmyard manure a nitrogen content of 0.6 % on a fresh mass basis was assumed, i.e. 33.000 kg farmyard manure fertilized per ha contained approximately 198 kg N$_t$ ha^{-1} (Table 2).

In 1998, there was a low fertilization level (50 kg N ha^{-1} at plot 1, 42 kg N ha^{-1} at plot 2 and 35 kg N ha^{-1} at plot 3) and a low level in pest management. The result of such a low management level is a high weed density and a low yield level at all three management variants.

Conclusions

Such a data set like the one described in this paper may be a valuable data base for the evaluation and validation of different agro-ecosystem approaches due to its 6-year record length. Additionally, the data set consists meteorological and soil hydrological data with a daily time step, a wide range of soil and crop values measured at each experimental plot and very detailed crop management information. On the other hand, the data set also has some deficiencies, in particular, there are time lacks in the Time Domain Reflectometry (TDR) and pressure head measurements, discrepancies in the frequency of the measurements between the first and the second half of the experimental time period. However, there are some measurements missing for a whole water and nitrogen balance calculation like surface water run-off, rooting depth and plant root densities, percolation and nitrogen leaching.

Experimental results, for example at plot 3, or in 1998 at all plots, are helpful data sets for agro-ecosystem model testing, especially under extreme conditions. The model testing results based on such data sets show the quality of models in the case of their application for conditions far from conditions used for model verification and show the range of applicability of tested models.

An accurate model evaluation requires thorough assessment of the quality of the data set itself. Here, this difficult task is complicated a little bit by the varying time resolution of the measurement and by the usual problems associated with measurements under field conditions. Hourly and daily measurements help greatly, but quality assessment of even these data requires experience and attention to detail. It is not possible to assess the quality of all given data in detail because several data collectors were acting during the 6-year period, and therefore, it is necessary to rely on the quality control provided by each of them.

However, the presence of independent sets of water content measurements from two different methods offers the possibility to compare the results and to make conclusions about the data quality. Here, parallel time series for water content were compared, from daily TDR measurements on one side and from more or less two-weekly augered gravimetric samples on the other side. In Fig. 7 water content time courses for the three soil layers 0–30 cm depth, 30–60 cm depth and 60–90 cm depth measured by TDR compared with the gravimetric measured water content values are shown for plot 1. Although, both measurement methods were realised correct, a deviation between the measurements was obtained in the same range for all soil layers, but both methods gave measurement results at the same water content level. The deviations between both methods were random and not systematic. The time periods of soil drying and soil wetting were identified properly in both cases. In comparison with the TDR method, the wetness and dryness pikes in the deeper soil layers were overemphasized by the gravimetric method in most cases (see Fig. 7). The fact, that the deeper the soil layer, the smoother the time courses of water content, is shown by both measurement methods correctly. The comparison results for the water content time courses between both measurement methods for plot 2 and for plot 3 were similar to those of plot 1. The comparison of water content measurements for all three plots between both methods is given in Fig. 8 for the three 30 cm soil layers, separately. Typical for all three soil layers was, that for water contents from 12–15 vol% and higher the gravimetric method gave higher measurement values in comparison with the TDR method. For water contents lower than 12–15 vol% it was opposite, i.e. the TDR method gave higher measurement values in comparison with the gravimetric method. Between both measurement methods, there was a deviation within −6.1–6.6 vol% (mean absolute deviation: 2.5 vol%) for the soil layer 0–30 cm depth, within −4.1–8.0 vol% (mean absolute deviation: 1.3 vol%) and within −4.8–6.2 vol% (mean absolute deviation: 1.7 vol%) for the soil layers 30–60 cm depth and 60–90 cm depth, respectively. It seems that there was a systematic difference in calibration of TDR method because the gravimetric method is a direct method and exact in its realization.

For analysing the actual water content dynamic over the whole time in detail, the TDR measurements were more suitable as basis for dynamic model verification. It is concluded that augered water content samples taken fortnightly gave only a very coarse and uncertain picture of the actual time course of soil water dynamic in the soil layers. These measurements are not adequate for validating soil-water flow models.

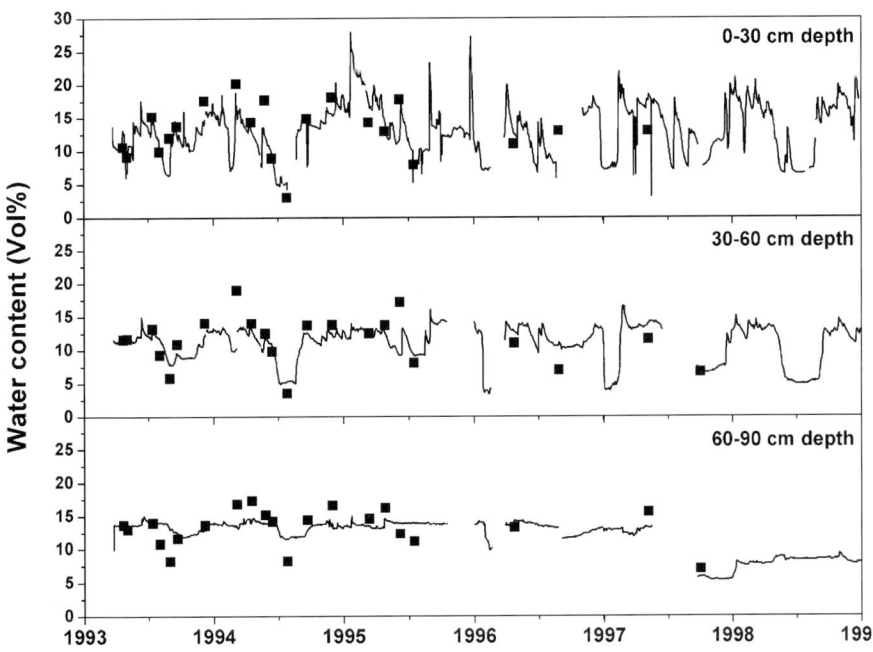

Fig. 7 Water content time courses for the soil layers 0–30 cm depth, 30–60 cm depth and 60–90 cm depth measured by TDR (full line) compared with gravimetric measured ones (full squares) for plot 1

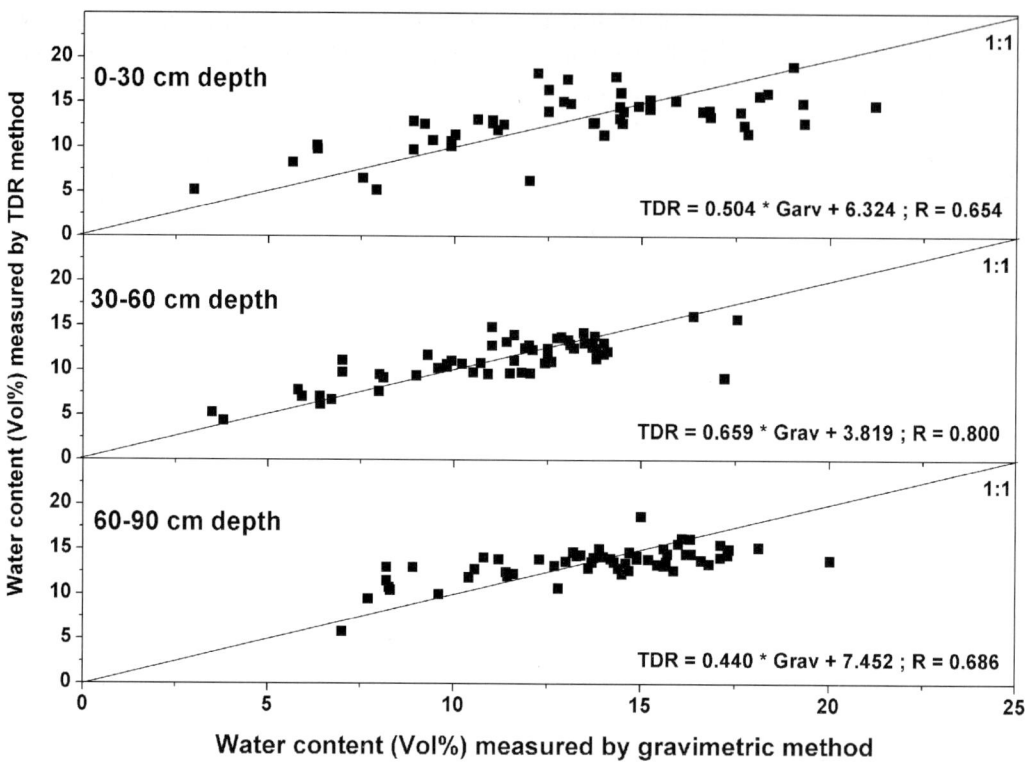

Fig. 8 Comparison of water content measured by TDR method on the one hand and measured by gravimetric method on the other hand for all three plots together, separate for the three 30 cm soil layers up to 90 cm depth

The comparison results between both measurement methods show, that model verification and model validation results depend on measurement methods and their accuracy in every case, i.e. the model accuracy after model verification is at the same level like the measurement accuracy. A combined determination of soil water contents with the TRIME-method and the observation of pressure heads with automatically recording tensiometers with a daily time resolution allows a more critical evaluation of the simulation quality of soil water balance models than the isolated use of one of those measurement methods.

Acknowledgements This contribution was supported by the German Federal Ministry of Consumer Protection, Food and Agriculture and the Ministry of Agriculture, Environmental Protection and Regional Planning of the Federal State of Brandenburg (Germany). We thank W. Hoehn, head of the Experimental Station Müncheberg up to 1999, for his support during realization of the field experiments, and M. Baehr and S. Dittmar for their efforts in the fields of equipment installation and service and handling of all experimental data. We also thank D. Schulz, head of the Central Laboratory, for realizing all necessary chemical analyses.

References

Anonymous (1997) Bestimmung von mineralischem (Nitrat-) Stickstoff in Bodenprofilen (N_{min}-Labormethode). In: Bassler R (ed) Handbuch der Landwirtschaftlichen Versuchs-und Untersuchungsmethodik (Methodenbuch) – 1. Band, 4. Auflage, VDLUFA-Verlag, Darmstadt, pp A6.1.4.1.1–A6.1.4.1.20

Hack H, Bleiholder H, Buhr L, Meier U, Schnock-Fricke U, Weber E, Witzenberger A (1992) Einheitliche Codierung der phänologischen Entwicklungsstadien mono-und di kotyler Pflanzen – Erweiterte BBCH-Skala, Allgemein.-Nachrichtenbl Deut Pflanzenschtzd 44:265–270

Kersebaum K.-C, Hecker M, Mirschel W, Wegehenkel M (2007) Modelling water and nutrition dynamics in soil-crop systems: a comparison of simulation models applied on a common data set. In: Kersebaum KC, Hecker J-M, Mirschel W, Wegehenkel M (eds) Modelling water and nutrient dynamics in soil-crop systems. Springer, Dordrecht, pp 1–17

Riegler H, Schiek O (1970) Automatisches Polarimeter für die Zuckerindustrie. Lebensmittel-Industrie 11/70, pp 1–2

Schindler U (1980) Ein Schnellverfahren zur Messung der Wasserleitfähigkeit im teilgesättigten Boden an Stechzylinderproben. Arch. Acker-, Pfl Boden 24:1–7

Strube W, Scholze D (1970) Erprobung des automatischen Polarimeters POLAMAT S. Lebensmittel-Industrie 11/70, pp 2–4

Wegehenkel M (1998) Zum Einsatz von TRIME-TDR zur Messung der Bodenfeuchte auf leichten Sandböden. Z Pfl Boden 161:577–582

Wegehenkel M (2004) The validation of soil water balance models using pressure head and soil water content data. Hydrological Processes 18 (in print).

Wehrmann J, Scharpf H.-C (1979). Der Mineralstickstoffgehalt des Bodens als Massstab für den Stickstoffdüngungsbedarf (Nmin-Methode). Plant Soil 52:109–126

Wenkel K.O, Mirschel W (eds) (1995) Agrarökosystemmodellierung - Grundlage für die Abschätzung von Auswirkungen möglicher Landnutzungs-und Klimaänderungen. ZALF-Bericht 24, pp 187

Zadoks J.C, Chang T.T, Konzak C.F (1974) A decimal code for growth stages of cereals. Weed Res 14:415–421

CHAPTER SEVENTEEN

Dynamics of water, carbon and nitrogen in an agricultural used Chernozem soil in Central Germany

Uwe Franko*, Martina Puhlmann, Katrin Kuka, Frank Böhme and Ines Merbach

Abstract The paper describes a data set from the experimental site at Bad Lauchstädt intended to be used by models for soil and plant dynamics. The data set includes the 100-year-old history of four treatments of one field of the well-known "Static Experiment" as well as more detailed information about crop development and soil moisture dynamics for a 4-year crop rotation from 1999 to 2002 that is identical with the crop rotation of the long-term experiment and a black fallow treatment, intended to provide basic data about soil moisture dynamics and nitrogen leaching without plant impact. Soil physical parameters sampled at four soil profiles, all of the same soil unit give an idea about the heterogeneity of the loess soil at this site. Climate data and especially their transformation to turnover conditions give strong indications for the impact of climate change on soil organic matter (SOM). Nitrogen deposition amounts to 47.3 kg/ha/year as an average value for the crop rotation of sugar beet, spring barley, potatoes and winter wheat. Soil water samples from ceramic suction cups show only very low concentrations of nitrogen below the root zone indicating that there is nearly no loss of the available nitrogen due to leaching.

Keywords Black fallow, Carbon and nitrogen dynamics, Climate change, Crop rotation, Long-term experiment, Soil moisture dynamics, Soil water quality

Introduction

The experimental site of Bad Lauchstädt is well known to the scientific community because of its long-term experiment that started in the beginning of the last century. The original aim of this experiment, the investigation of nutrient demands of crops and the effects of mineral fertilizer and manure for plant nutrition, are meanwhile less important comparing to the unique potential to investigate processes in the soil–plant system at the same site, but having a wide range of soil organic matter (SOM) content.

The recent development of modelling soil and crop processes resulted in a growing demand for more detailed experimental data that could not be supplied from the long-term experiment without serious disturbances. For this reason, an additional experimental field (the "Intensivmessfeld") has been established in 1997, which has the same crop rotation as the long-term experiment that has been equipped with numerous sensors in different depths providing data about

Uwe Franko, Martina Puhlmann, Katrin Kuka, Frank Böhme and Ines Merbach
UFZ-Centre for Environmental Research Leipzig-Halle, Department Soil Science, Theodor-Lieser-Str. 4, 06120 Halle, Germany
*Author for correspondence
tel.: +345-558-5432; fax: +345-558-5449;
e-mail: uwe.franko@ufz.de

soil moisture, soil temperature and soil water quality and with the opportunity to accomplish intermediate harvest actions. Additional to the traditional crop rotation of the long-term experiment, there have been established some alternatives representing different types of land use: industrial crops, organic farming, fallow and grassland.

In the soil–plant system there are processes with quite different timescale to be observed. Temperature and water dynamics, as well as the plant development show strong dynamics within months or even days, whereas the turnover of SOM and the subsequent nitrogen mineralization is much slower and has to be studied over years or even decades. Having these different timescales, it is a general problem in modelling to fit the fast processes as well as the slow ones. For this reason this data set contains besides the long-term data, a record of a 4-year crop rotation and a black fallow plot including information about plant development, soil temperature, soil moisture and soil water quality. Both experiments together provide a data set for the evaluation of soil or crop models for short-term and long-term dynamics. The long-term data include the main treatments of one out of the four main fields of the "Static Experiment" with the main driving variables, crop yields and soil carbon as the results, whereas the short-term data set contains additional data about soil water and temperature for one 4-year crop rotation as well as a black fallow treatment.

Material and methods

Long-term experiment

The long-term experiment "Static Experiment Bad Lauchstädt" has been established in autumn 1902 with two treatments of farmyard manure (FYM). In 1906, it has been extended to a third FYM treatment. Each of the FYM treatments is combined with treatments of mineral fertilizer applying nitrogen, phosphorus and potassium in various combinations. The crop rotation is sugar beet, spring barley, potatoes and winter wheat. The experiments consist of different fields in order to grow each crop in every year. A detailed description of the first 90 years of this experiment is given by Körschens (1994). A great part of the experimental data is accessible over the EURO-SOMNET database (Franko et al. 2002;

Smith et al. 2002). The data set here represents only one field (field "I-south") because the time course of SOM change is very similar at the other fields.

Site conditions

The reconstruction of weather data was difficult because from 1906 to 1955 there were available only aggregated climate data (pentades) about air temperature and precipitation. Beginning with 1956, daily observations of air temperature, sun shine duration and precipitation started that are still continued. As most models depend on daily climate data, it was decided to use the weather generator of the CANDY model (Oelschlägel 1995) in order to fill that gap. The generator has been parameterized with the available daily climate data. For each month, a set of daily data repeatedly has been generated until it was representing the observed pentade values.

Soil parameters are available from a general site description that has been published by (Körschens 1994). The data in Tables 1 and 2 give an overview about the basic soil physical parameters required for modelling. In order to take into account the impact of SOM on soil moisture dynamics Table 2 contains the hydrological parameters for the both treatments with the largest difference in organic carbon content.

Management

Most of the management actions on the field have been reconstructed from the annual reports of the experiment. In some cases, the actual dates of certain actions were not available and have been replaced by average values. The treatments selected for this paper are the control with no input of fertilizer and manure, the NPK treatment with mineral fertilizer input, the

Table 1 General soil physical parameters for the long-term experiment

Depth (cm)	Bulk density (g/cm^3)	Particle density (g/cm^3)	Field capacity (vol%)	Wilting point (vol%)
0–10	1.33	2.67	32.9	18.0
10–20	1.26	2.67	32.5	17.0
20–30	1.44	2.68	32.1	19.9
30–70	1.45	2.68	33.1	20.1
70–130	1.40	2.71	29.8	9.5
130–200	1.72	2.74	20.9	13.1

Table 2 Hydraulic parameters for the soil of the long-term experiment

Plot	Depth (cm)	pF = 0 (vol%)	pF = 1.8 (vol%)	pF = 2.5 (vol%)	pF = 3.0 (vol%)	pF = 3.4 (vol%)	pF = 3.8 (vol%)	pF = 4.2 (vol%)
FYM + NPK	0–40	51.5	43.9	32.6	21.8	20.6	18.6	17.5
	40–200	46.1	39.4	32.5	27.4	24.6	21.0	20.0
Control	0–40	49.3	38.5	28.9	21.5	20.8	17.5	16.8
	40–200	47.0	38.0	30.6	24.5	23.5	20.2	18.9

FYM treatment with a manure application of 300 dt/ha given to the row crops every second year, and the combination of manure and mineral fertilizer (FYM–NPK) with the same manure application as the FYM treatment. The input of mineral nitrogen changed over time parallel to the regional agricultural practice (see Fig. 1).

Crop rotation and black fallow

The "Intensivmessfeld" is situated close to the long-term experiment and generally has the same climate and soil conditions. This data set should provide more information about processes in the soil–plant system and for that reason it contains additional climate data and more detailed data about soil properties and crop development.

Site conditions

Climate data including temperature (daily values as maximum, minimum and average), rainfall, global radiation, humidity, soil temperature, wind speed and wind direction have been collected by an automatic meteorological station on a 10-min time resolution and have been aggregated to hourly and daily values. The total rainfall over this 4-year period represents the general site conditions, but the precipitation in 2002 was above average (Fig. 2).

In order to address the phenomenon of soil heterogeneity, the physical soil parameters for the "Intensivmessfeld" plots have been determined at four different soil profiles that were all very close (<50 m) to the experimental plots. The results for the most important soil physical properties required for modelling are represented in Table 3 with average and range. Particle density and hygroscopicity have been determined with five replications from a sample of disturbed soil, whereas six replications of undisturbed soil samples using soil sample rings were available for all other parameters.

Management

Management data shown in Table 4 include only information relevant for this special modelling purpose.

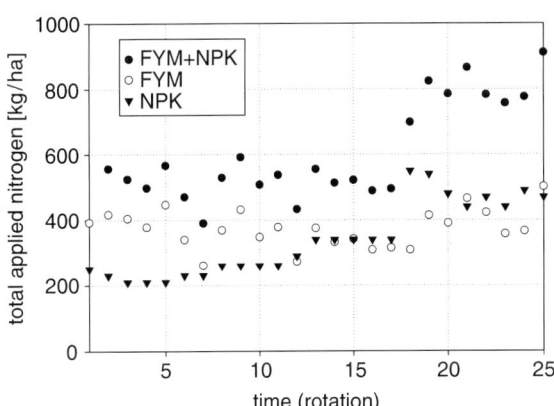

Fig. 1 Application of nitrogen during the history of the Static Experiment

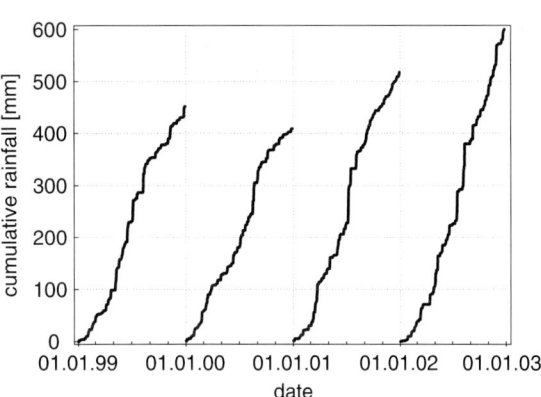

Fig. 2 Rainfall pattern for the 4-year period from 1998 to 2002

Table 3 Average and range for soil physical properties at the "Intensivmessfeld" observed at four different soil profiles

Soil profile	Depth (cm)	Particle density (g/cm³)	Hygroscopicity (M%)	Bulk density (g/cm³)	Water content pF 0 (vol%)	pF 1.8 (vol%)	pF 2.5 (vol%)	pF 3.0 (vol%)
1	20–24	2.56/**2.58**/2.62	5.53/**5.71**/6.01	1.17/**1.27**/1.42	43.90/**46.72**/51.70	39.70/**42.73**/45.70	27.10/**28.43**/31.00	20.80/**23.23**/27.70
	45–49	2.59/**2.60**/2.61	5.70/**5.81**/5.93	1.23/**1.34**/1.39	43.80/**46.33**/48.90	42.90/**44.62**/45.80	28.20/**31.45**/32.90	21.40/**24.83**/27.60
	115–119	2.66/**2.67**/2.67	2.82/**2.90**/2.96	1.50/**1.52**/1.54	42.60/**44.55**/45.80	42.40/**43.28**/44.30	32.90/**34.63**/36.50	14.50/**16.27**/19.30
	185–189	2.65/**2.66**/2.67	3.54/**3.59**/3.67	1.75/**1.77**/1.82	30.20/**30.97**/31.80	29.10/**29.98**/31.00	19.50/**20.28**/20.90	15.80/**16.58**/17.70
2	20–24	2.57/**2.58**/2.58	5.57/**5.75**/5.90	1.22/**1.35**/1.60	38.50/**45.02**/48.00	36.90/**38.70**/40.60	25.60/**28.50**/33.10	21.30/**25.53**/30.90
	45–49	2.59/**2.60**/2.61	5.91/**5.96**/6.06	1.23/**1.32**/1.40	43.50/**46.08**/49.60	41.60/**42.47**/43.40	26.10/**29.23**/31.40	24.30/**25.48**/26.20
	115–119	2.66/**2.67**/2.68	2.81/**2.90**/2.95	1.44/**1.47**/1.52	42.20/**44.25**/45.60	40.70/**41.83**/42.50	27.70/**29.27**/30.60	13.80/**17.00**/21.70
	185–189	2.64/**2.65**/2.66	3.35/**3.40**/3.46	1.77/**1.81**/1.85	30.00/**31.33**/32.60	27.80/**29.10**/29.80	17.00/**19.02**/19.80	14.40/**16.00**/16.80
3	20–24	2.55/**2.55**/2.56	5.85/**5.91**/6.00	1.29/**1.31**/1.33	47.20/**49.62**/51.50	41.20/**43.57**/45.50	27.60/**28.43**/29.30	23.70/**25.25**/26.60
	45–49	2.57/**2.59**/2.60	6.39/**6.47**/6.58	1.37/**1.41**/1.44	42.90/**43.95**/44.90	40.20/**40.82**/41.60	29.90/**30.90**/32.00	25.90/**26.98**/27.80
	115–119	2.66/**2.67**/2.69	2.79/**2.86**/2.91	1.49/**1.51**/1.52	43.40/**44.32**/45.90	41.70/**42.20**/42.70	27.80/**28.28**/29.00	15.50/**18.23**/25.30
	185–189	2.64/**2.65**/2.66	3.42/**3.52**/3.59	1.79/**1.81**/1.84	29.90/**30.77**/31.50	27.50/**28.18**/28.80	19.70/**19.85**/20.10	17.30/**17.63**/17.90
4	20–24	2.55/**2.56**/2.57	5.70/**5.80**/5.95	1.18/**1.21**/1.24	45.50/**49.77**/53.40	35.70/**39.38**/42.20	25.30/**26.48**/28.30	20.90/**22.73**/25.50
	45–49	2.58/**2.59**/2.61	6.26/**6.33**/6.38	1.37/**1.40**/1.42	40.10/**42.23**/46.50	40.10/**40.68**/41.20	30.60/**31.38**/32.40	26.30/**27.22**/28.30
	115–119	2.67/**2.67**/2.67	2.79/**2.88**/2.94	1.48/**1.50**/1.52	42.20/**42.85**/43.40	41.30/**41.85**/42.60	28.00/**29.63**/30.90	15.10/**16.82**/22.40
	185–189	2.64/**2.66**/2.68	3.45/**3.54**/3.64	1.73/**1.78**/1.83	27.10/**27.23**/27.60	26.50/**26.85**/27.20	18.90/**19.12**/19.40	16.70/**16.98**/17.40

Values given as minimum/**average**/maximum

Table 4 Management data of the crop rotation including the pre-experimental treatment

Date	Event
28.10.1996	Emergence: winter barley
13.03.1997	Mineral N fertilizer: 30 kg N/ha calcium ammonium nitrate
05.05.1997	Mineral N fertilizer: 30 kg N/ha calcium ammonium nitrate
25.07.1997	Harvest, crop residues removed
13.10.1997	Emergence: winter wheat
09.03.1998	Mineral N fertilizer: 40 kg N/ha calcium ammonium nitrate
22.04.1998	Mineral N fertilizer: 50 kg N/ha calcium ammonium nitrate
07.08.1998	Harvest, crop residues removed
08.08.1998	Organic manure: 40.45 dt FM/ha straw
18.01.1999	Soil tillage: 25 cm
30.03.1999	Soil tillage: 3 cm
16.04.1999	Emergence: sugar beet
03.05.1999	Mineral N fertilizer: 100 kg N/ha calcium ammonium nitrate
25.05.1999	Mineral N fertilizer: 80 kg N/ha calcium ammonium nitrate
12.10.1999	Harvest, crop residues removed
05.11.1999	Organic manure: 239.3 dt FM/ha sugar beet leaves
05.11.1999	Soil tillage: 25 cm
21.03.2000	Soil tillage: 6 cm
23.03.2000	Soil tillage: 8 cm
08.04.2000	Emergence: spring barley
25.04.2000	Mineral N fertilizer: 30 kg N/ha calcium ammonium nitrate
23.05.2000	Mineral N fertilizer: 20 kg N/ha calcium ammonium nitrate
31.07.2000	Harvest, crop res. removed
01.08.2000	Organic manure: 23.05 dt FM/ha straw
01.08.2000	Soil tillage: 8 cm
12.09.2000	Soil tillage: 8 cm
03.11.2000	Soil tillage: 25 cm
19.04.2001	Soil tillage: 5 cm
16.05.2001	Mineral N fertilizer: 50 kg N/ha calcium ammonium nitrate
24.05.2001	Emergence: potatoes
05.06.2001	Mineral N fertilizer: 50 kg N/ha calcium ammonium nitrate
14.09.2001	Harvest, crop residues removed
27.09.2001	Soil tillage: 25 cm
20.11.2001	Emergence: winter wheat
18.03.2002	Mineral N fertilizer: 30 kg N/ha calcium ammonium nitrate
31.07.2002	Harvest, crop residues removed
02.08.2002	Organic manure: 48.54 dt FM/ha straw
02.08.2002	Soil tillage: 6 cm
22.08.2002	Soil tillage: 6 cm
18.09.2002	Soil tillage: 6 cm
28.10.2002	Soil tillage: 28 cm

Applications of pesticides and input of phosphorus and potassium followed general requirements according to agricultural rules. The black fallow treatment has been tilled from time to time to keep the soil free from weeds.

Crop data

Intermediate harvest actions provided data about biomass development. The plants were removed from a sub-plot of about 1 m² size without disturbance to the kernel plot (55 m²) that was reserved for main harvest. The fresh material has been weighed at the field with a subsequent determination of dry matter. Plant material was dried and analysed for its nutrient content. Results are represented in Tables 5 and 6.

Sensor data

Soil temperature and soil moisture data were collected by dataloggers using frequency domain reflectometry (FDR) sensors (Theta Probe type ML1 from Delta-T Devices) for soil moisture. Data were recorded on an hourly time base and aggregated to daily averages.

Table 5 Results of crop analysis. FM: fresh matter, DM: dry matter, N: nitrogen, C: carbon

Crop	Date	Sample	FM yield (dt/ha)	DM content (% of FM)	N content (% of DM)	C content (% in DM)
Sugar beet	17.05.1999	Leafs	5	10.11	4.49	–
	17.06.1999	Leafs	312	10.14	3.99	–
	17.06.1999	Beet	93.3	15.36	1.42	–
	12.10.1999	Leafs	239.3	20.18	2.08	36.7
	12.10.1999	Beet	735.1	27.12	0.89	–
Spring barley	16.05.2000	Shoot	109	15.33	3.11	–
	06.06.2000	Shoot	240	22	1.4	–
	31.07.2000	Grain	34.61	83.4	2.54	45.7
	31.07.2000	Straw	23.05	81.06	0.83	46
Potatoes	11.06.2001	Leafs	117.3	8.06	4.7	–
	25.06.2001	Leafs	180	10.03	3.96	–
	25.06.2001	Tuber	108	16.13	1.66	–
	14.09.2001	Tuber	549.2	23.21	1.29	42
Qinter wheat	30.04.2002	Shoot	178.4	15	3.41	–
	29.05.2002	Shoot	437.3	18.96	1.67	–
	31.07.2002	Grain	56.4	89.45	2	40.5
	31.07.2002	Straw	48.54	81.82	0.42	41

Table 6 Observation of plant development

Crop	Date	DC code
Sugar beet	30.03.1999	0
	16.04.1999	10
	12.05.1999	14
	17.05.1999	15
	07.06.1999	39
Spring barley	23.02.2000	0
	08.04.2000	10
	17.04.2000	13
	25.04.2000	21
	02.05.2000	22
	11.05.2000	31
	16.05.2000	31
	31.05.2000	47
	05.06.2000	57
	31.07.2000	92
Potatoes	26.04.2001	0
	24.05.2001	10
	28.05.2001	16
	05.06.2001	17
	11.06.2001	17
	15.06.2001	35
	25.06.2001	63
	29.06.2001	67
Winter wheat	23.10.2001	0
	20.11.2001	10
	05.02.2002	13
	18.03.2002	21
	25.03.2002	23
	30.04.2002	32
	21.05.2002	45
	29.05.2002	51
	31.05.2002	55
	04.06.2002	65
	19.06.2002	73
	11.07.2002	85
	18.07.2002	89
	23.07.2002	92
	31.07.2002	92

The sensors are horizontally installed in three depths (45, 90 and 170 cm) with three replications from a trench beside the plot that has been refilled afterwards. This type of installation requires only horizontal holes and should avoid the creation of artificial preferential flow routes in vertical direction. Measurements in the top soils have been accomplished by the same type of sensors, but in this case the sensors always had to be removed for soil tillage actions and were replaced immediately afterwards at the same spot. All FDR sensors have been calibrated off-site for the special soil conditions. Top soil sensors have been recalibrated using gravimetric soil moisture measurements.

Soil solution was collected in the same depths as the fixed sensors using three replications of ceramic suction cups connected with a vacuum unit that is controlled by tensiometers. The tension inside the suction cups has an offset of 50 hPa to meet the resistance of the ceramic material and an additional offset according to the installation depth in order to move the water to the collecting bottles at the soil surface. The water was collected over 4 weeks and then analysed. In times with risk of low temperatures only one collecting bottle of plastic material for each depth was used in order to prevent damage by freezing. Results of water quality analyses are shown in Tables 7 and 8.

Table 7 Soil solution data for crop rotation. Results for ammonia nitrogen (NH_4-N), nitrate nitrogen (NO_3-N) and acidity (pH) given as average/standard deviation with sample mode m: mixed sample, s: single values, n: number of replications

Depth (cm)	Date	Sample mode	n	NH_4-N (ppm)	NO_3-N (ppm)	pH
45	07.01.1999	M	1	0.01/–	100/–	7.4/–
	10.02.1999	M	1	0.05/–	93/–	5.179/–
	03.03.1999	M	1	0/–	84/–	8.159/–
	31.03.1999	M	1	0.02/–	94/–	7.404/–
	28.04.1999	S	3	0.01/0.01	74.67/32.52	7.73/0.08
	02.06.1999	S	3	0.00/0.00	65.67/31.07	7.20/0.20
90	07.01.1999	M	1	0/–	0.1/–	7.305/–
	10.02.1999	M	1	0/–	2.2/–	5.911/–
	03.03.1999	M	1	0.02/–	6.2/–	7.762/–
	31.03.1999	M	1	0.02/–	1.2/–	7.568/–
	28.04.1999	S	3	0.02/0.01	0.27/0.21	7.81/0.10
	02.06.1999	S	3	0.00/0.00	0.37/0.23	7.20/0.15
	30.06.1999	S	3	0.01/0.01	0.23/0.12	7.42/0.17
	25.08.1999	M	1	7.5/–	0.5/–	7.81/–
	31.05.2000	S	1	0.03/–	0.5/–	8.11/–
	25.04.2001	S	2	0.05/0.04	0.30/0.00	8.22/0.19
	30.05.2001	S	2	0.01/0.00	0.63/0.46	8.19/0.27
	31.07.2002	S	1	1.52/–	11/–	8.56/–
170	31.03.1999	M	1	0.02/–	0.4/–	7.487/–
	28.04.1999	S	3	0.01/0.01	0.17/0.29	7.89/0.06
	02.06.1999	S	3	0.00/0.00	0.43/0.29	7.49/0.07
	30.06.1999	S	3	0.06/0.02	0.57/0.06	7.57/0.09
	28.07.1999	S	3	0.14/0.16	0.80/0.66	7.54/0.18
	29.09.1999	M	1	1.88/–	0.2/–	7.19/–
	29.03.2000	M	1	0.01/–	0.3/–	7.85/–
	26.04.2000	S	3	0.00/0.00	0.58/0.03	8.56/0.12
	31.05.2000	S	2	0.05/0.02	0.50/0.00	8.36/0.10
	28.06.2000	S	3	0.03/0.02	0.62/0.03	8.15/0.10
	26.07.2000	S	3	0.06/0.11	0.50/0.00	8.52/0.06
	19.12.2000	M	1	0.18/–	0.5/–	8.23/–
	25.04.2001	M	1	0.56/–	0.5/–	8.282/–
	30.05.2001	S	3	0.03/0.02	0.50/0.00	8.52/0.25
	30.01.2002	M	1	0.04/–	0.2/–	8.1/–
	27.02.2002	M	1	0.45/–	0.2/–	7.74/–
	29.05.2002	S	2	0.06/0.06	0.10/0.00	8.54/0.13
	26.06.2002	S	3	0.13/0.07	0.50/0.00	8.41/0.19
	18.09.2002	S	2	0.27/0.13	0.50/0.00	8.10/0.32
	30.10.2002	M	1	0.04/–	0.5/–	8.39/–
	28.11.2002	M	1	0.06/–	0/–	7.96/–

Table 8 Soil solution data for black fallow treatment. Results for ammonia nitrogen (NH_4-N), nitrate nitrogen (NO_3-N) and acidity (pH) given as average/standard deviation with sample mode m: mixed sample, s: single values, n: number of replications

Depth (cm)	Date	Sample mode	n	NH_4-N (ppm)	NO_3-N (ppm)	pH
45	07.01.1999	M	1	0.2/–	10/–	7.862/–
	10.02.1999	M	1	0.03/–	69/–	6.684/–
	03.03.1999	M	1	0/–	62/–	8.154/–
	31.03.1999	M	1	0/–	28/–	7.926/–
	28.04.1999	S	3	0.01/0.01	29.77/22.98	8.02/0.18
	02.06.1999	S	3	0.18/0.21	20.50/20.79	7.68/0.15
	30.06.1999	S	3	0.50/0.66	19.90/17.94	7.75/0.23
	28.07.1999	S	3	2.81/2.54	14.92/9.30	7.60/0.47
	29.09.1999	S	3	2.85/1.79	9.13/10.65	8.09/0.05
	27.10.1999	M	1	11/–	20/–	8.17/–
	21.12.1999	M	1	13.6/–	20/–	8.3/–
	23.02.2000	M	1	3.50/–	26.50/–	8.86/–
	26.04.2000	S	2	0.00/0.00	53.50/54.45	8.15/0.01
	31.05.2000	S	3	0.00/0.00	3.75/ 4.60	8.12/0.25
	28.06.2000	S	1	4/–	128/–	8.13/–
	26.07.2000	S	3	15.13/14.12	71.67/25.97	8.12/0.24
	26.09.2000	S	1	5/–	130/–	7.76/–
	25.07.2001	S	1	5.9/–	76/–	8.16/–
	29.08.2001	S	1	0.37/–	21/–	7.69/–
	24.10.2001	S	1	0.26/–	15.5/–	7.54/–
	27.02.2002	M	1	0.61/–	27.75/–	8.07/–
	27.03.2002	M	1	0.06/–	1/–	7.7/–
	29.05.2002	S	1	0.06/–	14.8/–	7.69/–
	26.06.2002	S	1	0.09/–	28/–	7.45/–
	31.07.2002	S	1	0.01/–	28.5/–	7.22/–
	27.08.2002	S	1	0.03/–	35/–	7.31/–
90	07.01.1999	M	1	0/–	0.8/–	7.827/–
	10.02.1999	M	1	0/–	0.35/–	7.031/–
	03.03.1999	M	1	0.01/–	0.4/–	7.847/–
	31.03.1999	M	1	0.01/–	0.7/–	7.752/–
	28.04.1999	S	2	0.02/0.00	0.10/0.14	7.79/0.04
	02.06.1999	S	3	0.12/0.10	0.27/0.15	7.49/0.19
	30.06.1999	S	3	1.02/0.86	0.62/0.51	7.59/0.18
	28.07.1999	S	2	0.88/1.03	0.45/0.49	7.33/0.32
	25.08.1999	S	3	3.82/4.38	0.23/0.15	7.67/0.24
	29.09.1999	S	3	1.13/1.96	2.88/2.99	7.82/0.19
	27.10.1999	M	10	0.01/–	0.4/–	7.8/–
	27.11.1999	M	10	0/–	0.4/–	7.75/–
	21.12.1999	M	10	5.1/–	54/–	8.32/–
	27.01.2000	M	1	0.04/–	0.35/–	7.67/–
	23.02.2000	M	1	0.02/–	0.90/–	8.48/–
	29.03.2000	M	1	0.00/–	0.30/–	7.97/–
	26.04.2000	S	3	0.01/0.02	3.10/2.90	7.90/0.39
	31.05.2000	S	2	0.03/0.04	3.35/0.78	7.97/0.10
	28.06.2000	S	1	0.30/–	0.40/–	7.60/–
	26.07.2000	S	2	1.00/1.41	1.00/0.14	8.08/0.01
	30.08.2000	S	2	0.00/0.00	0.45/0.21	8.16/0.03
	26.09.2000	S	2	0.62/0.88	0.23/0.11	7.53/0.29
	25.10.2000	M	1	0.035/–	0.2/–	8.2/–
	29.11.2000	M	1	0.075/–	0.2/–	8.34/–
	19.12.2000	M	1	0.06/–	0/–	8.35/–
	31.01.2001	M	1	0.27/–	0.2/–	0/–
	07.03.2001	M	1	0.05/–	0.4/–	7.962/–
	28.03.2001	M	1	0.01/–	0.8/–	8.091/–
	25.04.2001	S	1	0/–	0.5/–	7.845/–
	30.05.2001	S	2	0.42/0.56	3.90/4.53	7.85/0.13

Table 8 Soil solution data for black fallow treatment. Results for ammonia nitrogen (NH$_4$-N), nitrate nitrogen (NO$_3$-N) and acidity (pH) given as average/standard deviation with sample mode m: mixed sample, s: single values, n: number of replications—Cont'd

Depth (cm)	Date	Sample mode	n	NH$_4$-N (ppm)	NO$_3$-N (ppm)	pH
	27.06.2001	S	1	2.89/–	0.75/–	7.95/–
	25.07.2001	S	1	0.48/–	0.2/–	7.7/–
	29.08.2001	S	1	1.17/–	1.1/–	7.59/–
	26.09.2001	S	1	4.3/–	0.2/–	8.18/–
	29.05.2002	S	2	0.04/0.04	2.75/2.19	7.58/0.23
	26.06.2002	S	1	1.99/–	0.8/–	7.33/–
	31.07.2002	S	1	0.57/–	0.6/–	7.61/–
	27.08.2002	S	2	0.20/0.15	3.20/1.84	7.62/0.33
	18.09.2002	S	2	0.11/0.16	5.45/4.74	7.51/0.04
	28.11.2002	M	1	0.01/–	5.6/–	7.72/–
170	07.01.1999	M	1	0/–	0.6/–	8.063/–
	10.02.1999	M	1	0.01/–	0.4/–	7.402/–
	03.03.1999	M	1	0.03/–	0.3/–	8.371/–
	31.03.1999	M	1	0.02/–	0.4/–	8.121/–
	28.04.1999	S	2	0.01/0.01	0.45/0.07	8.02/0.03
	02.06.1999	S	3	0.14/0.21	0.30/0.26	7.61/0.09
	30.06.1999	S	3	0.94/0.82	0.30/0.20	7.76/0.19
	28.07.1999	S	2	3.01/4.09	0.43/0.04	7.53/0.19
	25.08.1999	S	3	2.81/2.75	0.25/0.26	8.22/0.35
	29.09.1999	S	3	2.10/1.90	0.57/0.16	7.94/0.10
	27.10.1999	M	1	0/–	0.3/–	7.82/–
	27.11.1999	M	1	0.01/–	0.5/–	8.07/–
	21.12.1999	M	1	0.01/–	0.3/–	8.34/–
	27.01.2000	M	1	0.03/–	0.70/–	7.85/–
	23.02.2000	M	1	0.01/–	1.00/–	8.66/–
	29.03.2000	M	1	0.00/–	0.30/–	8.20/–
	26.04.2000	S	3	0.01/0.02	1.27/1.33	8.36/0.06
	31.05.2000	S	3	0.01/0.01	0.40/0.10	8.23/0.04
	28.06.2000	S	3	2.20/2.60	1.37/1.42	7.51/0.26
	26.07.2000	S	3	1.06/1.44	0.70/0.44	8.20/0.15
	30.08.2000	S	3	0.17/0.29	1.00/0.26	8.18/0.37
	26.09.2000	S	3	0.38/0.66	0.83/0.67	7.65/0.30
	25.10.2000	M	1	0.03/–	0.4/–	8.28/–
	29.11.2000	M	1	0.05/–	0.4/–	8.5/–
	19.12.2000	M	1	0.055/–	0.5/–	8.56/–
	31.01.2001	M	1	0.03/–	1.0/–	8.168/–
	07.03.2001	M	1	0.04/–	0.4/–	8.18/–
	28.03.2001	M	1	0.07/–	0.4/–	8.25/–
	25.04.2001	S	3	0.02/0.02	0.97/1.07	8.22/0.19
	30.05.2001	S	3	0.10/0.14	0.23/0.21	8.35/0.22
	27.06.2001	S	3	0.39/0.59	0.27/0.15	8.36/0.28
	25.07.2001	S	3	0.62/1.05	0.50/0.40	8.41/0.11
	29.08.2001	S	3	0.39/0.53	0.90/1.06	8.38/0.06
	26.09.2001	S	3	0.47/0.52	1.13/1.18	8.29/0.39
	24.10.2001	S	2	0.29/0.36	1.85/1.34	8.54/0.18
	28.11.2001	M	1	0.01/–	0.45/–	8.28/–
	30.01.2002	M	1	0.015/–	0.2/–	8.43/–
	27.02.2002	M	1	0.02/–	0.2/–	8.17/–
	27.03.2002	M	1	0.05/–	0.2/–	8.08/–
	29.05.2002	S	2	0.02/0.01	1.25/1.20	8.17/0.11
	26.06.2002	S	3	0.57/0.92	0.47/0.15	8.27/0.08
	31.07.2002	S	3	0.15/0.12	0.57/0.50	8.07/0.10
	27.08.2002	S	3	0.54/0.50	1.47/1.76	8.05/0.02
	18.09.2002	S	3	0.25/0.35	1.00/1.22	7.96/0.09
	30.10.2002	M	1	0.07/–	0.3/–	8.33/–
	28.11.2002	M	1	0/–	0.4/–	7.85/–

Results

Nitrogen deposition from atmosphere has been proved to be a significant source for plant nutrition. Measurements of total nitrogen deposition, carried out with the integrated total nitrogen input (ITNI) system, show that the annual rate of nitrogen that can be gathered from atmosphere may be up to 65 kg/ha (Weigel et al. 2000). This amount is resulting from different sources: bulk deposition, gaseous deposition and direct uptake by plant. Direct measurements from Böhme et al. (2002, 2003) for different plants indicate a crop-specific deposition rate, but because of the high variability between years it seems useful to apply one average rate of 190 g N/ha/vegetation day for all crops. Results from Mehlert (1996) suggest that the rate of bulk deposition has to be increased by 28% to account for the organic compounds that are not included in the standard method. Gaseous (7 kg/ha) and the corrected bulk deposition (27.3 kg/ha) together lead to an annual deposition of 34.3 kg/ha on black fallow soil. The total deposition, calculated using the vegetation time for each crop and the above given deposition rate for the fallow time, amounts to an average rate of 47.3 kg/ha for the whole crop rotation.

The crop yields (see Fig. 3) in the long-term experiment showed variations caused by the special conditions of the individual growing seasons and a trend that is partly a result of improved agricultural techniques, especially plant protection and more productive varieties, but is also driven by the increase of fertilizer input especially during the last 40 years.

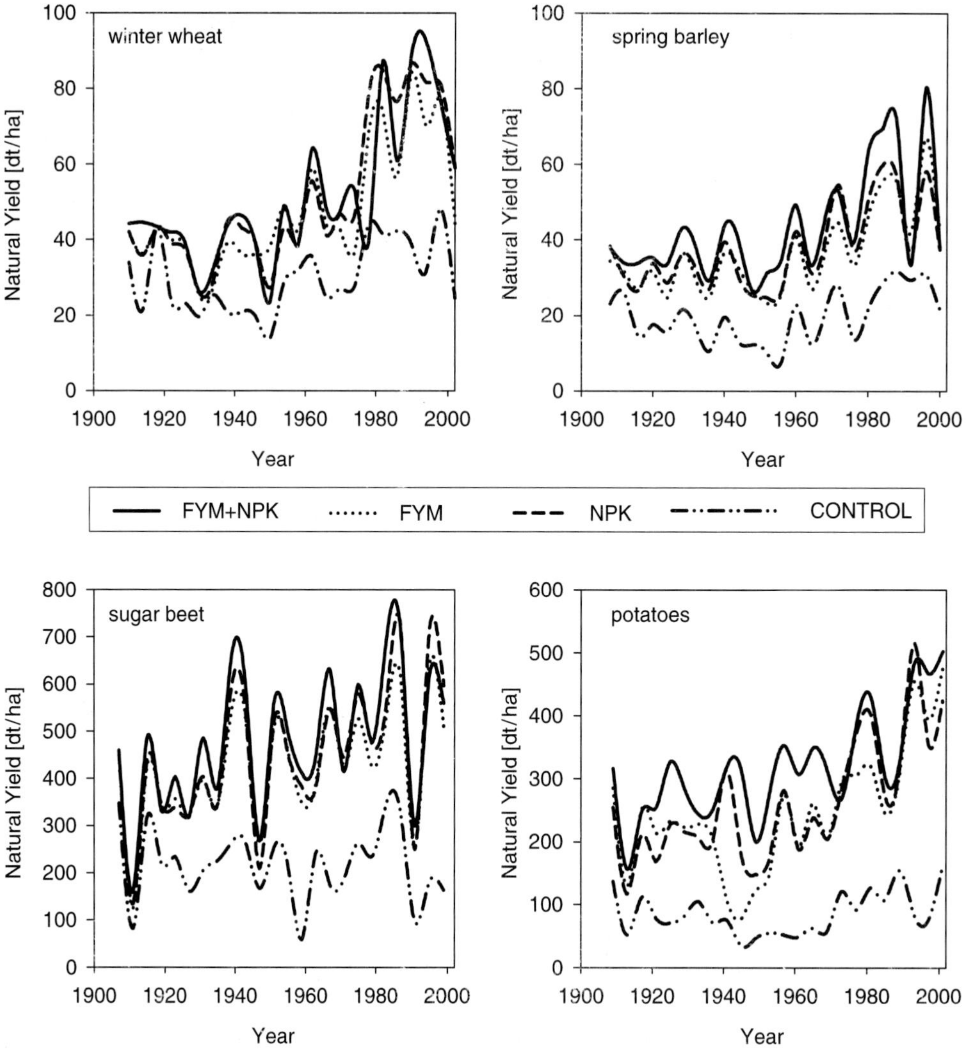

Fig. 3 Dynamic of crop yields during the history of the Static Experiment (smoothed lines)

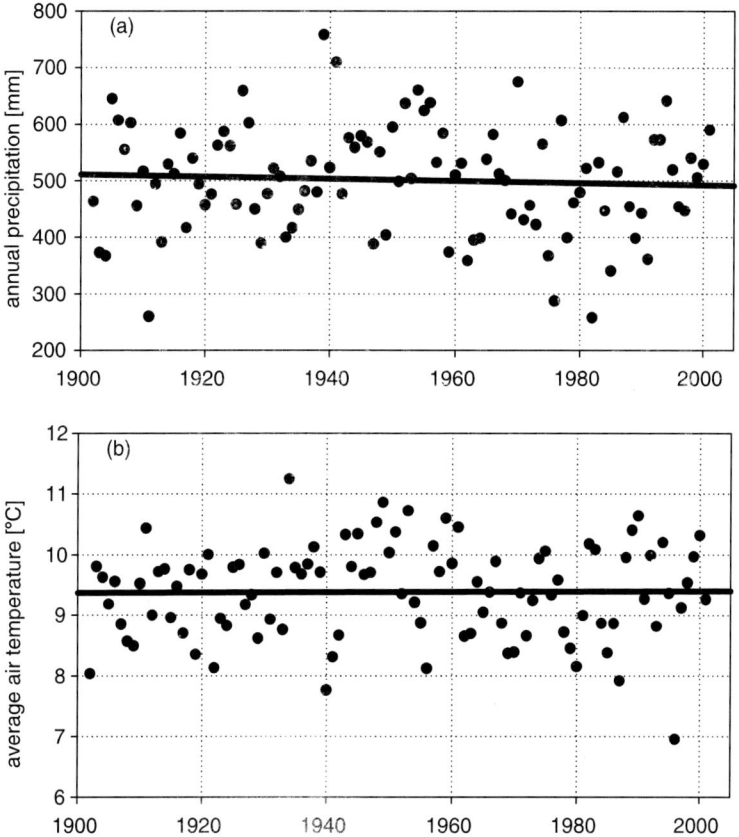

Fig. 4 Climate change impact to turnover conditions in Bad Lauchstädt: annual precipitation (a), annual average of air temperature (b)

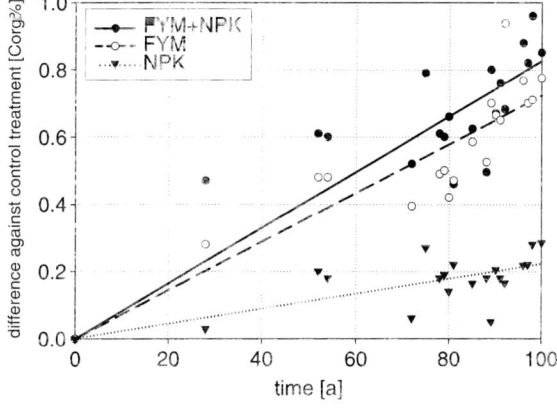

Fig. 5 Changes of soil organic matter storage related to the control treatment. Increment rates: 0.0022 (NPK), 0.0072 (FYM,) and 0.0083 (FYM+NPK)

Climate change may also have an influence on the experimental results. Fig. 4a and b represents the development of precipitation and air temperature where the temperature shows a little positive trend (0.03° in 100 years) and precipitation has a negative trend with an average reduction of 19 mm during the century.

The impact of the different treatments to the organic carbon content becomes better visible, if the difference between a given treatment and the control plot is calculated (see Fig. 5). It is obvious that the experiment is not yet in its steady state because of the still ongoing differentiation between the treatments.

Yields of all fertilized treatments of the long-term experiment are comparable to the "Intensivmessfeld" plot (see Fig. 6). In all plots, there was an untypical low yield of winter wheat in 2002 because of relative high biomass losses during the winter due to mice.

In the 4-year crop rotation there is a total nitrogen uptake of 516 kg/ha from the field. With a mineral fertilizer input of 360 kg/ha there was a gain of 39 kg/ha nitrogen/year that was taken from atmospheric deposition and mineralization of initial SOM. Observed changes of SOM are documented in Table 9. The mineralization rate for this short period of time can not be calculated from C_{org} and N_t observation because of their variability. Only measurements of hot water-soluble carbon on the cropped plot showed a significant reduction of this pool over the time indicating that there was net mineralization.

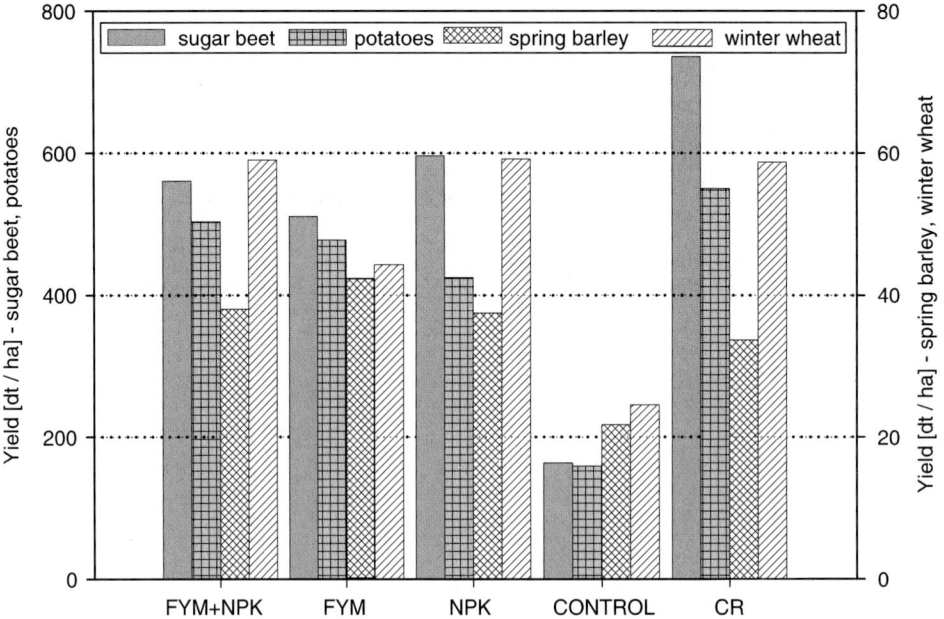

Fig. 6 Comparison of crop yields in the long term experiment (FYM+NPK, FYM, NPK, CONTROL) and the four year crop rotation (CR)

Table 9 Results of soil sampling in 0–20 cm with nitrate nitrogen (N-NO$_3$), ammonium nitrogen (N-NH$_4$), total nitrogen (N_t), organic carbon (C_{org}) and hot water-soluble carbon (C_{hws})

Treatment	Date	N-NO$_3$ (ppm)	N-NH$_4$ (ppm)	C_{org} (%)	C_{hws} (ppm)	N_t (%)
Crop rotation	13.08.1998	8.18	4.58	1.85		0.17
	04.03.1999	3.76	0.63			
	23.09.1999	1.37	0.55	2.06		0.18
	06.03.2000	5.09	0			
	27.10.2000	10.81	0.73	2.15	513.0	0.19
	07.03.2001	10.11	0.89			
	23.10.2001	10.01	0.39	1.98	410.1	0.15
	04.03.2002	6.56	0.3			
	09.10.2002	7.7	0.73	2.04	348.4	0.18
Black fallow	01.01.1999			(2.03)*		
	04.03.1999	3.43	2.06			
	23.09.1999	19.81	0.47	2.02	269.9	0.16
	06.03.2000	5.29	1.15			
	01.11.2000	21.28	0.6	1.91	368.3	0.16
	07.03.2001	8.8	0.77			
	20.11.2001	12.49	0.75	1.79	345.0	0.19
	04.03.2002	6.03	0.711			
	11.10.2002	14.76	0.37	1.88	235.7	0.16

*Initial value for black fallow: estimated from previous treatment

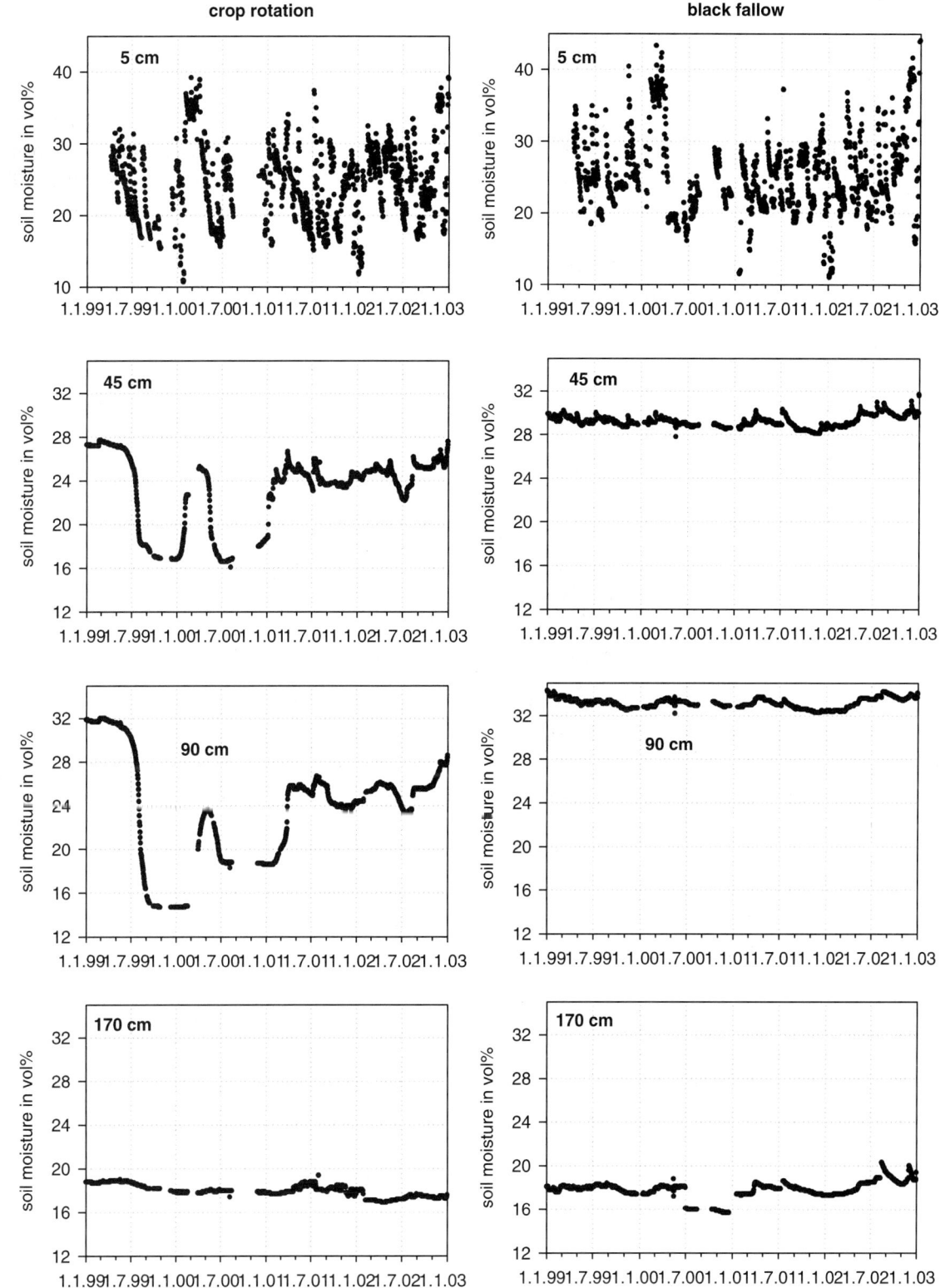

Fig. 7 Soil moisture dynamics of cropped and fallow soil

Soil moisture measurements (see Fig. 7) at the different depths showed the expected differences between cropped soil and black fallow. Rooting activity was clearly indicated up to a depth of 90 cm, whereas in the bottom layer (170 cm) there was only a small trend of reducing soil moisture that might result from capillary rise. At the bottom layer of the black fallow treatment, there appeared a small increase of soil moisture that may indicate a slow refilling of water capacity in that depth.

The absence of significant soil moisture dynamics in the bottom layers of both plots suggests that there is only a small water flux to groundwater. This way also a small amount of nitrogen leaching should appear. The results of water quality sampling support this hypothesis. The average nitrogen concentration in soil water at 170 cm depth was 0.60 ppm in the crop rotation and 0.99 ppm for the black fallow treatment.

Discussion

Long-term experiments raise often questions about the value of the most recent results compared to the accumulated knowledge during the long history of the experiment. Despite of the high variability of organic carbon measurements, the long-term data set gives the opportunity to quantify trends of SOM. In contradiction to previous indications that the steady state of this experiment has already been reached (Körschens and Mahn 1995), it seems now obvious that the differentiation between the investigated treatments has been not yet finished. The ongoing decrease of SOM in the control plot is partly caused by the improved turnover conditions indicated by the biologic active time. The other treatments take advantage from the increasing crop yields. But as the relative C_{org} trend of the NPK treatment (against control) is lower than the rate of carbon decrease at the control treatment, it is very likely that the yield increase is not sufficient to stabilize the humus content at the current level.

The crop rotation data set provides more detailed information about the soil processes at the site of Bad Lauchstädt. The soil moisture measurements together with the measurements of soil water quality give a clear indication that losses of nutrients due to leaching are very low under the site conditions of Bad Lauchstädt. This is in a good agreement with the results of the nitrogen balance of the 4-year crop rotation and previously published results from the long-term experiment (Körschens and Mahn 1995).

Soil physical parameters sampled at the four soil profiles showed that there has to be considered a remarkable heterogeneity within that loess soil at Bad Lauchstädt. Gaps in the balance calculations have to be closed using modelling techniques. For this purpose it should be advantageous to validate the model parts concerning SOM turnover with the long-term data and calibrate submodels for crop development and soil moisture with the data sets from the "Intensivmessfeld" plots.

References

Böhme F, Russow R, Neue H-U (2002) Airborne nitrogen input at four locations in the german state of Saxony-Anhalt – Measurements using the 15N-based ITNI-system. Isotopes Environ Health Stud 38:95–102

Böhme F, Merbach I, Weigel A, Russow R (2003) Effect of crop type and crop growth on atmospheric nitrogen deposition. J Plant Nutr Soil Sci 166:601–605

Franko U, Schramm G, Rodionova V, Körschens M, Smith P, Coleman K, Romanenkov V, Shevtsova L (2002) EuroSOMNET – a database for long-term experiments on soil organic matter in Europe. Comput Electron Agric 33:233–239

Körschens M (1994) Der Statische Düngungsversuch nach 90 Jahren. B.G. Teubner Verlagsgesellschaft, Stuttgart-Leipzig

Körschens M, Mahn G (1995) Strategien zur Regeneration belasteter Agrarökosysteme des mitteldeutschen Schwarzerdegebietes. B.G. Teubner Verlagsgesellschaft, Stuttgart-Leipzig, pp 168–202

Mehlert S (1996) Untersuchungen zur atmosphärischen Stickstoffdeposition und zur Nitratverlagerung. Ph.D. Thesis, University of Hamburg, UFZ-Bericht, p 22.

Oelschlägel B (1995) A method for downscaling global climate model calculations by a statistical weather generator. Ecol Model 82:199–204

Smith P, Falloon P, Körschens M, Shevtsova L, Franko U, Romanenkov V, Coleman K, Rodionova V, Smith JU, Schramm G (2002) EuroSOMNET – a European Database of long-term experiments on soil organic matter: the WWW-metadatabase. J Agric Sci 138:123–134

Weigel A, Russow R, Körschens M (2000) Quantification of airborne N-input in long-term field experiments and its validation through measurements using 15N isotope dilution. J Plant Nutr Soil Sci 163:261–265

CHAPTER EIGHTEEN

The lysimeter station at Berlin-Dahlem

Heiko Diestel, Thomas Zenker, Reinhild Schwartengraeber and Marco Schmidt

Abstract The lysimeter station at Berlin-Dahlem consists of 12 weighable lysimeters in each of which all components of the soil water balance can be measured. A permanent record of the weight of the water entering or leaving the monolith at its upper surface (precipitation and evapotranspiration), as well as at its bottom plane (percolation and capillary rise) is obtained. A permanent record of the changes in the water content of the monolith is an additional data output in using this procedure. Properly equipped lysimeters are, indeed, adequate physical models of the field situation and give insights into the in situ – situation which can not be obtained with other methods, not even with field measurements. This refers to the dynamics of water and solutes in the soil and to the reactions of the components of the soil water balance to the climate, as well as to the validation and calibration of mathematical models, or to the possibility of comparing the behaviour of different soil types and different vegetation under identical climatic conditions.

Heiko Diestel, Thomas Zenker, Reinhild Schwartengraeber and Marco Schmidt
Department of Applied Hydrology, Resource Protection, Irrigation and Drainage
Technical University of Berlin
Albrecht-Thaer-Weg 2, 14195 Berlin. Tel.: (+4930) 314 71 220; fax: (+4930) 314 71 228; e-mail: heiko.diestel@tu-berlin.de

Keywords Capillary rise, Evapotranspiration, Groundwater recharge, Hydrologic variants, lysimeters, Monoliths

Lysimeter investigations

The term "lysimeter" is sometimes used to describe disturbed in situ or transported soil samples containing some measurement instruments. In the context of the work discussed here, "lysimeter" refers to a large soil sample installed in such a way, that it is possible to keep a permanent record of the weight of the water entering or leaving the monolith at its upper surface (precipitation and evapotranspiration), as well as at its bottom plane (percolation and capillary rise). A permanent record of the changes in the water content of the monolith is an additional data output in using this procedure. Depending on the question to be investigated, the soil sample may be a properly extracted and transported, undisturbed monolith or may consist of a filling, which takes place at the station if artificial systems are to be investigated. As compared to laboratory studies or to measurements at field sites, lysimeter investigations offer a number of advantages. These benefits must be weighed against shortcomings or problems which are specific to lysimeter installations. Experiences after a number of years of work at the Berlin-Dahlem station and evaluations of the relevant literature allow some conclusions on lysimeter investigations. Properly equipped lysimeters are, indeed, adequate physical models of the field situation and give insights into the in situ – situation, which can not be obtained with other methods, not even with field measurements. This refers, i.s., to the dynamics of water and solutes in the soil and to the reactions of the components of the soil water balance to the climate, as well as to the validation and calibration of mathematical models, or to the possibility of comparing the behaviour of different soil types and different vegetation under identical climatic conditions. It still is difficult to segregate "lysimeter effects" on the dynamics of water in soils from the "natural" soil hydrological behaviour in situ. The same is true for the quantification of the climatological "undisturbedness" of a lysimeter site and of a meteorological station, as well as of the degree, to which the oasis effect can be avoided.

If research questions on the behaviour of cultivated soils are investigated, an important deficiency of lysimeter investigations is the fact that the normal cultivation procedures cannot be carried out on the soils in the monoliths.

On the other hand, lysimeters allow undistorted physical simulations of artificial stratifications and layers such as constructed soil covers or sealings (Markwardt et al., 2003). A high sensitivity of the balance, on which the monolith stands may have the disadvantage of the disturbances by the wind pressure on the monolith (Zenker 2003, p. 147).

Modelling studies tend to confirm that the so-called "lysimeter effect" (commonly understood as a possible disturbance of the water dynamics by the bottom boundary layer) can be minimized or even avoided by proper installation procedures. From the soil hydrological point of view, it obviously represents a permissible technology to simulate a remote groundwater level by hanging water columns. However, monolith depths smaller than 3 m may lead to problems regarding to water uptake from the water supply system at the monolith bottom by the root system of some crops. Installation of soil moisture sensors may disturb the structure of the soil.

This WORKSHOP provided an excellent opportunity to study the outputs of results, as well as the advantages and shortcomings of lysimeter studies.

The lysimeter station at Berlin-Dahlem

The lysimeter station is operated by the Department of Applied Hydrology, Resource Protection, Irrigation and Drainage of the Technical University of Berlin, headed by the main author.

General description

The lysimeter station and the climatological station (Fig. 1) are situated at the same location in the southwest of the city of Berlin. (52°28′ N and 13°18′ E, 51 m above sea level) The mean air temperature is 9.3°C and the average precipitation is 545 mm per year. It has been described in previous publications (Diestel et al. 1993; Markwardt et al. 1993; Markwardt and Diestel 1994; Zenker 2003).

Extraction/construction

The lysimeter station at Berlin-Dahlem consists of 12 weighable lysimeters, in which all components of the soil water balance can be measured.

The stainless steel lysimeters have a surface of 1 m^2 and a depth of 150 cm. They are installed on even level with the ground surface and placed on stationary balances. These are accessible to the operating personnel in a basement.

The soil monoliths are extracted without disturbance of the natural soil by the method of Helfesrieder et al. (1989), see also Markwardt et al. (1993).

Weighing system

The stationary balances are connected with electronic balances, so that it is possible to register the change of weight of the monoliths with a sensitivity of 100 g, respectively 0,1 mm for evapotranspiration and precipitation.

A second weighing system measures the capillary rise, respectively the percolation with a sensitivity of 5 g. A magnetic switch valve directs these two flows from a reservoir for capillary rise, respectively into a reservoir for the percolate representing the groundwater recharge.

Fig. 1 Lysimeter area with meteorological station (lysimeter areas left unharvested for better visibility)

Hydrological variants

There are two hydrological variants with different ground-water levels. Some monoliths have a constant groundwater level at 135 cm depth (variant "shallow groundwater", see Fig. 3). In some lysimeters (variant "deep groundwater"), a groundwater level of 210 cm depth is simulated by a hanging water column (see also Fig. 2).

Deep groundwater (groundwater low) At the bottom of the lysimeter, 7 cm of the soil are taken out and a system of 86 suction cups is now embedded into a quartz flour 1 layer. Using hanging water columns with a length of 63 cm, a vacuum is applied. The groundwater level of 2.15 m is simulated. The constant water level is adjusted by a vessel of regulation of the groundwater level and a reservoir for groundwater recharge controlled by a magnetic switch (Fig. 2).

Shallow groundwater (groundwater high) At the bottom of the lysimeter, 15 cm of the soil are taken out, and two crossed pipes of steel are embedded in gravel. Through this cross water was accumulated in the gravel bed, so that the groundwater level results in 1.35 cm. The constant water level is also adjusted by a vessel of regulation of the groundwater level.

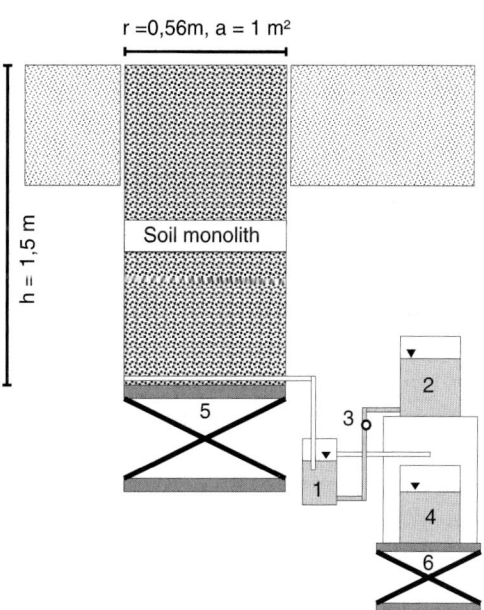

1 - Expansion tank for regulation of the groundwaterlevel
2 - Reservoir for groundwater recharge
3 - Magnetic switch
4 - Tank for percolat
5 - Scale for lysimeters
6 - Scale for percolat

Fig. 2 The weighing system: deep groundwater level

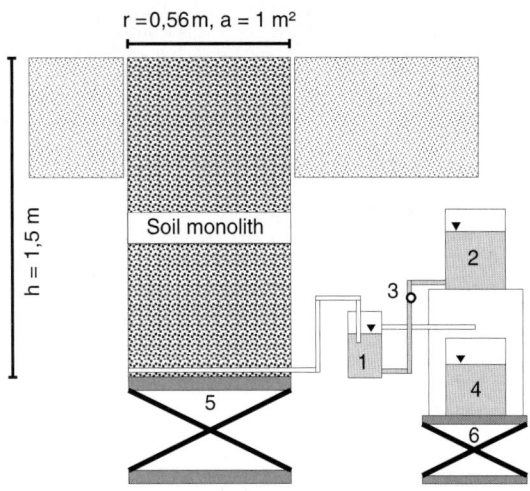

1 - Expansion tank for regulation of the groundwaterlevel
2 - Reservoir for groundwater recharge
3 - Magnetic switch
4 - Tank for percolat
5 - Scale for lysimeters
6 - Scale for percolat

Fig. 3 The weighing system: shallow groundwater level

The soil variants

From 1990 until 2004, the following soil types were installed in the monoliths:

Podzol (Podsol) from Wildeshausen (State of Lower Saxony) and Cambisol (Braunerde) from Weckesheim (State of Hessen). From 1992 until 2001: Stagnic Luvisol (Parabraunerde-Pseudogley) from Parlow-Glambeck (State of Brandenburg). These variants were replaced in 2001 by material from the restoration layer of a waste disposal site.

Tables 1 and 2 show the most important data on the soil variants:

Parameters measured at the meteorological station

The following parameters are measured at the meteorological station

Parameters		Units
Precipitation	0 m/1 m (electronically and by hand)	[mm]
Air temperature	2 m	[°C]
Soil temperature	0/0.1/0.2/0.3 m	[°C]
Air humidity	2 m	[%]
Wind speed	2 m/1 m	[m s^{-1}]
Global radiation	1 m	[w m^{-2}]
Diffuse radiation	1 m	[w m^{-2}]
Net radiation	1 m	[w m^{-2}]

Details on the climatological situation during time phases subjected to modelling investigations with data from the Dahlem lysimeters by authors from other research institutions are given in the respective publications.

The data set

One of the reasons for selecting the data set of 1996–1998 for distribution in this workshop was the fact, that it covers a time phase with a continuous vegetation without harvests (excluding grass cuttings), fallow periods or crop rotation. This has the consequence, that the soil water balance is characterized by the small rooting depth and the specific evapotranspiration rates of the grass vegetation with lower fluctuations as it would be the case with deeper rooting field crops.

The climatological characteristics of these 3 years fluctuate between the relative extremes of 1997 and 1998. 1997 is, as a whole, a relatively dry year (65 mm below the long year precipitation average), 1998 a rather humid year (32 mm above the long year average), both years had above-average temperatures. The deficits of the climatological water balance of this relatively dry site vary between 29 mm for 1996 and 264 mm for 1997 (Zenker 2003, p. 35).

The weights of the lysimeter balance carrying the monoliths (change of water content, precipitation and evapotranspiration) as well as those of the second balance (percolation and capillary rise) are registered electronically at 15-min intervals.

The data for all these years and for all variants are accessible as weight differences for each 15-min interval.

The meteorological data are registered electronically at 1-min intervals and averaged to 15-min interval values. These match the balance data. These data are accessible as well.

Partially automatically, partially by hand, errors in this data set were corrected within defined tolerances

Table 1 Physical parameters of the soil variations

Wildeshausen

Lysimeter 1–4   Lys 1–2 groundwater low   Lys 3–4 groundwater high

Podsol (KA 4)   Haplic Podzol (FAO)

Horizon	Depth (cm)	Soil class	Sand (%)	Silt (%)	Clay (%)	Humus (%)	PV (vol%)	pF 1.8 (vol%)	pF 2.5 (vol%)	pF 4.2 (vol%)	Ksat (mm/day)
Ap	0–40	Su2	81.5	14.9	3.6	4.0	42.7	23.0	16.5	5.0	1400
Bsh 1	40–60	Su2	80.4	15.1	4.5	1.3	38.6	16.8	10.7	2.3	2210
Bsh 2	60–150	Ss	87.5	9.0	3.5	0.5	36.3	25.3	16.6	4.3	490

Weckesheim

Lysimeter 9–12   Lys 9–10 groundwater high   Lys 11–12 groundwater low

Braunerde (KA 4)   Eutric Cambisol (FAO)

Horizon	Depth (cm)	Soil class	Sand (%)	Silt (%)	Clay (%)	Humus (%)	PV (vol%)	pF 1.8 (vol%)	pF 2.5 (vol%)	pF 4.2 (vol%)	Ksat (mm/day)
Ap	0–20	Tu4	3.7	66.9	29.4	2.8	36.0	34.5	34.0	21.5	255
Bv 1	20–60	Tu4	4.5	68.8	26.7	1.0	42.0	35.2	33.7	19.7	33
Bv 2	60–150	Ut4	6.2	74.5	19.3	0.4	35.0	32.5	30.5	15.5	48

Parlow-Glambeck

Lysimeter 5–8   Lys 5–8 groundwater low

Parabraunerde-Pseudogley (KA4)   Stagnic Luvisol (FAO)

Horizon	Depth (cm)	Soil class	Sand (%)	Silt (%)	Clay (%)	Humus (%)	PV (vol%)	pF 1.8 (vol%)	pF 2.5 (vol%)	pF 4.2 (vol%)	Ksat (mm/day)
Ap/Al	0–40	Su3	62.2	29.8	8.0	1.6	35.9	21.6	14.8	9	2173
Bt-Sw	40–90	Ls4	52.7	28.7	19.6	0.5	31.1	23.1	16.7	15.1	16.4
Cc-Sd	90–150	Sl4	52.5	33.7	13.8	0.2	39.4	30.1	22.6	13.2	1.3

(soil classes according to Bodenkundliche Kartieranleitung (1995) and FAO–Unesco soil map of the world (1988))

Table 2 Vegetation on the lysimeters

Year	Lysimeter	Vegetation
1990	1–4; 8–12	Winter barley
1991	1–4; 8–12	Winter barley
	1–4; 8–12	Autumn-sown rye
1992	5–8	Perenial rye grass
	1–4, 8–12	Autumn-sown rye
	1–4, 8–12	Rape
1993	5–8	Perennial rye grass
	1–4; 8–12	Rape
1994	All	Maize
1995	All	Spring wheat
	All	Winter barley
1996	All	Winter barley
	All	Lawn
1997	All	Lawn
1998	All	Lawn
1999	All	Lawn
2000	All	Alfalfa
2001	All	Alfalfa
2002	All	Grass*
2003	All	Grass*
2004	All	Grass*

*Typical grass for a waste disposal site

based on a software specifically designed for this purpose. Weight changes due to grass mowing were distributed over the time period back to the preceding cut, extreme values caused by miscellaneous factors were adjusted to the mean value of adjoining data. In this context, special attention had to be given to the influence of the wind on the lysimeter weights. As the deviations caused by the wind pressure onto the monoliths depend on the wind speed, it is problematic to design a systematic error correction procedure. At the Dahlem installation, the measuring frequency of the balances subjected to the wind effect was increased and the correction was carried out by a calculation of averages (Zenker 2003).

For the precipitation data, only those errors due to the electronic recording equipment were corrected. Systematic errors specific to the Hellmann rain gage were not taken into account, but can be assumed to lead to an underestimation of approximately 5–7% of the actual precipitation.

Daily sums and daily averages were computed based on the measurements carried out in intervals of 15 min.

In the data set, the following abbreviations are used: "GW-low" for a groundwater depth of 210 cm, "GW-high" for a groundwater depth of 135 cm. The soil type Podzol is designated as "sand", the soil type Stagnic Luvisol (Parabraunerde-Pseudogley) as "loam" and the soil type Cambisol (Braunerde) as "silt". Thus, as an example, the designation "silt, GW-low" is used for the Cambisol with a groundwater level at 210 cm.

This data set provides an enormous amount of information on the reactions of the soil–plant–water system to the dynamics of climate. The scientific community is invited to benefit from this data pool.

Possible range of research topics

In Figs. 4 and 5, precipitation, soil water deficit against field capacity and percolation are represented for the soil variant silt and for the years 1996–1998. Evaporation can be deduced from these parameters.

The distance to the groundwater is 2.10 m for the data shown in Fig. 4 and 1.35 m for those in Fig. 5. It becomes evident, that a soil water deficit arises through evaporation during summer, this deficit being lower for the high groundwater level, dependent on the groundwater uptake into the root zone, than for the low groundwater level.

Percolation phases can be identified for both variants in connection with a full soil water reservoir. There is no significant difference between the distribution of percolation of the two variants. The groundwater level only has a small effect on the percolation quantity with 423 mm for the low groundwater level against 475 mm for the high groundwater level.

Investigations conducted on the basis of the datasets from the lysimeter station in Berlin-Dahlem included work on the relations of plant evaporation with soil properties, groundwater depth, dynamics of percolation and climatic water balance, as well as on the water balance of waste disposal covering layers. Model validations related to these research questions were carried out.

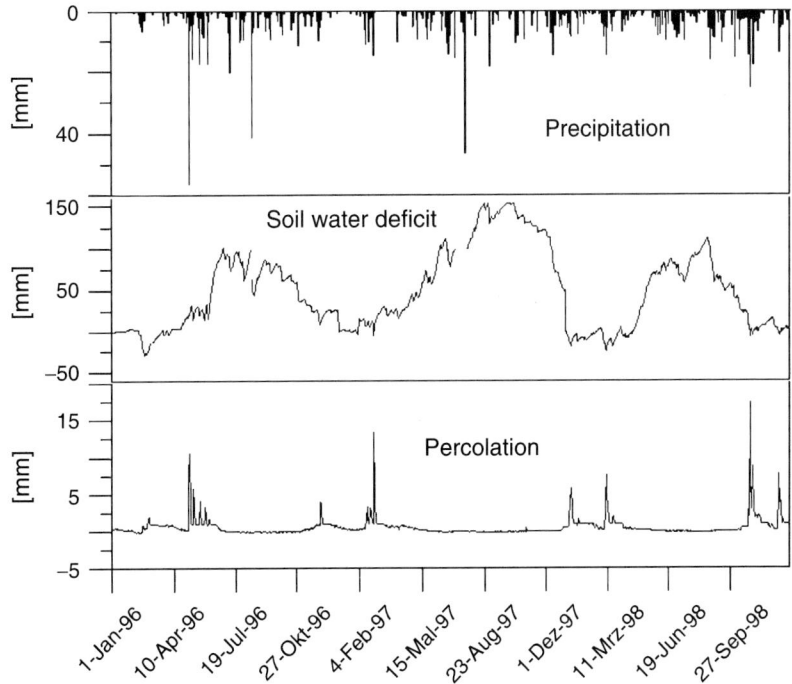

Fig. 4 Measured water balance components for the silt variant with low groundwater level (please note: increases in soil moisture are shown as "negative soil water deficits")

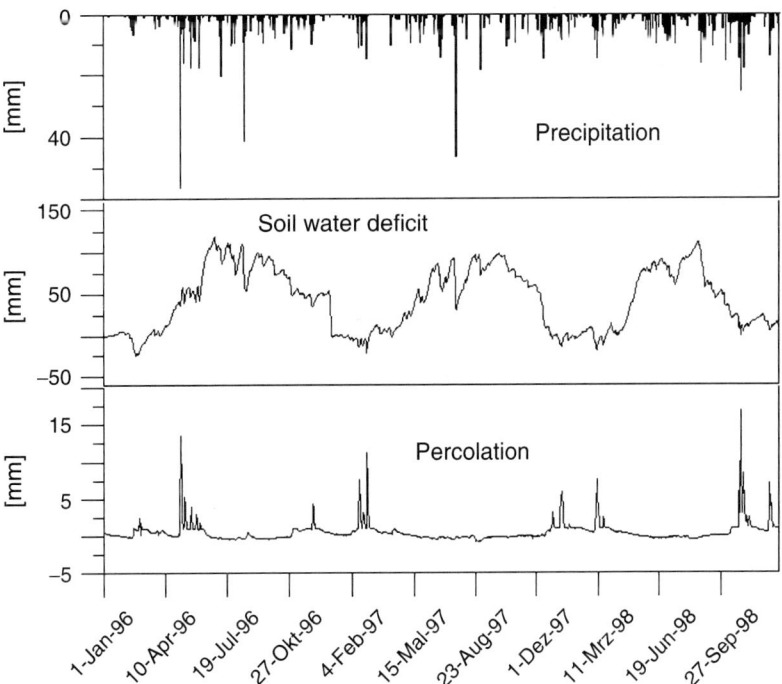

Fig. 5 Measured water balance components for the silt variant with high groundwater level (please note: increases in soil moisture are shown as "negative soil water deficits")

References

Diestel H, Markwardt N, Moede J (1993) Experimentelle Untersuchungen sowie Modellentwicklungen zur Verlagerung von Pflanzenschutzmitteln in der ungesättigten Bodenzone. Bodenökologie und Bodengenese, Heft 10. ISSN 0939-7787

Bodenkunde AG (1995) Bodenkundliche Kartieranleitung. Bundesanstalt für Geowissenschaften und Rohstoffe und den Geologischen Landesämtern in der Bundesrepublik Deutschland 4.Aufl, Hannover. 392 S

FAO – Unesco (1988) Soil map of the world. Food and Agriculture Organization of the United Nations, Rome 119 S

Helfesrieder K, Bartilla C, Ploeg RR van der (1989) Bau eines wägbaren Unterdrucklysimeters zur Erfassung des Herbizidaustrages unter Feldbedingungen. Mitteilungen der Deutschen Bodenkundlichen Gesellschaft 59/1:181–184. ISSN 0343-107X

Markwardt N, Diestel H, Schmidt M (1993) The technical characteristics and the applicability of weighing lysimeters. Mitteilungen der Deutschen Bodenkundlichen Gesellschaft 71:149–152. ISSN 0343-107X

Markwardt N, Diestel H (1994) Die Auswirkungen der extremen Sommerwitterung des Jahres 1992 auf die Grundwasserneubildung. Potsdam Institute for Climate Impact Research: PIK-Reports. N 2 :501–524

Markwardt N, Bender D, Diestel H (2003) Alternative Oberflächenabdichtungssysteme unter Berücksichtigung der unterschiedlichen klimatischen Bedingungen. Symp.Umwelttechnik 2003, Bauhaus Universität Weimar. Schriftenreihe Geotechnik, Heft 10:91–106

Zenker T (2003) Verdunstungswiderstände und Gras-Referenzverdunstung. Lysimeteruntersuchungen zum Penman-Monteith-Ansatz im Berliner Raum. Dissertation. Technical University of Berlin. http://edocs.tu-berlin.de/diss/2003/zenker_thomas.pdf

Index

actual evapotranspiration (AET) 6-8, 19-23, 27, 30, 32, 39, 40, 44, 47, 53, 54, 56, 61, 63, 69, 184, 193, 210
agro-ecosystem model 1, 14, 37, 40, 41, 59-61, 67, 71, 129, 130, 147, 219
AGROSIM 2-7, 10-13, 16, 38, 59-72, 75
AGROTOOL 2-4, 75, 79, 84-86
air temperature 20, 29, 38-40, 52, 54, 64, 66, 76, 70, 86, 108, 111, 116, 117, 123, 125, 127, 132, 135, 149, 158, 165-167, 169, 211, 220, 221, 223, 237, 238, 246, 255, 260, 262
AMBAV 2-4, 19-26
ammonium content 174, 176, 177, 223, 228-230, 239
ammonium nitrate lime 99, 116, 226
ammonium sorption 115
ammonium urea solution 116, 142, 226
ANIMO 3, 4, 111-115, 117, 121, 124
assimilation pool 63

Bad Lauchstädt 2, 3, 14-16, 27-29, 31, 36-38, 40-48, 59-61, 66-76, 79, 83, 85, 130-132, 135-144, 147, 148, 150, 151, 154-159, 197, 198, 295, 20-211, 213-217, 245, 246, 255, 258
BBCH code 54, 61, 232
Berlin-Dahlem 2, 3, 15, 27-30, 36-39, 41, 46-48, 259, 260, 264
BOWET 64, 66
bulk density 27, 29, 41, 47, 53, 79, 80, 83, 133, 135, 149, 167, 186, 199, 223, 231, 248

C/N ratio/C:N ratio 3, 101, 132, 149, 150, 165, 166, 191, 193, 194, 211
calibration 1, 2, 4, 15, 37, 43, 53, 59, 66-69, 71, 80, 91, 93, 96-99, 101-109, 111, 118, 129, 135, 136, 148, 150, 151, 158, 161, 163, 167-171, 176, 186, 187, 194, 198, 199, 227, 241, 249, 260
CANDY 2-5, 7-14, 209-212, 214, 216, 246
capacity model 64
capillary conductivity 27-29, 31
capillary rise 20, 27-30, 32, 36, 93-95, 97, 99, 101, 108, 132, 149, 258, 259, 260, 262
carbon content 53, 95, 150, 156, 157, 166, 186, 199, 201, 203, 212, 213, 223, 235, 236, 246, 250, 255
carbon pools/C pools 209, 210, 212, 216
carbon sequestration 190, 198, 204, 205
carbon turnover 3, 96, 131-133, 144, 198, 209
catchment modelling 130
CERES 2-14, 38, 43, 53, 75, 161-169, 173, 174, 176, 177, 183-193
climate change 60, 71, 162, 197, 198, 205, 219, 238, 245, 255
CO_2 assimilation 38, 77
CO_2 concentration 61, 63, 64, 66, 197-199, 205
coefficient of determination/R^2 10, 54, 59, 66-72, 117, 134, 152, 161, 168-173
coefficient of residual mass (CRM) 117, 118, 122-124
COST 2, 15, 159
critical N concentration 149

crop biomass 13, 37, 39, 55, 149, 153, 154, 156, 158, 169, 170, 177, 189, 190, 198
crop cover 2, 19, 64, 76, 94, 155, 185
crop data 21, 23
crop development/development stage 20, 24, 28, 52, 61, 63, 65, 94, 113, 127, 149, 164, 169, 170, 185, 187, 188, 220, 245, 247, 258
crop growth 1, 2, 4, 5, 13-16, 27, 37-39, 41, 44, 47, 48, 51-53, 59-61, 63, 67, 68, 71, 88, 112, 113, 129-131, 138, 139, 144, 147-154, 158, 159, 162, 164, 165, 167, 174, 183-188, 190-194, 197, 219, 220, 232, 237, 239
crop height 52, 113, 210, 211
crop residues 93, 107, 108, 121, 149, 154, 158, 165, 166, 194, 197, 202, 204, 211, 249
crop rotation 2, 12, 16, 34, 35, 38, 42-45, 59-61, 66-69, 71, 72, 92, 108, 111, 129-131, 133, 136-142, 144, 147, 150, 151, 157, 183, 184, 187, 194, 211, 219-221, 223, 231, 239, 240, 245-247, 249, 251, 254, 256-258
crop varieties 52, 155, 197-199
cultivar 64, 66, 71, 161-164, 167, 168, 170, 171, 221, 224, 226, 240

DARCY 28, 231
decimal code 61, 68, 232-234
decision support 1, 9, 164
denitrification 3, 4, 7, 9, 39, 40, 45, 93, 97, 101, 106, 107, 115, 132, 148, 149, 154, 158, 165, 166, 190, 211, 213
deposition 64, 66, 93, 97, 101, 106, 108, 116, 121, 122, 142, 154, 190, 197-199, 201, 205, 211, 245, 254, 256

ear 64, 172, 232-236
energy balance 51-53
EPIC 75, 132, 139
erosion 51, 129, 131
evaporation 2, 19, 20, 32, 35, 39, 40, 52, 55, 66, 75-77, 94, 113, 118, 119, 132, 149, 164-166, 184, 185, 188, 190, 191, 193, 220, 227, 264
experimental design 61, 220-222
EXPERT-N 2-9, 183-185
extensive management 185, 219-221
exudation 63

fallow 2, 32, 33, 35, 36, 131, 136, 139, 141, 142, 147, 150, 151, 154, 156-158, 212, 245-247, 252-254, 256-258
farmyard manure (FYM) 100, 142-144, 147, 150, 151, 154-157, 192, 194, 197-203, 221, 226, 240, 246, 247, 254-256
FASSET 2-14, 16, 197, 198, 204, 205
fertilisation/fertilization 12, 51, 60, 64, 79, 92, 112, 114, 129, 130, 133, 139, 141-144, 149-151, 155, 166, 192, 199, 204, 205, 211, 219-222, 240
fertiliser recommendation/fertilizer recommendation 1, 148
field capacity 20, 21, 23, 25, 29, 41, 53, 67, 70, 79, 86, 93, 97, 133, 137, 149, 152, 153, 164, 211, 212, 239, 264
frequency domain reflectometry (FDR) 16, 41, 43, 68, 70, 71, 84, 126, 137, 150, 156, 158, 249, 251

generic crop growth model 51, 183
genetic coefficients 162, 168
grain filling 61, 63, 64, 168, 171, 172, 187, 238
grass 2, 79, 86, 113, 223, 231, 262, 264
gravimetric metho/gravimetry 21, 22, 192, 223, 241, 242
groundwater/ground water 27-29, 36-38, 44, 46, 47, 93, 94, 99, 101, 108, 111-114, 131, 132, 136-138, 149, 163, 183, 220, 258-265

harvest index 132, 139, 155, 167, 204
HAUDE 4, 148-150, 155
heat capacity 64, 76, 114, 120
heat stress 56
HERMES 2-16, 147-150, 157-159
hydraulic conductivity 19, 29, 37, 39-41, 43, 48, 53, 79, 112, 117, 118, 122, 169, 186, 211, 231
hydrotopes 131
HYPRES 53
hysteresis 111, 117, 123, 125, 127

immobile water 2
immobilisation/immobilization 3, 4, 157, 158, 165, 166, 190, 191, 194, 211
Index of agreement (IA) 9-14, 37, 41, 43, 47, 59, 67, 68, 72, 101-106, 129, 133-136, 138-143, 147, 152-156, 158, 185-187, 191-193, 210, 212, 217
inert carbon 209-211, 213, 215, 216
Initialisation/initialization 133, 147, 149-151, 157, 158, 168, 184, 209, 213, 214, 216

Index

inorganic fertiliser/inorganic fertilizer 107, 142, 185, 226
intensive management 66, 67, 219, 220, 240
interception 6, 19, 39, 54-56, 64, 113, 118, 119, 198, 204, 210
irrigation 19, 20, 24-27, 29, 39, 60, 93, 97, 108, 116, 119, 149, 164, 166, 167, 220, 222, 226, 239, 240, 260

leaf area index (LAI) 20-23, 40, 52, 61, 64-66, 113, 116, 131, 132, 164, 165, 167, 177, 187, 188, 191, 193
leaf water potential 76, 77
long term experiment 143, 147, 150, 151, 155-158, 167, 197, 198, 215, 216, 245-247, 254, 256, 258
lysimeter 2, 16, 27-30, 36-39, 41, 46-48, 79, 86, 88, 131, 132, 136-138, 259-264

MAKKINK 4, 94
management practice 37, 61, 71, 112, 130, 133, 139, 144, 220-222, 224, 226
mean absolute deviation (MABS) 59, 65-68, 72, 241
mean bias error (MBE) 9-14, 59, 66-68, 72, 151-153, 155, 156, 158, 210, 212, 214, 215
meteorological data/weather data 20, 24, 25, 29, 52, 53, 61, 64, 80, 86, 101, 108, 112, 115, 116, 128, 148-150, 164, 169, 184, 210, 211, 220, 235, 237, 246, 262
microbial biomass 198
mineral nitrogen/inorganic nitrogen 5, 8, 11-16, 64, 91-93, 97, 98, 100-104, 106-108, 111, 112, 116, 117, 122, 123, 127, 132, 144, 147, 152-154, 157, 158, 210, 211, 214-217, 247
mineralisation/mineralization 3, 4, 7, 9, 39, 44, 45, 61, 64, 91, 93, 95-98, 122, 127, 132, 133, 148-150, 157, 158, 165, 166, 183, 185, 190-194, 210, 211, 213, 239, 246, 256
mobile water 2
model accuracy 59, 67, 68, 70, 72, 242
model comparison 144, 157, 220-222
model performance 1, 2, 9-16, 36, 43, 92, 102, 103, 109, 111, 122, 127, 133, 136, 138, 140, 141, 143, 147, 148, 151, 153, 156, 159, 161, 176, 177, 194
model verification 168, 219, 240-242
modelling efficiency (ME) 9-14, 67, 106, 118, 120-124, 152, 153, 155, 156, 158, 186, 200, 201, 205

modular simulation 183
monolith 47, 85-87, 259-262, 264
Müncheberg 1-16, 19, 20, 22-26, 28, 38, 53-56, 59-61, 66-72, 79, 80, 92, 96, 98, 99, 100, 109, 111-113, 118, 124, 130, 131, 135-143, 147, 148, 150-154, 158, 163, 174, 178, 185, 187, 194, 209-211, 213-216, 219, 220, 227, 237, 239
mustard 225

NDICEA 2-4, 13-15, 91-93, 98-102, 106-109
nitrate concentration 40, 97, 111, 117, 122, 124-127, 139, 142, 158, 174, 231
nitrate content 38, 43, 44, 47, 133, 142, 144, 161, 174, 176, 193, 223, 229, 230, 239
nitrate leaching 36, 123-126, 154, 183, 211
nitrogen balance 2, 36-39, 44, 63, 78, 106, 107, 121, 122, 124, 154, 158, 164, 183, 188, 240, 258
nitrogen fixation 98
nitrogen leaching/N leaching 7, 9, 27, 36, 44, 45, 64, 98, 101, 106, 121, 148, 154, 155, 158, 188, 190, 213, 231, 240, 245
nitrogen pools/N pools 132, 147, 149, 150, 154, 158, 198
nitrogen stress 47, 59, 61, 63, 64, 149, 155
nitrogen uptake/N uptake 5, 7, 9, 11-16, 44, 45, 47, 59, 64, 67-71, 77, 97, 98, 100-103, 106, 108, 115, 132, 147-149, 153, 154, 156-158, 166, 185-192, 194, 198, 209, 211

oil radish 116, 151, 154, 163, 224
ontogenesis 40, 59-70, 232-234, 238
OPUS 2-4, 16, 36-47, 71
organic farming 4, 92, 246
organic management 219-221, 239, 240
organic matter 14, 92, 93, 95-99, 101, 106-108, 114-116, 118, 127, 129, 130, 132, 133, 144, 147, 149, 150, 153, 154, 156-159, 165, 166, 183, 186, 190, 191, 193, 194, 198, 209-214, 216, 217, 240
organic nitrogen 91-93, 95, 96, 108, 132, 149, 150, 157, 158
oxygen 39, 97, 115, 132, 166, 212, 234

partitioning 2, 52, 64, 65, 77, 78, 94, 132, 149, 151, 164, 209
pedotransfer function (PTF) 19, 27-32, 36, 48, 53, 79, 80, 83, 85, 86, 133, 137
PENMAN 4, 38

PENMAN-MONTEITH 4, 19, 76, 113, 132
percolation 6-9, 21, 22, 27-30, 32, 35, 36, 61,
    64, 66, 93, 132, 188, 190, 193, 227, 239,
    240, 259, 260, 262, 264, 265
performance indicator 1, 10-14, 91, 153, 156
pH 39, 93, 95, 96, 115, 166, 167, 169, 223,
    227, 251-253
phenology 24, 161, 167
phosphorus 4, 39, 40, 114, 115, 131, 139, 164,
    166, 246, 249
photosynthesis 2, 52, 54, 61, 63, 64, 65-77,
    149, 151, 163
photosynthetic rate 61, 62
physical protection 209, 213, 214, 216
pore space class 209, 212, 213, 217
potatoes 20, 39, 44, 45, 77, 79, 83-86, 132, 139,
    140, 142, 147, 149, 151, 155, 164, 198-202,
    204, 205, 245, 246, 249, 250, 254, 256
potential evapotranspiration (PET) 4, 20, 32, 35,
    38-40, 47, 61, 63, 64, 76, 113, 132, 148-150,
    184, 185, 193
preferential flow 47, 109, 251
pressure head 37, 38, 40, 41, 43, 48, 111-113,
    116, 117, 119-121, 125, 127, 220, 227-229,
    239, 240, 242
PRIESTLEY-TAYLOR 4, 132, 165

radiation 20, 29, 38, 40, 52-56, 60-64, 66, 76, 80,
    86, 101, 116-118, 131, 139, 149, 150, 164-167,
    169, 199, 201, 211, 223, 237, 247, 262
radiation use efficiency 199, 201
respiration 38, 61, 63, 65, 66, 77
Richards equation 19, 40, 48, 112, 113, 184, 185, 193
root distribution 75, 149, 164, 194
root mean square error (RMSE) 9-14, 37, 41, 43,
    47, 54, 59, 66-68, 72, 91, 101-107, 117-124,
    133-136, 138, 140-144, 152, 153, 155-158,
    161, 168-170, 172, 173, 185, 186, 200, 201,
    204, 210, 212, 214, 216
root mortality 63, 64
rooting depth 20, 21, 39, 52, 59, 63-66, 71, 93,
    94, 108, 109, 113, 136, 148, 152, 164, 210,
    239, 240, 262
run-off 43, 44, 51, 115, 118, 119, 132, 164,
    174, 184, 185, 238, 240

sample correlation coefficient 134
seepage 115

senescence 40, 64, 65
shoot to root ratio 63, 75, 78
SIMWASER 2-4, 27-29, 32, 36
slope 38, 40, 51-56, 69, 156, 167, 174, 200,
    202, 204, 220
soil cover 39, 52, 113, 210, 260
soil data 1, 28, 80, 113, 132, 150, 152-154, 169,
    211, 222
soil layer 28, 29, 36, 37, 40, 41, 53, 59, 61, 64,
    66-68, 70, 71, 79, 82, 83, 96, 92-94, 97, 98,
    102-105, 107, 108, 117-119, 122, 123, 132,
    139, 149, 161, 164, 166-169, 173-178, 222,
    228, 231, 239, 241, 242
soil organic carbon (SOC) 129, 132, 142-144,
    147, 209-213, 216
soil organic matter (SOM) 92, 93, 95, 96, 98, 99,
    101, 106-108, 129, 130, 132, 133, 144, 149,
    150, 156, 158, 159, 183, 190, 191, 193, 194,
    197, 198, 209-214, 216, 217, 245, 246, 255,
    256, 258
soil profile 27-29, 31, 39, 40, 45, 98, 103, 104,
    116, 125, 131-133, 149, 150, 152, 158, 164,
    165, 185, 223, 239, 245, 247, 248, 258
soil temperature 52, 61, 64, 76, 93, 101, 111, 112,
    114, 120, 121, 127, 129, 130, 132, 135, 136,
    144, 165, 166, 210, 223, 231, 237, 246, 247,
    249, 262
soil texture 19, 29, 41, 47, 80, 86, 96, 97, 108,
    114, 120, 148-150, 152, 157, 167, 174,
    209, 210, 212, 213, 217
soil water characteristic/water retention curve 19,
    81, 84, 86, 87, 101, 193
soil water pressure 40, 113, 119-121, 199, 228
SOMNET 246
SPASS 3, 4, 9-14, 183-194
spring barley 45, 79, 83-86, 139, 140, 151, 155,
    198-202, 204, 205, 245, 246, 249, 250, 254, 256
STAMINA 2-6, 10-12, 51, 52, 56
stomata 24, 28, 63, 76
storage organ 11-13, 15, 55, 147, 149, 151, 153,
    158, 167, 186, 189, 223, 232-236
SUCROS 3, 4, 9-14, 52, 53, 72, 149, 162, 183-194
suction cups 27, 112, 122, 158, 245, 251, 261
sugar beet 20-22, 25, 39, 45, 51, 53, 54, 56, 59-61,
    64-66, 68-71, 102, 106, 107, 116, 139, 140,
    142, 149, 151-153, 155, 163, 183-185, 189-192,
    198, 220, 221, 224-226, 229, 232, 234, 238-240,
    245, 246, 249, 250, 254, 256

sunshine duration 149, 150, 237, 238
SWAP 2-5, 7-14, 43, 111-114, 147, 120, 124, 162
SWIM 2-5, 7-16, 129-133, 135, 138, 139, 142, 144, 145

tensiometer 41, 43, 112, 136, 227, 228, 231, 239, 242, 251
terrain 51, 52, 56
thermal conductivity 114, 120
THESEUS 2-4, 7, 8, 16, 37-48, 71
tillage 51, 93, 96, 112, 149, 167, 185, 211, 213, 217, 221, 222, 224, 226, 227, 249, 251
tiller 136, 223-226, 232-234
time domain reflectometry (TDR) 10, 15, 21, 22, 41, 43, 112, 119, 120, 136, 137, 192, 220, 223, 227, 240-242
time series 112, 168, 184, 240
total nitrogen 96, 97, 100, 102, 149, 155, 223, 254, 256
transpiration 19, 20, 26, 28, 29, 38, 39, 42, 45, 52, 63-66, 75-77, 119, 132, 137, 149, 165, 174, 188, 190, 191, 193, 227, 239
TURC-WENDLING 4, 148, 150

uncertainty analysis 111, 116-118, 122, 125, 132

validation 1, 37, 38, 41, 43, 48, 60, 64, 66, 68, 91, 130-132, 144, 148, 161, 163, 168-170, 172, 177, 219, 227, 240, 242, 259, 260, 264
vernalisation/vernalization 163, 171, 187
volatilisation/volatilization 4, 97, 115, 165, 176, 190

WASMOD 2-4, 6-9
water balance 2, 6, 19, 20, 22, 27, 29, 32, 36, 38, 39, 47, 48, 52, 53, 60, 64, 95, 118, 119, 132, 136-138, 158, 162, 164, 165, 183, 188, 190, 192-194, 220, 227, 231, 259, 260, 262, 264, 265

water content 4, 5, 10, 13, 20-22, 24, 26, 29-31, 33, 34, 37-43, 45, 48, 52, 53, 61, 64, 66, 79-86, 93, 97, 112, 114, 132, 135-138, 150, 152, 153, 156, 158, 164, 166, 174-177, 186, 191-194, 199, 219, 220, 223, 227-231, 239-242, 248, 259, 262
water pressure 40, 113, 119-121, 199, 223, 227-229, 239
water stress 45, 47, 51, 53, 61, 63-65, 113, 116, 136, 198, 239
water use efficiency 53
weather scenario 118, 124, 126
weed 71, 99, 102, 104, 167, 221, 239, 240, 249
wilting point 20, 40, 53, 79, 97, 164, 211, 212, 239
wind speed 20, 29, 38, 52, 76, 80, 86, 117, 118, 237, 247, 262, 264
winter barley 20, 21, 24, 59-62, 66, 68, 69, 71, 100, 102, 112, 116, 139, 140, 151, 161, 163, 168, 183, 184, 187-189, 220, 224, 225, 249, 264
winter cereals 25, 59, 61, 64, 65 67, 68, 238, 240
winter rye 21, 22, 59-62, 66, 68, 69, 71, 79, 81, 82, 100, 112, 116, 139, 140, 151, 163, 183, 184, 189, 220, 224, 225, 238, 239
winter wheat 20-22, 24-26, 38, 44, 45, 53-55, 59-62, 66, 68-71, 79, 81-85, 100, 102, 112, 116, 139, 140, 142, 147, 151, 152, 155, 161, 163, 183, 184, 188, 189, 192, 198-202, 204, 205, 220, 224-226, 238, 240, 245, 246, 249, 250, 254, 256
WOFOST 2, 38, 39, 75, 113, 162

yield 11, 13, 14, 27, 37-39, 44, 45, 47, 48, 51-56, 59-61, 67-70, 81, 83, 84, 91, 99-101, 107, 109, 129-132, 138-140, 142, 144, 147, 149, 155, 156, 161, 162, 167-169, 171-173, 176, 177, 186, 187, 197-205, 210, 211, 219, 237-240, 246, 250, 254, 256, 258